Pantanal & Cerrado

National Parks:

1 Pantanal Matogrossense
2 Chapada dos Guimarães
3 Serra da Bodoquena
4 Ilha Grande
5 Emas
6 Araguaia
7 Brasília
8 Chapada dos Veadeiros
9 Serra da Canastra
10 Nascentes do Rio Parnaíba
11 Grande Sertão Veredas
12 Cavernas do Peruaçu
13 Sempre-Vivas
14 Serra do Cipó
15 Serra das Confusões
16 Chapada das Mesas

Advisory Committee

Martha Argel (WCS)
Juan Mazar Barnett
(Seriema Tours)
Dennis Driesmans Beyer
Dante Buzzetti
Braulio Carlos
(Pantanal Bird Club)
Pedro F. Develey
(SAVE Brasil/Birdlife)
Neiva Guedes (Projeto Arara Azul)
John A. Gwynne (WCS)

Alexine Keuroghlian (WCS)
Katherine Lemcke (WCS)
Vincent Kurt Lo (IBAMA)
André De Luca
(SAVE Brasil/Birdlife)
Fernanda Marques (WCS-Brasil)
Fernanda Melo
(Caiman Ecological Refuge)
Leonardo Vianna Mohr
(ICMBio-Ministério do Meio Ambiente)
Maria Antonietta Pivatto
(OrnitoBr)

Wandir Ribeiro
Rômulo Ribon
(Universidade Federal de Ouro Preto)
Robert S. Ridgely
(World Land Trust)
Fernando C. Straube
(Hori Consultoria Ambiental)
Eduardo Martins Venticinque
(WCS)
Carlos Yamashita
(IBAMA)

Artists

Michael DiGiorgio
Dale Dyer
John A. Gwynne

Thomas R. Schultz
Guy Tudor
Andrew C. Vallely

Sophie Webb
Jan Wilczur

Photographers

Maria Allen
Amazon Araguaia Lodge
Jenny Bowman
Dante Buzzetti
Luciano Candisani

Braulio Carlos
Fabio Colombini
Mario Friedlander
Adriano Gambarini
Daniel de Granville

Peter Milko
Haroldo Palo, Jr.
João Quental
Rômulo Ribon
Cassiano Zaparoli

WILDLIFE CONSERVATION SOCIETY
BIRDS OF BRAZIL
THE PANTANAL & CERRADO OF CENTRAL BRAZIL

WCS Project Director: **John A. Gwynne**

Senior Author: **Robert S. Ridgely**

Art Director: **Guy Tudor**

WCS Project Coordinator
and Translator: **Martha Argel**

Species Maps: Robert S. Ridgely,
Maria Allen, Terry Clarke

Book Design and Production: Terry Clarke

Comstock Publishing Associates
a division of Cornell University Press
Ithaca and London

Copyright © 2010 by Wildlife Conservation Society

Species accounts and associated maps on pages 30–314 copyright © 2010 by
Robert S. Ridgely
   (Recommended citation: Ridgely, Robert S. Species Accounts. In
   Birds of Brazil: The Pantanal and Cerrado of Central Brazil, by
   John A. Gwynne et al. Ithaca: Cornell University Press, 2010.)

Illustrations by Guy Tudor from *The Birds of South America, Volume 1: The
Oscine Passerines*, copyright © 1989; *The Birds of South America, Volume 2:
The Suboscine Passerines*, copyright © 1994; and *Field Guide to South
American Songbirds*, copyright © 2009, by Robert S. Ridgely and Guy Tudor.
Used by permission of the illustrator and the University of Texas Press.

All rights reserved. Except for brief quotations in a review, this book, or
parts thereof, must not be reproduced in any form without permission in
writing from the publisher. For information, address Cornell University Press,
Sage House, 512 East State Street, Ithaca, New York 14850.

First published 2010 by Cornell University Press
First printing, Cornell Paperbacks, 2010

Editor: Marybeth Sollins

Printed in China

Library of Congress Cataloging-in-Publication Data

Wildlife Conservation Society birds of Brazil / WCS project director, John
A. Gwynne ; senior author, Robert S. Ridgely.
   v. cm.
   Includes bibliographical references and index.
   Contents: v. 1. The Pantanal and Cerrado of Central Brazil
   ISBN 978-0-8014-7646-4 (v. 1 : pbk. : alk. paper)
   1. Birds--Brazil--Identification. I. Gwynne, John A. II. Ridgely, Robert S.,
   1946- III. Wildlife Conservation Society (New York, N.Y.) IV. Title: Birds
   of Brazil.

QL689.B8W53 2010
598.0981--dc22

2010013388

Cornell University Press strives to use environmentally responsible suppliers
and materials to the fullest extent possible in the publishing of its books.
Such materials include vegetable-based, low-VOC inks and acid-free papers
that are recycled, totally chlorine-free, or partly composed of nonwood
fibers. For further information, visit our website at
www.cornellpress.cornell.edu.

Paperback printing   10 9 8 7 6 5 4 3 2 1

Projeto Aves do Brasil

gratefully acknowledges the vision and generosity of donors who have made this project possible:

Sam and Nora Wolcott

Blue Moon Fund

Beneficia Foundation

The World Bank

John D. and Catherine T. MacArthur Foundation

James Large

The Penates Foundation

Abigail Congdon and Joe Azrack

Graham Arader

Whitney Simonds

Edith McBean

Adeline and Ted Kurz

Mikel Folcarelli

Loring McAlpin

Dane Nichols

Anonymous Donors

This publication has been made possible with funding from
the World Bank and the John D. and Catherine T. MacArthur Foundation

Márcio Ayres
1954-2003

This project was born of the inspiration of the late Dr. José Márcio Ayres, the Wildlife Conservation Society's visionary field scientist, primatologist, photographer, and longtime Brazil program director. Márcio understood that collaboration is the key to the success of sustainable conservation initiatives for conservationists and local populations alike. Few others have been as directly responsible for protecting so much of our planet's wild places as Márcio, who helped to create the Amazon's model Mamirauá Sustainable Development Reserve which in turn enabled the creation of Amanã Reserve and Jaú National Park. Together these reserves form the most extensive protected rainforest area in the world. Subsequent state reserves such as Piagaçu-Purus added further protection and additional sources of support for local people.

Márcio also recognized the need to connect the people of Brazil with its birds, and asked us to create easy-to-use, accessible field guides that would provide information about local birds and inspire the next generation of Brazil's conservationists. Márcio died young: *Birds of Brazil* is only a small part of his legacy. We can only imagine what more he might have achieved for Brazil and for the world.

Contents

Acknowledgments	viii
Introduction	x
Birding in Brazil	xii
Pantanal & Cerrado: Protecting Wild Brazil	1
Species Accounts	30
References	316
Index	317

Acknowledgments

Márcio Ayres simply asked for a good field guide to the birds of Brazil. But the challenge of Brazil's size and the number of its species required us to think about multiple volumes, with each compact enough for easy field use and affordable to most Brazilians. Determined to carry on Marcio's mission, we conceived of a series of regional guides to be published under the auspices of the Wildlife Conservation Society (WCS). Art and text could also be used for educational posters and articles in Brazil to stimulate interest in birding. Each volume would contain a well-illustrated section about habitats and conservation. And there would be an associated website. The dream of creating such a series became real with the combination of Robert Ridgely's encyclopedic field knowledge of Neotropical birds and Guy Tudor's generous offer to allow the re-use of passerine illustrations from their *Birds of South America* series.

As one of the Neotropics' foremost ornithologists, Bob brings to this volume a panoramic overview of South American birds, supplementing his evocative species accounts with an extensive map database, digitized into maps by Maria Allen. The species accounts have been supported with field checking by many in Brazil, including Martha Argel, Juan Mazar Barnett, Dante Buzzetti, and Braulio Carlos. Guy Tudor, the preeminent artist of Neotropical birds whose unparalleled illustrations of passerines form the core of this volume, recruited and lent support to other illustrators—Michael DiGiorgio, Dale Dyer, Thomas R. Schultz, Andrew Vallely, Sophie Webb, Jan Wilczur, and myself—who painted hundreds of Brazilian nonpasserines included here. Working with expert digital artist Terry Clarke, Guy and Bob also supervised the modification of some illustrations to depict the proper local races for central Brazil. Terry also designed and produced the pages of this volume, a tough task due to the mandate of compactness for field use. The team has been lucky to have Marybeth Sollins, who edited this volume and provided counsel surpassing normal editorial responsibilities.

More than a dozen scientists, most of them Brazilian, served as advisors, lending their combined expertise to shape this volume's conservation chapter, inspired by Bob Ridgely's earlier innovations in that arena. Representing conservation organizations, universities, tour operators, private organizations, and Brazilian government departments, they include Juan Mazar Barnett, Dennis Beyer, Dante Buzzetti, Braulio Carlos, Pedro Develey, Neiva Guedes, Vincent Lo, André De Luca, Fernanda Melo, Leonardo Vianna Mohr, Tietta Pivatto, Wandir Ribeiro, Rômulo Ribon, Fernando C. Straube, Carlos Yamashita, and Martha Argel, Alexine Keuroghlian, Kat Lemcke, Fernanda Marques and Eduardo Ventincinque of WCS, assisted by Ana Rosa Cavalcante, Terry Clarke, Giselle Cury, Paula Dip, Chris and Lu Simonds, and Flavio Zuffellato.

Many accomplished nature photographers have contributed to this volume: Maria Allen, Jenny Bowman, Dante Buzzetti, Luciano Candisani, Braulio Carlos, Fabio Colombini, Mario Friedlander, Adriano Gambarini, Daniel de Granville, Peter Milko, Haroldo Palo Jr., João Quental, Rômulo Ribon, Cassiano Zaparoli, and the photographers of the Amazon Araguaia Lodge. We are also grateful to Dante Buzzetti for his knowledge and for the collection of field sounds that is part of this project. The critical support of IBAMA and Instituto Chico Mendes de Conservação da Biodiversidade (both from Ministry of Environment) and the sharing of information from Conservation International, EcoRotas Turismo (from Alto Paraíso de Goiás), Panthera Foundation, Projeto Arara Azul, and Projeto Pato Mergulhão have been important contributions. Giselle Cury assisted the WCS team to begin creation of the website, and Humberto Moura Neto provided advice and guidance. Tietta Pivatto worked with Martha Argel to start guide-training workshops in the Pantanal to help create local jobs in birder tourism.

Martha Argel's local knowledge of Brazilian ornithology, her energy, and networking connections were invaluable to the coordination of this project in Brazil. She also helped shape and translate all texts into Portuguese. As Chief Creative Officer/Vice President for Design at WCS, I steered the project, found support for its many parts, oversaw design and finances, and coordinated the work of those cited above as well as publishers, legal advisors, development experts, publicists, event planners, and many others, all of whom we thank wholeheartedly.

WCS-Brazil is grateful to its many supporters. The generous participation of significant donors who saw the need to make information about Brazil's especially rich birdlife more available to Brazilians and the world secured the project's success. John Mitchell and Sharon Pitcairn Forsyth's vision enabled Beneficia Foundation's lead gift to launch this project, followed by Kathy MacKinnon's World Bank support.

Sam and Nora Wolcott have been the project's major supporters, showing great confidence in our team's effort toward global conservation. Our dream of making field guides affordable to Brazilians, a major innovation toward conservation, was made possible by the creative insight of Diane Edgerton Miller and Adrian Forsyth of the Blue Moon Fund. WCS Trustee Edith McBean kindly took on this project, contributing her characteristic creativity toward finding future supporters, ably assisted by Graham Arader. Trustee Jim Large enabled the inclusion of downloadable sounds to the website. Joe Azrack, Abigail Congdon, Mikel Folcarelli, Addie and Ted Kurz, Loring McAlpin, Sandra Montrone, Dane Nichols, Whitney Simonds, and anonymous donors helped bring the volume to publication.

This project owes its existence to the kindness of its many friends, including Ana Rita Alves of Instituto de Desenvolvimento Sustentável Mamirauá; Jim Brumm; John Danilovich, United States Ambassador to Brazil, and Irene Danilovich; the Panthera Foundation's Tom Kaplan, Alan Rabinowitz, and Luke Hunter; Israel Klabin and Tom Lovejoy; Coca-Cola Brasil; Muriel Matalon; and Antônio Patriota, Brazil's Ambassador to the United States. Claudia Bauer, Jean Black, Mark Boyer, André Carvalhaes, Andrew Farnsworth, Tim Hirsh, Bruno Lima, Fernando Pacheco, Giulia Moon, Roger Pasquier, Mary and Mike Tannen, Marc Weinberger, and Bret Whitney are among the friends who have offered insight. The American Museum of Natural History kindly allowed access to its reference collection, with Joel Cracraft, George Barrowclough, Paul Sweet, and Christine Blake generously providing their time and knowledge. The University of Texas Press has graciously assisted this project in the use of artwork from its *Birds of South America* series.

Heidi Lovette of Cornell University Press and Peter Milko of Editora Horizonte have provided information, hard work, and dedication to ensure the allocation of energy and resources to this project. Senior staff at WCS–Jean Boubli, Avecita Chicchon, Bill Conway, Bob Cook, Jim Large, Richard Lattis, Kent Redford, John Robinson, CEO Steve Sanderson, and Andrew Taber–all understood the need for this publication and enabled time to be spent on its creation. WCS's Sylvia Alexander, Márcia Arosteguy, Eileen Cruz-Minnis, Carlos Fajardo, Zach Feris, Valeria Guimarães, Evelyn Junge, Kat Lemcke, Thays Nicolella, Anne Rice, Alexandra Rojas, Steve Sautner, and Pat White capably charted this project's trajectory forward at WCS. WCS designers Sharon Kramer-Lowe, Gail Shube, Kimio Honda, and the late Georges Dremeaux were among those who contributed. And not least, I thank my long tolerant, ever enthusiastic partner Mikel.

This first volume of the Wildlife Conservation Society Field Guide Series has been a team effort, with all its members already engaged in other projects. As such it has also been a labor of love. We believe fervently that an easy-to-use field guide, accessible to the Brazilian people, will highlight the critical importance of conservation and lead to an enhanced awareness of Brazil's extraordinary natural heritage.

John A. Gwynne

Introduction

The Wildlife Conservation Society (WCS) is the United States' oldest international conservation science organization, with a long history of study in Brazil—in the Pantanal since the 1970s and in the Amazon with Márcio Ayres in more recent decades. In this first volume of *Birds of Brazil*, a series of regional field guides to the birds of Brazil published under the auspices of WCS in two editions (English and Portuguese), we are fortunate to bring together the work of one of the world's most knowledgeable ornithologists and some of its best bird illustrators, along with evocative landscape images from some of Brazil's most renowned photographers. Conceived as more than just a series of field guides, this project is the centerpiece of a broader mission—to spark widespread interest and initiatives in conservation among the people of Brazil and create a pathway toward that goal by providing comprehensive information about the extraordinarily rich birdlife of Brazil, region by region. This will help facilitate Brazilian birding, making bird identification seductively easy and linking basic conservation information to habitats and critical regional issues for bird protection. A secondary goal is to introduce Brazilian birds to the birders of the world and by doing so stimulate ecotourism in Brazil's parks and wild places, ultimately benefitting local economies and creating jobs through tourism.

Brazil's vast size and wealth of bird species make a single all-Brazil field guide too bulky for easy portability in the field. The volumes in this series, however, in which each region has been roughly delineated according to simplified IBGE (Instituto Brasileiro de Geografia e Estatística) vegetation maps, will be convenient for field use and will provide comprehensive, easy-to-use information for identifying species, habitat, behavior, and songs. Few people are aware that with 1825 species currently known, and more being named almost continuously, Brazil is one of the planet's richest centers of bird diversity. Among all the continents, South America is home to the largest number of bird species, almost 60% of which are found in Brazil. Brazil also has the largest number of endemic (234) and threatened (116) species of any nation in the New World—more than one-quarter of South America's threatened birds. These numbers demand an especially urgent imperative for conservation and for the development of widespread public interest in birds, which has been shown in other areas of the world to inspire the growth of a culture of conservation.

Among the benefits of creating regional guides is the inclusion of local habitat information as a focal point for the presentation of local conservation issues. This volume's conservation section, "Protecting Wild Brazil," links Brazil's extraordinary biodiversity to its wealth of habitats and presents current issues of conservation concern. One of the challenges of the 21st century is to achieve economic and social development within a framework of environmental protection. More effective safeguards are needed in Brazil to protect increasingly scarce fresh waters, the vegetation needed to buffer the effects of global warming, and Brazil's wealth of biodiversity. It is to that end that a conservation section is included here and projected to appear in future volumes of the series.

This volume focuses on the birds of two large regions in central Brazil—the Pantanal and the Cerrado. The Pantanal is already famous among birders worldwide for supporting one of the globe's most exuberant spectacles of inland waterbirds. But the relatively unknown wild Cerrado, which once represented a quarter of all Brazil's lands and is still one of the world's most important savannas with respect to biodiversity, is now gravely threatened by progressive habitat loss.

To assist birders in central Brazil, this volume focuses on those species most likely found in the region, eliminating the distraction of information about species not found there. Many species that are transitional between the Cerrado and adjacent biomes also appear in this volume but will be given more attention in subsequent volumes. Nearly all of central Brazil's resident birds, birds of key conservation concern, and most regular migrants are illustrated and treated here in individual species accounts along with some associated species. Green areas on species range maps denote each species' normal range; tan denotes the ranges of southern birds occurring here in austral fall and winter (approximately April-August); blue denotes the ranges of migrants from North America (approximately October-March). Colored dots indicate isolated locations where the species is known to have occurred outside its normal range or, in some cases, the only location where it has occurred. The color key and legend found inside this volume's front cover refer to the status (endangered, threatened, etc.) of bird species of high conservation concern in Brazil, according to the official status lists of the Ministry of Environment.

The goal of this volume (and this series) is to make bird identification easy in regions whose sheer numbers of species, let alone similarities between them, present an extraordinary challenge to most birders. The illustrations, carefully researched to facilitate comparisons between similar species, are all rendered to showcase the diagnostic field marks mentioned in the text that describes each species. The design of the volume further simplifies identification, with species accounts and range maps facing illustrations. The inclusion in the species accounts of succinct but carefully crafted and evocative descriptions of behavior, habitat, and voice further enhances ease of use.

Our dream is to inspire a nation of potential birders who will delight in the beauty of Brazil's abundant birdlife and become interested in helping to protect Brazil's vibrant ecosystems and natural heritage. It is our hope that this volume will provide the first sparks.

John A. Gwynne and Martha Argel

Brazil has one of the world's largest, and finest, avifaunas. Ornithologists and birders have long recognized this, but as yet all too few of Brazil's citizens do. That's changing, and it's my hope that this book, and this series, will serve to accelerate that process. Get out and go birding and, while you're at it, work to protect Brazil's fabulous birds.

Robert S. Ridgely

Birding in Brazil

Brazil boasts enormous numbers of bird species—740 in this volume alone. This extraordinarily rich celebration of birdlife proffers an enticing invitation for foreign visitors from bird-impoverished lands. There are many new species and families to learn, some of which are confusingly similar (such as woodcreepers and furnariids). Birders should visit diverse habitats—grasslands, forests, savannas, wetlands—as each has its own distinctive birds. Before visiting, try to learn the characteristic birds of each habitat.

Visiting Brazil

Brazil is large. Geographically it is the fifth largest nation on the planet—larger than Western Europe—and covers approximately the area of the continental United States. Travel distances are vast, requiring advance planning of itineraries. Major airports are usually located outside urban centers so it is easy to avoid the challenges of unfamiliar cities and go directly to the very friendly countryside where the birds are. Familiar car rental companies are easily found at many airports; Brazilians drive on the right (as in the United States), and road conditions are generally good in eastern and southern Brazil. Many Pantanal roads are inundated and impassable in flood season (dates vary locally). Seek information before traveling to any destination, and inform your hotel staff where you are going.

Birding is good at all times of the year in Brazil but perhaps best during spring (September-October), the advent of the breeding season and also a dry time in the Pantanal. Temperatures are cooler during the southern winter (June-August). In central Brazil, winter temperatures rarely drop below 10° C (50° F) but sometimes rise above 30° C (into the high 80s F) on sunny days. Summer is hot (over 40° C, or 100°-plus F, in Mato Grosso during January and February). Bear in mind that this is vacation time in Brazil, with many people traveling and many flights, tours, and hotels booked in advance.

Brazilians speak Portuguese, and are helpful, friendly, and full of fun. Visiting is easier if you can speak a few words in Portuguese. For the most part, English is only spoken in major urban centers. Brazil has a reciprocal visa policy; if your country usually requires Brazilians to obtain a visa (e.g., United States, Australia, New Zealand), a visa is required for you to enter Brazil. Yellow fever vaccination is required; polio and tetanus shots are recommended. If you plan to enter a known malarial or dengue zone, consult a physician to obtain preventive medication. Wear insect repellent and use sun block.

The currency of Brazil is the *real* (R$). US dollars and euros are not accepted: one must change money upon arrival in Brazil. Not all ATMs in Brazil are international, even at airports. Credit cards (especially Visa and Mastercard) are widely accepted in urban hotels and restaurants.

Harpy Eagle

Birding the Pantanal

A vast seasonally flooded wetland, the Pantanal is renowned among birders around the world as an easy place for good views of large numbers of waterbirds—egrets, storks, ibises, cormorants, ducks—as well as raptors and several species of parrots, including Hyacinth and Golden-collared Macaw and local Blaze-winged Parakeets. Other spectacular species include Jabiru, Maguari and Wood Storks, Southern Screamer, Bare-faced Curassow, Common Piping Guan, Limpkin, Great Black Hawk, and Black-collared Hawk. Scarlet-headed Blackbird, Yellow-billed Cardinal, Helmeted Manakin, Long-tailed Ground Dove, and American Pygmy Kingfisher are often well seen, too. Most Pantanal specialties are found here year round, with the larger concentrations of waterbirds occurring toward the end of the low-water season, between June and November. The wet season's tropical rains begin in November, with daily showers of one to two hours' duration. The northern Pantanal floods once a year, usually beginning in January, while the southern Pantanal tends to receive flood waters two months later, often flooding twice.

The main gateway to the northern Pantanal is Cuiabá. The easiest way to explore the area is to travel along the Transpantaneira road south from Poconé. Visitors to the southern Pantanal arrive via Campo Grande or Corumbá airports. Many *fazendas* (cattle ranches) and *pousadas* (inns) offer lodging and facilities for birdwatching, and can be discovered on the Internet.

Birding the Cerrado

The wild woodlands and grasslands of the Cerrado are known for dry-area specialties, with Gray-backed Tachuri and Dwarf Tinamou among the 30-plus endemics found only here. Each Cerrado habitat has its own distinctive bird fauna. In addition, many nocturnal mammals such as Giant Anteater, Maned Wolf, and Crab-eating Fox are fairly easy to find in the wild Cerrado.

Within the vast agricultural lands of the southern Cerrado there are several jewels. Brasília National Park (NP), only 10 kilometers from the capital, is inhabited by the Greater Rhea, Bare-faced Curassow, Brasilia Tapaculo, and Curl-crested Jay. At wonderful, remote Emas NP, Cock-tailed Tyrants and many other grassland specialties are likely. Even a White-winged Nightjar or Crowned Eagle is possible. When the rains begin, millions of Emas wildflowers come into bloom.

Chapada dos Guimarães NP, near Cuiabá, has dramatic rock formations and waterfalls as well as Fiery-capped and Helmeted Manakins, Blue-winged Macaw, and Dot-eared Coquette. Serra da Canastra NP boasts the very local Brazilian Merganser, Stripe-breasted Starthroat, and Gray-backed Tachuri. Serra do Cipó NP has the Cipó Canastero, Hyacinth Visorbearer, and White-naped Jay. The beautiful Bodoquena range and Bonito, with many birds, good habitat, and lodgings, can be reached via Campo Grande airport.

The less disturbed (to date) northern Cerrado has many birding destinations along the Araguaia River, such as São Félix do Araguaia, just across the river from seasonally-flooded Ilha do Bananal. Here a rich mix of habitats boasts Scarlet-throated Tanager, Bananal Antbird, and Crimson-fronted Cardinal. Kaempfer's Woodpecker and many other northern Cerrado species inspire birders to visit the dry northern Cerrado where new destinations emerge, such as Chapada das Mesas, near Carolina, Maranhão, and the Jalapão region, in the eastern Tocantins.

Airports

Some distances by road:

São Paulo to Brasília
...................................1,011 km

São Paulo to Cuiabá
...................................1,643 km

São Paulo to Emas NP
...................................1,000 km

São Paulo to Bonito
...................................1,201 km

Brasília to Cuiabá
...................................1,147 km

Brasília to Emas NP
......................................680 km

Brasília to Campo Grande
...................................1,192 km

Cuiabá to Emas NP
......................................500 km

Cuiabá to São Félix do Araguaia
...................................1,159 km

Cuiabá to Campo Grande
......................................694 km

Cuiabá to Cáceres
......................................221 km

Cuiabá to Poconé
......................................108 km

Campo Grande to Bonito
......................................312 km

Dot-eared Coquette

PANTANAL & CERRADO
Protecting Wild Brazil

PANTANAL & CERRADO
Protecting Wild Brazil

The Pantanal is a seasonally flooded wetland covering more than 80,000 square miles. Except for small extensions into Paraguay and Bolivia, the Pantanal is a Brazilian landform. When the Paraguay River and its tributaries flood their banks, the region becomes one of the world's most extensive wetlands. This rich floodplain supports one of the world's largest inland concentrations of waterbirds. The global significance of the Pantanal has been acknowledged under the Ramsar Convention and by its designation as a UNESCO World Heritage Site.

After the Amazon, the Cerrado encompasses Brazil's largest biome—740,100 square miles of Brazil, also with small extensions into Paraguay and Bolivia. The Cerrado plateau is a major source of water for the three main basins in Brazil: the Amazon in the north; the Paraná in the south and west; and the São Francisco in the east. Occupying fully one-quarter of Brazil, the Cerrado is a distinctive Brazilian habitat as well as one of the planet's biologically richest savannas. Once underappreciated as wasteland, the Cerrado is beginning to be known as an important biome in its own right. To date, approximately 694 bird species have been found in the southern and central Cerrado, of which 51 are endemic to Brazil and 33 near endemic. Plant diversity is even higher: 4,400 of the Cerrado's plant species are found only here.

Brazil's high bird diversity results from its wealth of different habitats, each of which supports a different array of birds. Species do not exist randomly. Over time they have evolved in specific environments to which they have adapted. There are many subhabitats in both the Pantanal and Cerrado, each with its own landscape and composition of birds. This is why these places are so rich for birds and birders.

Nature, Cows, and Crops

In recent decades, two very different approaches to land use have emerged in central Brazil. Cattle ranches dominate the landscape of the Pantanal, but many of them support the recovery of several previously over-exploited species. Hyacinth Macaw, Jacaré Caiman, Giant Otter, and even Jaguar populations are coming back in some areas because of the far-sighted work of certain private landowners, science researchers, and environmental agencies.

Most of the Cerrado, however, has been cleared and plowed. A generation ago the wild Cerrado seemed limitless. Now more than three-quarters of it has been converted to some form of agricultural use. To protect the last remaining wild Cerrado, it is incumbent upon government and private entities to set aside and protect a significant fraction of what remains. Much less than 5% of the Cerrado receives legal protection at present. Many of these protected "islands" may already be too small to prevent losing many of their rarer species. Most are too isolated for birds to disperse between them. The more fragmented they become, the less they can sustain populations of many of Brazil's most special bird species.

PREVIOUS PAGE PHOTO: D. DE GRANVILLE

Pantanal Waters Derive from the Cerrado

The Pantanal is a mosaic of habitats created by water, but few people realize that it would not exist but for the abundant waters that flow into it annually from the southern Cerrado. Without this flooding the Pantanal would be covered by more grassland, with an aspect not dissimilar from the Cerrado. Each wet season, Cerrado rainwater pours from Brazil's central plateau (the *planalto*) down the Paraguay, Cuiabá, Piquiri, and Taquari rivers, and many other smaller affluents. As their levels rise, these rivers inundate the Pantanal's grasslands and gallery forests with nutrient-rich waters, creating a vast temporary wetland.

Flood timing varies. The rainy season usually begins in November, but it may take as many as two months for Cerrado rainwater to begin to reach the northern Pantanal. Flooding progresses slowly southward, requiring fully four months to reach the southern Pantanal, where highest flood levels can occur during what is actually the peak of the local dry season. The gradual movement of water provides the Pantanal with an influx of nutrients that insure a lush growth of annual plants; it also supports a surge of fish consumed by local inhabitants (*pantaneiros*) and waterbirds alike.

Heavy herbicide and fertilizer use, and erosion in the dense agricultural zones of the Cerrado, are beginning to have a negative effect on water quality in the Pantanal. The reduction of chemical input and the preservation of natural vegetation next to streams, which act as natural filters, are both critical to protect the Pantanal.

The Cerrado is the Pantanal's water source.

In dry months, thousands of fish-eating birds feast at the Pantanal's shrinking pools.

PANTANAL
Brazil's Wildlife Spectacle

The sprawling Pantanal boasts some of the world's most impressive seasonal concentrations of wading birds. With approximately 450 other species it is one of the world's famed birding sites, where it is easy to see the Toco Toucan, Hyacinth and Golden-collared Macaws, Bare-faced Curassow, and a host of others. Oftentimes birds are literally everywhere.

In the flooding period (*cheia*), the Pantanal's waterbirds disperse, feeding in scattered flocks. When flood waters recede, lagoons shrink and become isolated as small ponds. Enormous numbers of fish can then become trapped in the drying shallow pools. This concentration of fish provides a bonanza for waterbirds, sometimes in staggering numbers. The star of many such waterbird aggregations is the huge Jabiru, standing tall among the restless legion of hungry egrets, cormorants, spoonbills and other storks. Five kingfisher species join the feast.

The Pantanal is increasingly important for tourism. Most of its lands are privately owned cattle ranches (fazendas). Many visitors drive slowly south from Poconé along the Transpantaneira road. Stopping at different habitats will reveal a constantly changing panorama of motion and color comprising a wide range of bird species. On a good day, with a good guide, even in the mid-day heat, one can see upwards of 100 species and the sky may be full of birds.

PHOTO: JENNY BOWMAN

The Pantanal's wildlife spectacle is not limited to birds. Its relatively open landscape facilitates seeing some of South America's larger land animals, including Jacaré Caiman, Capybara, Yellow Anaconda, and Marsh Deer. Brazilian Tapir, peccaries (both Collared and White-lipped), Giant Otter (ever increasing and not hard to see in some areas), and now even Jaguar can be occasionally encountered.

Brazilians have reason to be proud of the Pantanal's exuberant wildlife. But long-term protection of its nature spectacles will require monitoring as new challenges emerge in the 21st century. Pantaneiro cattle ranching has been a good example of sustainable land use, combining cultural tradition and environmental protection, but ranching has been ecologically problematic when stocking rates have been too high. Preservation of the Pantanal's semi-wild landscapes requires unmodified native grassland and woodland habitats, the maintenance of Cerrado water quality, and annual flooding from the slow-flowing Paraguay River. It is the intertwining of habitats–forested islands next to marshes, grasslands (*campos*) next to ponds–that creates the Pantanal's rich biodiversity.

> The Pantanal has the greatest concentration of fauna in the Americas.... People outside Brazil know only the Amazon.
> Dr. Maria Tereza Jorge Pádua, former Director of Brazil's National Parks, Getty International Conservationist

Caimans bask in large groups.

PHOTO: J. QUENTAL

PHOTO: C. ZAPAROLI

PANTANAL
Seasonal Grasslands

The annually flooded grasslands of the Pantanal change radically during the year, alternating between dry grassy fields and vast areas that are flooded for two to three months. The flooding is not uniform. The northern Pantanal is often flooded while the south is still dry. This variability argues for widespread conservation of the Pantanal. Some birds, such as the lovely Azure Gallinule, engage in local movements to take advantage of this flooding; they may totally vacate the Pantanal during the dry season.

As flooding ends, soil is exposed and dormant grass seeds begin to sprout. Wetlands turn into fields where a different assemblage of birds now forages alongside cows. Greater Rheas, Red-legged Seriemas, Whistling Herons, and Buff-necked Ibis scout the grass for food, as do Yellowish Pipits and many other smaller birds. Atop termite mounds Burrowing Owls perch sleepily and White-rumped Monjitas scan for passing insects. Near ranch buildings flocks of Black-hooded and Monk Parakeets sometimes drop to the ground to forage for seeds and fallen fruit.

At the end of the dry season grasses mature and seeds ripen, attracting up to nine species of migratory seedeaters, some of them at risk of extinction. Jabirus, Roseate Spoonbills, Wood and Maguari Storks, and several egret species gather at the ever-shrinking pools to feed on stranded fish. Then, after some weeks of drought, floodwaters from the Cerrado reappear and the cycle begins anew.

Muscovy Duck, Plumbeous Ibis, and Jabiru

PHOTO: H. PALO JR.

PANTANAL
Marshes

Unlike seasonally flooded grasslands, marshes, ponds, and rivers retain water year-round. Many bird species occur in both habitats; some live only in permanently wet places. Least Bitterns, rare Pinnated Bitterns, Scarlet-headed and Unicolored Blackbirds, Southern Yellowthroats, and the Black-capped Donacobius especially favor marshes. Yellow-billed Cardinals abound in fringing shrubbery and open areas. Crakes specialize to habitat: Gray-breasteds in high wet grass near lakes; Ash-throateds in marsh edges far from open water; Rufous-sideds in marshes with dense vegetation; and Russet-crowneds in woodland edges.

Crested Doraditos remain poorly known or are overlooked, favoring large expanses of wet grass. White-headed Marsh-Tyrants, Black-collared Hawks, and Snail Kites perch at the water's edge in low shrubs. Purple Gallinules and Rufescent Tiger Herons stalk the edges of thick aquatic vegetation. Ringed Teals, Southern Screamers, and noisy flocks of Black-bellied and White-faced Whistling Ducks frequent lagoons where Wattled Jacanas and Black-backed Water Tyrants walk on floating vegetation.

Some marsh birds (e.g., screamers, herons, and ibis) take advantage of seasonal high waters by moving into flooding campos. Capybaras, Marsh Deer, and anacondas sometimes move, too, but the key point is that it is the marshes that are the year-round mainstay for the Pantanal's waterbird abundance.

Black-capped Donacobius, Scarlet-headed Blackbird, and Wattled Jacana

7

PHOTO: F. COLOMBINI

PANTANAL
Forested Islands

Small forested islands (*capões*) on slightly elevated patches of ground are always above water. These groves, scattered across the landscape, range from approximately 5 to 300 feet across. A large fig tree or *manduvi* typically rises in the center of each *capão*, with smaller trees in its shade and *acuri* palms around it. Hyacinth Macaws nest in this habitat within hollows of the largest manduvi trees and feed on the seeds of the palms. Jabirus use the same big trees for building their immense stick nests, and sometimes are joined by a nesting colony of noisy Monk Parakeets. The capões' trees also provide nesting sites for colonies of Wood Storks, spoonbills, cormorants, and Anhingas. Other birds typical of capões are the Great Rufous Woodcreeper, Pale-crested and Golden-green Woodpeckers, Blue-crowned Motmot, and Golden-collared Macaw.

In dry periods, forested islands function as stepping-stone refuges across open grassland for shy birds and mammals. Undulated Tinamou, Gray-crested Cacholote, Great Antshrike, Fawn-breasted Wren, Pearly-vented Tody-tyrant, Narrow-billed Woodcreeper, and White-lored Spinetail are capão specialists. In flood periods capões also become crucial as high-water refuges for terrestrial mammals such as the Giant Anteater, Marsh and Gray Brocket Deer, Paca, and several species of small armadillos. Here many palms and trees produce fruits consumed by monkeys, agoutis, and squirrels.

Golden-collared Macaw, Pale-crested Woodpecker, and Great Rufous Woodcreeper

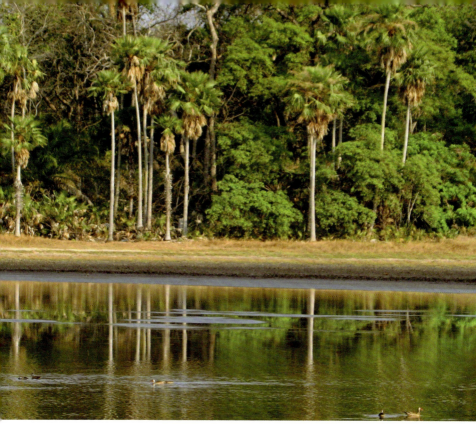

PHOTO: C. ZAPAROLI

PANTANAL
Forests

The Pantanal supports two major forest types. Gallery forest is found along rivers; where the water table is high enough there is evergreen foliage. Gallery forest plays an important role in impeding erosion, as tree roots hold the soil against onrushing water. Breeding colonies of storks, herons, cormorants, and other waterbirds are sometimes found in riverside trees. Giant Otters can be locally numerous along rivers, having increased in recent decades due to the cessation of most hunting.

Gallery forest birds include Brazilian specialties such as the Mato Grosso Antbird and White-lored Spinetail, both near-endemics, as well as Bare-faced Currasow, Ashy-headed Greenlet, Large-billed Antwren, Blue-crowned Trogon, and Plain Inezia. Certain monkeys are also restricted to these forests, with Black-and-gold Howler Monkeys being most common.

Dry forest is seasonally semideciduous, located on higher ground above flood levels, and often has vegetation more typical of the Cerrado. Dry forest supports the Chestnut-bellied Guan and Planalto Slaty Antshrike, both endemic to Brazil, as well as Collared Forest Falcon, Rufous-tailed Jacamar, and the Rufous Casiornis. Higher ground also provides refuge for forest mammals–Jaguars and smaller wild cats, brocket deer, and tapirs–that feed in the grasslands. Today, high-ground forest is becoming scarce due to wood extraction, periodic fires, and trampling by cattle taking refuge there during the high-water season.

Mato Grosso Antbirds, Blue-crowned Trogon, and Red-billed Scythebill

The Pantanal is famed for flocks of the fabulous Hyacinth Macaw: these macaws and cattle-ranching can

PANTANAL
A Conservation Success

Once there were many more Hyacinth Macaws in the Pantanal, but by the 1980s their population had declined to only about 1500 individuals, this after thousands had been trapped and sold as cage birds. Today, however, there are more than 5,000 (most of Brazil's population), an exciting recovery of one of the world's most spectacular birds. The Hyacinth Macaw Project (Projeto Arara Azul), which began in the 1990s, has helped to more than triple the species' local population in fifteen years. In the Pantanal the Hyacinth Macaw eats the hard seeds within the fruits of acuri and *bocaiúva* palms. Flocks feed in the palm trees but they also forage on the ground for seeds, some of which have passed through cattle stomachs, the cows having digested the fleshy outer fruits and dropped the hard, inner seeds.

Cattle and Hyacinth Macaws can coexist, but cows become a problem when they eat or trample young bocaiúva palms and manduvi seedlings. Cattle ranchers who want to increase local populations of this iconic bird need to plant seedlings within fenced enclosures to prevent cows from damaging them.

Protect palms for macaws.

Palm seeds are prime food.

coexist in harmony here. PHOTO: L. CANDISANI

Hyacinth Macaw populations are also limited by the natural scarcity of their preferred nesting trees. Seventy percent nest in manduvis that are more than eighty years old and have flat-bottomed hollows large enough to accommodate macaws and their nest. Old manduvis are fragile and vulnerable to strong winds if exposed by forest clearing. Keeping smaller trees around old manduvis can create a protective barrier. Installing man-made nest boxes is a proven way to increase nest-site availability.

Pantanal ecotourists can join Projeto Arara Azul staff in season to see macaw chicks in their natural nests or nest boxes. Among the project's field bases are Refúgio Ecológico Caiman, Pousada Xaraés, and Pousada Araraúna (check the Internet for current information). An increasing number of fazendas welcome tourists in Mato Grosso do Sul and Mato Grosso state. Smaller lodges are also springing up to serve the growing market, both domestic and international.

The restoration of the Hyacinth Macaw population in the Pantanal is a notable Brazilian environmental success story that can be repeated with other species in central Brazil. Similar projects are starting for Golden-collared Macaws and Turquoise-fronted Amazons. Other successes have been the Wildlife Conservation Society-Brazil's Jaguar Conservation Program and an ongoing Pantanal/Cerrado initiative, both of which are parts of a larger program partnering with landowners to promote sustainable ranch management practices that are profitable alternatives to deforestation and habitat conversion.

Macaws nest in manduvis.

Nest boxes increase numbers.

There is an urgent need to protect more of the Cerrado's wild places, a unique heritage for Brazil.

CERRADO
Beautiful but Unknown

Though barely known to foreigners or even many Brazilians, the Cerrado is a national treasure. It has one of the greatest plant biodiversities of any savanna in the world and, next to the Amazon and Mata Atlântica, boasts one of Brazil's richest bird biomes. Among the bird species endemic here are the Collared Crescentchest, Sharp-tailed Grass Tyrant, Campo Miner, and Coal-crested and Blue Finches.

There are approximately 10,000 plant species in the Cerrado (with even more plant families than in the Amazon), and the Cerrado supports the greatest variety of large land mammals in South America, once having been a stronghold for many of their populations. Maned Wolf, Giant Anteater, Pampas Deer, White-lipped Peccary, Puma, Jaguar, Brazilian Tapir, and even the rare Giant Armadillo can still be seen in and around protected areas. For a short period at the onset of the rains, the Cerrado's termitaria glow at night with the mysterious light of bio-luminescent glowworms. But all will vanish if the march of industrial agriculture is permitted to continue across all of what was once the wild Cerrado.

PHOTO: JENNY BOWMAN

The Cerrado's Maned Wolf

PHOTO: F. COLOMBINI

The Cerrado's hidden treasures

PHOTO: A. GAMBARINI

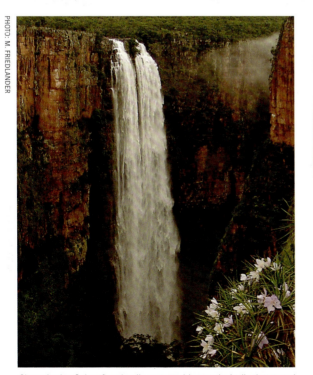

PHOTO: M. FRIEDLANDER

The subtle but powerful beauty of the Cerrado landscape—where the wild cries of the seriema greet the dawn and silence envelops vast mid-day vistas of gnarled trees, rolling grasslands, cliffs, and cascading waterfalls—is uniquely Brazilian.

Chapada dos Guimarães: legally protected but ecologically threatened

PHOTO: P. MILKO

CERRADO
Campo

The Cerrado supports three subtly different types of grassland: open grassland (*campo limpo*); grassland with tall shrubs and scattered small trees (*campo sujo*); and wet grassland (*campo úmido*), which occurs on permanently saturated soils. An additional specialized habitat, rocky grassland (*campo rupestre*), is found in certain higher elevation areas with shallow soil and very rocky terrain. Each habitat has distinctive plants; many also have specialized bird species.

Many birds are typical of campo limpo, including the Greater Rhea, Cock-tailed Tyrant, Campo Miner (especially in recently burned areas), Coal-crested Finch, Sharp-tailed Grass Tyrant, Ocellated Crake, Black-masked Finch, and Bearded Tachuri. Among campo sujo specialties are the Cinereous Warbling Finch, Dwarf Tinamou, and Collared Crescentchest. Widespread grassland species include Red-winged Tinamou and Burrowing Owl. The very rare Great-billed Seedfinch favors shrubbery at the edge of campo úmido, with Lesser Grassfinch out in the grass itself. Campo rupestre is home to the Blue Finch and Horned Sungem.

Natural (cattle-free) campos support a different assemblage of grass species, which are naturally adapted to live with occasional fire. Most campos can be lightly grazed, but when ranchers want to increase production they typically plant African grasses, which eliminate the indigenous grasses, wildflowers, and birds.

White-winged Nightjar, Coal-crested Finch, and Red-winged Tinamou.

PHOTO: F. COLOMBINI

CERRADO
Wooded Savanna

Wooded savanna or "true" cerrado is very open woodland where the low tree crowns seldom touch but emerge from a groundcover of various grasses with scattered small bushes. This type of savanna results from the mixing of two distinct floras–one herbaceous from the grasslands, and the other of tree species from dry forest. Tree trunks are often charred black from past fires. Many areas support wildflowers, at times in spectacular display, especially after the first rains.

Throughout the Cerrado sporadic natural fire has shaped a mosaic of campos, savanna, and dry forest. In areas with frequent fires, wooded savanna tends to evolve into grassy campos. In the absence of fire, trees can grow taller and denser as the wooded savanna becomes dry forest (cerradão). The species of birds change as these landscapes change. Some species require the interface of two habitats: seriemas patrol open campos but nest in the trees of adjacent wooded savannas.

A rich bird fauna occurs here, where it is common to see Blue-and-yellow Macaw, Toco Toucan, Red-legged Seriema, Curl-crested Jay, Yellow-faced Parrot, Peach-fronted Parakeet, and White-vented Violetear. Other specialties are Chapada Suiriri, White-rumped and Shrike-like Tanagers, Black-throated Saltator, Plumbeous Seedeater, and the huge but increasingly rare Crowned Solitary Eagle. Although hard to see in the dense and tall grass, ground-inhabiting birds including Spotted and Lesser Nothuras and Dwarf Tinamou sometimes reward the determined birder.

Toco Toucan, Blue-and-yellow Macaw, and Curl-crested Jay

PHOTO: D. BUZZETTI

CERRADO
Dry Forest

Two types of dry forest occur in the Cerrado. Cerradão forest forms where lack of fire allows trees to grow closer together, with intertwined crowns. This low forest with nearly continuous canopy intergrades with more open, wooded savanna, and the two share many tree species. In areas with more fertile soils, a taller deciduous forest (*mata seca*) grows, with its own unique tree species.

Cerrado dry forests share some bird species with adjacent biomes–Amazonian to the north, *caatinga* to the northeast, and the Atlantic forest to the east and south–but they also harbor their own bird species. Dry forests are fast becoming one of the Cerrado's rarer ecosystems.

Some cerradão forest specialties are the Chestnut-eared Araçari, Caatinga Puffbird, Undulated Tinamou, Rufous-winged Antshrike, Narrow-billed Woodcreeper, Campo Suiriri, and Southern Antpipit. Cinnamon and Black-faced Tanagers are both common at Serra da Canastra. The central Cerrado's deciduous forests support interesting avian endemics. Some (e.g., Pfrimer's Parakeet, which only occurs around Chapada dos Veadeiros) have a very restricted range. Other tall-forest specialties are Henna-capped and Planalto Foliage-Gleaners, Reiser's Tyrannulet, and Moustached Woodcreeper. Red-and-green Macaws visit both types of forest for feeding.

Red-and-green Macaw, Laughing Falcon, and White-eared Puffbird

PHOTO: M. ALLEN

CERRADO
Gallery Forest

Gallery forests closely follow river courses, forming green, biologically rich corridors that traverse drier habitats. Typically they are edged by palm groves (characterized by one species of palm, *Mauritia flexuosa*, known as the *buriti*), in turn edged by a zone of wet grassland that becomes wooded savanna as the terrain rises away from the river. Typical gallery forest birds include Undulated Tinamou, Bare-faced Currasow, Red-shouldered Macaw, Rufous-tailed Jacamar, Planalto and Cinnamon-throated Hermits, Planalto Foliage-gleaner, Helmeted Manakin, and White-striped Warbler. The rare and endemic Brasilia Tapaculo can be found very locally, creeping close to the ground.

Species characteristic of the palm groves are Red-bellied Macaw, Neotropical Palm Swift, Sulphury Flycatcher, and the unique Point-tailed Palmcreeper. Semipermanently flooded forest (*bosque de várzea*) also occurs very locally; here a few of the critically endangered Cone-billed Tanagers persist (fewer than 250 apparently survive in scattered pockets of remnant habitat).

Gallery forests and palm groves are so ecologically interconnected with the adjacent campo habitats that some forest bird species will disappear if the forest remains but the surrounding savanna is cleared. For instance, Blue-and-yellow Macaws breed in the palm groves but regularly forage in nearby savannas. Gallery forests occupy only a tiny fraction of the Cerrado's total area but they stand out for their richness.

Red-bellied Macaw, Helmeted Manakin, and Point-tailed Palmcreeper

Dramatic rock formations provide habitat for specialty birds and plants. PHOTO: D. BUZZETTI

CERRADO
Stone Gardens

Campo rupestre is a distinctively Brazilian habitat, found above 2,900 feet on some mountaintops and plateaus in the Espinhaço Range in central Brazil. Existing in this striking landmass of stony outcrops and scattered rivulets is an exceptional flora. Although only a small number of areas have been inventoried, it is estimated that approximately 3,000 plant species grow here, of which 30 % are endemic. Some of this unique low vegetation grows on almost bare rock with adaptations for water retention.

There are numerous endemics among the fauna here, with many mammals, amphibians, reptiles, and fish unique to campo rupestre. There are specialties among the birds, too, including the Gray-backed Tachuri, Cipó Canastero, Horned Sungem, Blue Finch, and Pale-throated Serra Finch. Some of the best sites for finding these birds are Serra do Cipó, the Canastra and Caraça ranges in Minas Gerais, and Chapada dos Veadeiros in Goiás.

Horned Sungem and Blue Finch

Found only in campo rupestre

PHOTO: R. RIBON

Intermittent fires shape the Cerrado landscape. PHOTO: F. COLOMBINI

CERRADO
Birds of Fire

Occasional fires are natural in the Cerrado. Sparked by lightning, they prevent trees from invading open grasslands and tend to favor native grasses and wildflowers which have protective underground root systems and bulbs. Without fire, campos evolve into wooded savannas, and wooded savannas become cerradão dry forests.

Cerrado birds have adapted to occasional fire, with some species even favoring recently burned areas; others need long grass that has not been burned for years. Specialists of burned areas often appear quickly and almost miraculously after a fire has raced through. Long grass harbors the Cock-tailed Tyrant, Sharp-tailed Grass Tyrant, Collared Crescentchest, Ocellated Crake, Dwarf Tinamou, Lesser Nothura, and Plumbeous Seedeater. These birds must go elsewhere when fire occurs. Specialists of burned areas include the Coal-crested Finch, Campo Miner, and the very rare White-winged Nightjar along with generalists such as Greater Rhea and White-tailed Hawk.

Nowadays anthropogenic fire occurs much more frequently than is natural in most areas, and these fires are almost invariably now set deliberately. Too frequent fire, establishment of crop monocultures that reduce large areas of natural vegetation to fragments, and the introduction of nonnative grasses that overpower native grasses and wildflowers endanger much of the Cerrado's unique flora and specialist birds.

Cock-tailed Tyrant

PHOTO: D. BUZZETTI

Wildflowers sprout after fires.

Brazilian Mergansers can persist only in crystal-clear waters. PHOTO: D. BUZZETTI

CERRADO
Icon of Clear Water

The near-endemic Brazilian Merganser is one of the Western Hemisphere's most critically endangered birds, a living symbol of healthy Cerrado water. This fish-eating duck requires clear-water streams and thus is a sensitive indicator of habitat degradation: with disturbance, the Cerrado's red soils erode quickly, contaminating the water. A small population of Brazilian Mergansers persists in and around Serra da Canastra National Park, where a protection project employs strategies such as conservation and replanting of gallery forests (preventing agriculture from reaching the water's edge); controlling erosion and siltation; prohibiting recreation in prime merganser habitat; controlling sewage and pesticide pollution; and preventing dam building and forest fires.

Brazilian Mergansers occur only in the São Francisco, Tocantins, Paraná, and Doce watersheds with very few known from Argentina and Paraguay. Among the places they persist are Jalapão, Chapada dos Veadeiros, the Tibagi River and its tributaries, and Itacolomi State Park (Minas Gerais).

In a world where clean fresh water is becoming increasingly scarce, we must make all possible efforts to protect water sources and restore native waterside vegetation. This should be a national priority: not only would it permit merganser populations to increase, but people would benefit, too.

PHOTO: F. COLOMBINI

Typical merganser habitat

Fewer than 250 mergansers survive.

The Araguaia River – one of the northern Cerrado's birding frontiers. PHOTO: AMAZON ARAGUAIA LODGE

PANTANAL & CERRADO
Many Places for Birding

National parks such as Emas and Serra da Canastra are superb for birding, but birders should also look for birds outside parks, especially in places where some wild habitat remains. With a rising interest in birding in Brazil and more information available on the Internet, new birding spots are constantly being discovered. In turn, ecolodges in new places are supporting the needs of birders.

Several localities recognize the business opportunities of nature tourism. A few fazendas in the Pantanal welcome guests. Nearby, atop the Serra da Bodoquena plateau, Bonito, Jardim, and Bodoquena are known for pristine clear waters. Bird sightings here include the Hyacinth Macaw, Fasciated Tiger Heron, Chaco Chachalaca, and Buff-bellied Hermit.

Birders are increasingly interested in sites in the northern Cerrado, an area particularly rich in bird species due to its mix of Cerrado and Amazonian vegetation. Much of this area is hardly explored, presenting exciting potential for discovery. The Araguaia River provides access points, with nearby fishing camps already accommodating birders.

Today travel is the world's largest business. Harboring so much wildlife, Brazil has huge untapped potential for nature tourism and can play a leadership role in developing sustainable sites that also generate local employment, benefit communities, and protect environments. Birders are often pioneers, finding the best areas for ecotourism.

PHOTO: D. DE GRANVILLE

Fish watching at Bonito

Too much wild Cerrado is being cleared for agriculture.

CERRADO
Trouble in Paradise

The Cerrado is Brazil's most endangered large ecosystem, more threatened than the Amazon. Unlike the rainforest, its destruction has gone mostly unnoticed.

Imagine a future without the wild Cerrado. No rheas would graze in golden grasslands. Seriemas would no longer greet the dawn with their wild cries. Macaws would have no places to nest and find food (already many forage along railroad tracks for soy spilled from trains). Migrating seedeaters and nomadic specialty "birds of fire" would have trouble locating the few remaining grasslands.

It is not too late to preserve some of what remains of central Brazil's wild lands. Protection requires more support for existing Cerrado parks, more support for actual land acquisition, and more local community awareness. Government, private landowners, and the agribusiness sector must act in tandem to combine development with the creation of conservation zones such as wild corridors and protected wetlands, in compliance with existing Brazilian law.

The Pantanal has witnessed the increase of several large native animals' populations in recent years but–during the same period–declining water quality, waste from ethanol plants, overgrazing, expansion of alien grasses, and *cordilheira* deforestation have become problematic. Proposed river diversion projects here will create inevitable and severe negative impacts on wildlife.

Living off spilled soybeans

Rheas also need some wild habitat.

PHOTO: D. DE GRANVILLE

1950 – Cerrado is abundant

2009 – Cerrado becomes fragmented

2030 – Cerrado almost gone

SOURCE: CONSERVATION INTERNATIONAL

Rapid protection needed now for the Cerrado

Formerly wild, today about 80% of the Cerrado has been cleared for large-scale agriculture. If recent rates of destruction continue, almost no native Cerrado habitat will exist in twenty years outside a few protected areas, all of them small. Less than 5% of the Cerrado presently receives protection. Protected areas are unconnected and are too isolated to insure survival of the Cerrado's rich biodiversity for the long term.

Ultimately, the decision to save the environment must come from the human heart.

From "Humanity and Ecology," © 1988, The Office of His Holiness the Dalai Lama

Too-frequent fires upset ecology.

Protection of national parks like Chapada dos Guimarães requires ongoing management and support.

PANTANAL & CERRADO
Support Parks

Parks protect biodiversity and maintain the myriad complex interactions between species, from microorganisms to great trees and vertebrates. Parks protect the purity of watersheds, preserve species of potential value to people, and provide sites for leisure opportunities. Parks inspire us by nurturing our sense of wonder and beauty; they remind us of our responsibilities as stewards of our planet to care for all forms of life. Parks are for everyone–poor and rich. They enhance the quality of our lives in an increasingly urbanized world.

Brazil's existing parks struggle with insufficient budgets, poor infrastructure, and the need for greater government support. To become more than just parks "on paper," they must connect with their greater constituencies, find innovative links with local communities, and explore new methods of protection. Some can't even afford staff to enable them to be open to the public (though visitors could generate needed revenue) and to train existing staff to enforce laws. Many parks even have unresolved land titles within their borders.

Central Brazil's parks–Emas, Serra da Canastra, Chapada dos Guimarães, Chapada das Mesas, Serra do Cipó, Chapada dos Veadeiros, Brasília, Pantanal, and Araguaia–should be among the most famous in the world, but they are not. They are filled with magnificent landscapes and abundant wildlife, including rare birds. But they need our support to fulfill their potential–a potential as yet unmet.

There is a desperate need to expand the boundaries of Emas National Park to protect the habitat of the Cone-billed Tanager. Unknown for seventy years after initial discovery in the 1930s, fewer than 250 Cone-billed Tanagers currently survive in only two places—one of them partially protected at the edge of Emas, the other at risk of being flooded by a hydroelectric scheme on the Juruena River, Mato Grosso.

PHOTO: J. QUENTAL

Chapada das Mesas National Park

Many places like the spectacular and biologically diverse Parnaíba River basin in Piauí are not legally protected.

PANTANAL & CERRADO
Protect More Land

Although the Pantanal comprises many habitats, it has only one national park. Only 4% of the Pantanal, and less than 5% of the Cerrado, is protected. Although many of Brazil's legally protected areas appear to be large, they are not large enough to support more than a few individuals of certain of their scarcer species. Emas National Park (at 1000km^2) appears to be enormous, but it is now so surrounded by agricultural lands that it may be inadequate to sustain a population of Crowned Solitary Eagles, Maned Wolves, Brazilian Tapirs, or Giant Armadillos.

Parks alone will not provide enough land to protect all of Brazil's species. Individual landowners can help, for they have the option of preserving some of their properties as private natural heritage reserves (RPPNs). The RPPN program plays a strategic role in Brazil's conservation efforts. Unused land is not useless land: where large enough, it can support a wide range of species and can help to connect Brazilian citizens to their natural heritage.

One way to expand habitat needed by wide-ranging birds and mammals is to create ecosystem corridors, mosaics of adjacent public and privately owned lands. Corridors that connect parks and other large wild reserves are especially valuable, enlarging protected areas and allowing species to roam over wider areas. More government incentives are needed to create partnerships that promote such conservation initiatives.

Unknown since its discovery in 1926, Kaempfer's Woodpecker was recently found again near Goiatins, Tocantins. Its habitat, dry forest mixed with bamboo, is threatened by agricultural and highway development. Perhaps only 250 birds exist. As of 2007, only 23 actual individuals were known.

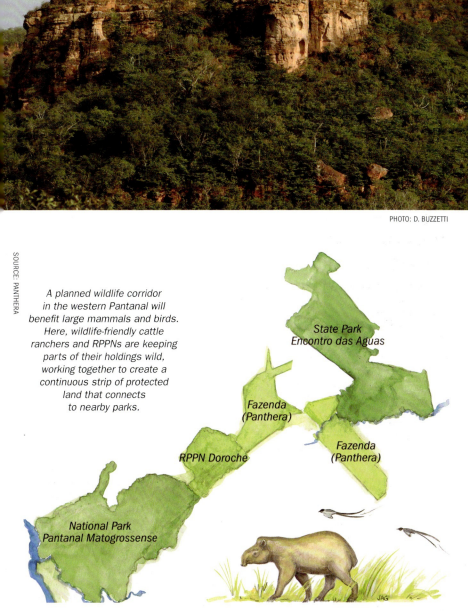

PHOTO: D. BUZZETTI

SOURCE: PANTHERA

A planned wildlife corridor in the western Pantanal will benefit large mammals and birds. Here, wildlife-friendly cattle ranchers and RPPNs are keeping parts of their holdings wild, working together to create a continuous strip of protected land that connects to nearby parks.

State Park Encontro das Aguas

Fazenda (Panthera)

Fazenda (Panthera)

RPPN Doroche

National Park Pantanal Matogrossense

Abundant birdlife and wild landscapes cannot be taken for granted. We must take action to preserve both.

CONSERVATION
A Priority for Brazil

In 1992, Brazil assumed a global leadership position at the United Nations Conference on Environment and Development in Rio de Janeiro and brought together world leaders to envision sustainable development—a pattern of land use that aims to meet human needs while preserving the environment for future generations. Brazil has sound environmental laws, but there is a need for more recognition and enforcement of the statutes to protect Brazil's watersheds and natural heritage.

If recent rates of agricultural expansion continue, the environmental health of central Brazil is at risk of irreversible damage. As global warming and poor agricultural practices create drying trends, water will become increasingly scarce and valuable. The unregulated flow of silt, chemicals, and sewage into streams and rivers already results in the contamination of water sources for towns and wild habitats downstream of pollution points. Native vegetation, which functions as a natural filter around wetlands and at headwaters, and which prevents erosion of steep slopes, is presently subject to legal protection. Government incentives are needed to raise public awareness of the importance of watershed protection and encourage its practice. Collaboration between the government and private sectors (especially agribusiness and cattle ranching) is key to the successful creation and maintenance of innovative land-use models in concert with more protection of remaining undeveloped or wild land.

PHOTO: D. BUZZETTI

The Cerrado's distinctive plant communities have evolved through time on low-nutrient, sandy soils and subject to the effects of natural fires. Natural vegetation can regenerate with time in cattle pastures but, once plowed and limed, soil chemistry changes and native seeds are lost, making it nearly impossible to restore native Cerrado vegetation.

PHOTO: F. COLOMBINI

The slogan *Think Globally, Act Locally* expresses the idea that the initiative for solving the world's complex and interrelated problems–environmental, economic, and cultural–begins at home. The concept rings as true for Brazil as elsewhere. Individuals can work toward the health and protection of the environment by making informed political and personal choices. Action at the local level is the first step. Learn about local, regional, and–by extension–national environmental problems. Become informed about the positions of government representatives and their political parties on such issues. Vote for candidates who promote protection of the environment by supporting incentives for landowners (such as tax advantages for the preservation of watersheds, special habitats, and ecosystems), endorse the creation and expansion of existing parks, and legislate for air- and water-quality protection. Join and become active in organizations that work in environmental education, restoration of damaged habitats, regulation of hunting and trapping of animals, and promote scientific research and conservation initiatives.

On a personal level, there are many opportunities to make choices according to ecologically sound principles and practices, from reducing consumption of water, electricity, and other resources to using public transportation, and choosing biodegradable products. Whether one's approach to conservation is political or personal, individual actions can contribute to the health of the planet.

Central Brazil's Crowned Solitary Eagles need large areas of wild habitat to survive.

SERIEMAS (Cariamidae) are large, long-legged, long-tailed birds of cerrado habitats, with raptor-like heads.

RED-LEGGED SERIEMA *Cariama cristata*   89-94 cm | 35-37"

Fairly common and widespread in cerrado, campos, and agricultural areas (where especially in large pastures). This *large, long-legged*, mostly terrestrial bird can adapt to disturbed conditions and has spread into some formerly forested areas. As it is not found in areas that flood regularly, very local in Pantanal. *Heavy hooked bill and long legs reddish; conspicuous long-feathered frontal crest often looks "messy."* Iris pale brown; one of few birds with eyelashes. *Brownish gray above* with faint darker vermiculations; *grayish white below* with narrow dark streaking; outer tail feathers broadly white-tipped. Often looks solidly brownish due to staining from dust and soil. In flight shows *boldly black-and-white banded flight feathers*. Unmistakable; a signature bird of the Cerrado. Usually seen singly or in pairs, it strides through open terrain in search of prey (large insects, rodents, lizards). Reputed to eat many snakes, but does not eat carrion. Sometimes makes quick flights to perch on fence posts and in low trees (where it nests), slipping to the ground if disturbed. Walks away to distance itself from disturbances, running if pressed. Often one of the first bird sounds heard at dawn, the seriema's far-carrying loud yelping call, given by both sexes, sometimes in a duet, is a long series of strident notes that gradually fade away: "kyup-kyúp-kyup-kyup-kyupkyup-kyup-kyo-kyo-kyo-kyo."

RHEAS (Rheidae) are ratites, an ancient flightless group that includes the ostriches, emus, and kiwis found in various areas in the Old World. The two rheas, whose fossil ancestors lived more than 40 million years ago, are South America's largest birds.

GREATER RHEA *Rhea americana*   145-160 cm | 57-63"

Locally fairly common in campos, cerrado, and open agricultural areas (mainly in large pastures), but only where not excessively persecuted and extirpated where human population density is high. By far our largest and heaviest bird; flightless. Very long legs grayish, unfeathered; iris pale. *Mostly gray*, somewhat paler below, with feathers of upperparts soft and floppy; *neck very long*, its lower part blackish in ♂, more or less uniform in considerably *smaller* ♀. Lacks tail. Unmistakable; huge. Often in small groups, but sometimes in larger ones composed mainly of younger birds. The rhea walks slowly, feeding as it goes on a variety of plants, insects, and small vertebrates. Rather wary, it tends to keep considerable distance between itself and observers, moving away almost imperceptibly as it is approached. Startled at close range, it lowers its neck and abruptly and rapidly zigzags away, raising its wings and shuffling its feathers. ♂♂ incubate and care for young alone, with a harem of several ♀♀ laying eggs in a large nest, usually well hidden in tall grass. Rarely vocalizes, but breeding ♂ does make a deep booming sound, sounding almost like the roar of a large mammal. One of the iconic birds of Brazil's Cerrado, the rhea is now sadly much reduced in overall numbers and range; a large population is found in the national park that carries its name, Emas NP.

RED-LEGGED SERIEMA
SERIEMA

GREATER RHEA
EMA

TINAMOUS (Tinamidae) are plump, *terrestrial* birds of an ancient Neotropical order, well represented in our area. Though *vocalizing* frequently, tinamous are usually *hard to see*, remaining in dense cover.

GRAY TINAMOU *Tinamus tao* — 43-46 cm | 17-18"
Uncommon and local in forest in *NW of our area*. Bill yellowish below; legs gray-blue. *Crown and neck blackish with freckled black-and-white throat and stripe down neck. Dark olivaceous gray above with blackish barring and vermiculations throughout*. Below paler grayish brown; crissum rufescent. The largest forest tinamou in our area; confusion unlikely. The Gray Tinamou is a rarely seen, solitary denizen of remote forests where it is terrestrial except while roosting at night on low branches. Not very vocal. ♂'s song, given at long intervals, an abrupt, slightly tremulous single hoot, "hooooo." Sometimes vocalizes at night.

CRYPTURELLUS tinamous are plump forest or woodland (*not* grassland) birds. All our species are *plain*, mostly brown and gray. Heard more than seen, with *easily recognized songs*. They rarely fly.

UNDULATED TINAMOU *Crypturellus undulatus* — 25.5-26.5 cm | 10-10.5"
Common and widespread in várzea, gallery and deciduous forest, woodland, and sometimes even coming out into shrubby clearings. The only numerous tinamou in Pantanal. Iris pale brown; bill grayish; legs olive gray. Crown dusky; above dull grayish brown with blackish vermiculations; below paler with some dark barring on flanks and crissum. *Lacks obvious field marks; usually identified by voice.* Little Tinamou is smaller, rich rufous below (not mostly gray). Though shy, Undulated is somewhat more easily seen than other forest-based tinamous. Song a distinctive series of whistled notes with characteristic cadence, "whooh, whoh-hoah?"

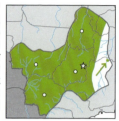

LITTLE TINAMOU *Crypturellus soui* — 21.5-23 cm | 8.5-9"
Locally common in *secondary habitats*, woodland, and overgrown clearings in *NW of our area*. *Small and plain*. Bill pale greenish to horn below; legs olive. *Crown blackish, sides of head grayish; rich rufous brown above, ochraceous-rufous below. Throat whitish*. Some birds are blacker above, duller below; ♀♀ average brighter. Small-billed Tinamou has a grayer breast, reddish bill. Extremely furtive, rarely emerging from dense thickets. Either freezes or, more often, scurries away when closely approached. Hard to see even where frequency of song indicates many are around. Primary song a series of clear, tremulous whistles, each slightly higher-pitched than the previous, gradually increasing in volume, ending abruptly. Also gives a slurred call that rises and falls, "pee-ee-ee yer-r-r."

Brown Tinamou (*C. obsoletus*) occurs locally in forest in SE of our area, e.g., at Serra da Canastra. Larger and even harder to see than Little Tinamou. Gray throat; obvious dark flank barring. Voice, an often heard series of 6-8 whistled notes, rises in pitch with a long pause after first note; it resembles a police whistle like the Tataupa's, but does not descend.

TATAUPA TINAMOU *Crypturellus tataupa* — 23-24 cm | 9-9.5"
Fairly common in semideciduous woodland and scrub in S and E of our area. Bill coral pink; legs dull purplish red. Head, neck, and breast gray, throat whitish. Above rufescent brown; belly pale grayish, flanks white with black scaling. Small-billed Tinamou is smaller, has dark-tipped bill, orangey eye (not dark). Little Tinamou is mainly rich rufous, has grayish legs, very different voice. Hard to see, Tataupa is often heard. Song a descending series of well-separated whistled notes (like police whistle), the first few longer, slightly rising, then gradually accelerating: "drreeyp? dreeéyp? dreey dreey dreey dri dri dri-dri-dri-dri-dru." Also gives a shorter version "dreeéyp? dreey-dri-dri-dri-dru." Small-billed's main song is much longer, more musical.

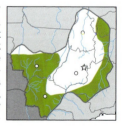

SMALL-BILLED TINAMOU *Crypturellus parvirostris* — 21-22 cm | 8.25-8.75"
Fairly common and widespread in shrubby, grassy areas. Bill reddish; iris orangey; legs pink. Similar plumage to Tataupa Tinamou but with dark-tipped bill. Shorter legs impart a "dumpier" impression. Both are very difficult to see, though frequently heard (sometimes together) in favorable habitats. Rarely can be seen scurrying across a road or feeding after dawn at the edge of a little-traveled road. Song, given mainly at dawn and dusk, starts with hesitant clear whistled notes on the same pitch (often with a long pause after the first), then accelerates and rises into a series of piping notes before it quickly descends into a series of "churrs." During the day gives a much shorter song of two long trilled or muffled notes, the second trailing away.

GRAY TINAMOU
AZULONA

UNDULATED TINAMOU
JAÓ

LITTLE TINAMOU
TURURIM

TATAUPA TINAMOU
INHAMBU-CHINTÃ

SMALL-BILLED TINAMOU
INHAMBU-CHORORÓ

DWARF TINAMOU *Taoniscus nanus*   14-15 cm | 5.5-6"

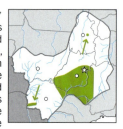

🇧🇷
VU *Very local* in grassy cerrado and *campo sujo*. Recorded now mainly in Emas NP and Brasília NP grasslands. Believed rare, this *tiny, very inconspicuous* bird may just be hard to find. Iris yellow; legs dull yellowish. Above rufous brown, barred and scaled whitish, paler below; *breast and flanks barred blackish*. In flight primaries uniformly blackish (*no paler barring* as in nothuras). Recognize this *smallest tinamou* by size, but beware other tinamous' chicks! Cf. Ocellated Crake, also very small. Seen singly and in well-separated pairs, almost always in tall grass. May hide and roost in armadillo holes. Very occasionally one will emerge from grass to feed at roadside soon after dawn. Distinctive song so insect-like that it is often overlooked, a succession of high-pitched cricket-like "trrree-tii" and "tii" notes with little pattern.

NOTHURAS are midsized, *mainly brownish* tinamous. *They sometimes flush when disturbed.* Widespread in grassy areas, but not found in Pantanal, likely due to its seasonal flooding.

LESSER NOTHURA *Nothura minor*   18-19 cm | 7-7.5"

VU *Rare and local* in cerrado and campo sujo in S of our region; mostly recorded at Emas and Serra da Canastra NPs, and in Brasília region. Seems to require *substantial* areas of unmodified habitat. Iris yellow; legs yellowish. *Above rufescent brown* with blackish-and-white scaling and chevrons; crown blacker; below pale buff, *flanks rufescent*. Resembles the more numerous, widespread, and larger Spotted Nothura. Lesser's smaller size is hard to judge, but it is more rufescent overall, especially on lower underparts (beware Spotteds stained by reddish soil). In flight the pale barring on Lesser's primaries is less evident. *Only Spotted ranges in agricultural areas.* Apparently often hides and escapes fires in armadillo holes. Song, given in late afternoon or at dusk, a series of very high-pitched piping notes at varying paces.

SPOTTED NOTHURA *Nothura maculosa*   24-26 cm | 9.5-10.25"

Fairly common in campos, cerrado, and pastures in S and E of our area; with agricultural expansion seems to be increasing and expanding range. Iris yellow; legs yellowish. *Above brown* with blackish-and-white scaling and chevrons; crown blacker. *Below buff* with blackish streaking and barring. Some are paler, others darker (sometimes merely stained by reddish soil). In flight primaries narrowly barred yellowish buff. A uniform brownish tinamou of open country, Spotted Nothura is the most often seen of its family (along with Red-winged Tinamou). Cf. much more local Lesser Nothura and (in NE of our area) White-bellied Nothura. Usually seen singly, often in fields or along roads; sometimes quite tame. Flushed birds burst up underfoot and fly off, scaling away on set wings after rapid wingbeats. Song a high-pitched musical trill, sometimes given by several birds at once. Singing birds may break into an accelerating series of well-enunciated piping notes that ends with several faster descending notes.

WHITE-BELLIED NOTHURA *Nothura boraquira*   27-28 cm | 10.5-11"

Uncommon in caatinga scrub and adjacent pastures in far NE of our area. Iris yellow; *legs bright yellow*. Above resembles Spotted Nothura. Below buffy whitish, foreneck streaked and sides and flanks barred blackish; *midbelly whitish*. Spotted is larger, markedly buffier below, with less brightly colored legs. Behavior similar; the two may occur together locally, though White-bellied prefers less altered terrain. Song a sweet "pseeu pii," or simply "pseee," repeated many times with a leisurely cadence, very different from Spotted's.

RED-WINGED TINAMOU *Rhynchotus rufescens*   40-42 cm | 15.75-16.5"

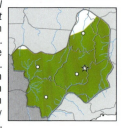

Widespread and often fairly common in campos, cerrado, and agricultural areas. Large, long-necked. Iris yellow; legs dusky-olive. *Head, neck, and breast pale rufous*; blackish crown and white throat; above brown with intricate pattern of black-and-white scaling. Lower underparts brownish, flanks scaled black. In flight shows *large area of rufous in outer primaries*. An elegant, attractive tinamou, Red-winged is *much larger than the nothuras with which it occurs*. Usually in tall grass, and heard much more often than seen, yet can be seen regularly on roadsides and in fields, especially in early morning. Prefers to run when alarmed; adept at hiding, even where vegetation is sparse; flushes with a roar of wings when pressed, often flying several hundred meters. The easily recognized song of the *perdiz* is one of the classic sounds of the Cerrado, a simple, far-carrying musical phrase: "plee-dip, deer, deer." Though given mainly at dawn and dusk, it can sometimes be heard throughout the day.

DWARF TINAMOU
INHAMBU-CARAPÉ

LESSER NOTHURA
CODORNA-MINEIRA

SPOTTED NOTHURA
CODORNA-COMUM

WHITE-BELLIED NOTHURA
CODORNA-DO-NORDESTE

RED-WINGED TINAMOU
PERDIZ

35

CHACHALACAS, GUANS, & CURRASOWS (Cracidae) are large birds, superficially pheasant-like but plainer. They range in forest or woodland. Most are arboreal, and many are heavily hunted.

CHACO CHACHALACA Ortalis canicollis   53-58 cm | 21-23"
Locally common in Pantanal in *deciduous and gallery woodland and nearby open areas*. Ranges both on the ground and in trees. *Bare face and dewlap pinkish*. Head and neck dull gray; above olive brown; tail dusky, *outer feathers broadly tipped rufous* (seen in flight). Breast brownish, paler on belly. *Dull-plumaged, the only chachalaca here*. Cf. the larger, more arboreal Chestnut-bellied Guan, often with it. When not persecuted can be very numerous, in flocks of 10-20 or more, and also become tame; where hunted much warier. Unmistakable *loud* raucous call, a repeated "cha-ka-ra-ká," is given by one bird, then in a chorus. Also gives various chortled or cackling alarm calls.

RUSTY-MARGINED GUAN Penelope superciliaris   61-69 cm | 24-27"
Uncommon and local in *more humid forest and woodland* (even where fragmented); largely absent from Pantanal. Bare face gray blue; red dewlap; legs pinkish. Above grayish brown; *conspicuous narrow cinnamon superciliary; wing feathers with narrow rufous edging*. Below gray brown, belly paler, breast with distinct whitish chevrons. Cf. Chestnut-bellied Guan. Found in pairs, less often small groups. Mostly arboreal; hops between branches and walks along larger horizontal limbs. Numbers reduced by hunting and forest clearance, but can survive in largely secondary habitats. Feeds on fruit of many tree species. Gives a variety of harsh honking calls.

Spix's Guan (*P. jacquacu*) occurs in forest in NW of our area, e.g., Serra das Araras. Larger than Rusty-margined, with more rufescent belly. Lacks rufous wing edging.

White-browed Guan (*P. jacucaca*) occurs in deciduous woodland around Peruaçu, N Minas Gerais. Also resembles Rusty-margined but has *prominent narrow white brow*, no rufous edging on wings.

CHESTNUT-BELLIED GUAN Penelope ochrogaster   71-76 cm | 28-30"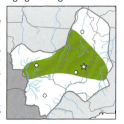
Uncommon and local in *deciduous forest, woodland, and gallery forest in N Pantanal and Rio Araguaia drainage*; formerly extended S to S Goiás and N Minas Gerais. *Highly colored* with bare face dusky, red dewlap. Above brown, more rufous on head and neck; *prominent silvery white brow*. Below rich rufous; breast feathers with narrow white chevrons. Rusty-margined lacks prominent brow, is much duller and darker below, favors more humid habitats; not known to range with Chestnut-bellied. Arboreal behavior similar; Chestnut-bellied is a more beautiful, elegant bird. Sometimes seen eating flowers of *Tabebuia* trees with Chaco Chachalacas and piping guans.

White-crested Guan (*P. pileata*), with *obvious white in crest* and *rich rufous foreparts*, is another forest-inhabiting guan, known from N end of Ilha do Bananal.

COMMON PIPING GUAN Pipile pipile   66-71 cm | 26-28"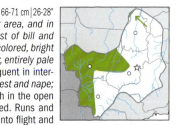
Fairly common in forest and gallery woodland in NW of our area, and in deciduous forest and woodland in S (mainly Pantanal). Most of bill and bare skin around eye pale blue; legs coral red. In N dewlap bicolored, bright red with dark blue chin; in S (illustrated), dewlap much longer, entirely pale blue. Sometimes considered two species but intergrades frequent in intermediate areas, e.g., N Pantanal. Mainly black. Shaggy white crest and nape; conspicuous *white patch on wing-coverts*. Small groups perch in the open especially in early morning. Tame where not overly persecuted. Runs and hops with agility between branches, when disturbed bursting into flight and then gliding in gradual descent. "Piping" is a series of 6-8 clear piercing whistled notes that gradually rise in pitch. Birds in display launch from perch with sharp wing-claps, then wing-rattle during long gliding flight.

BARE-FACED CURASSOW Crax fasciolata   86-94 cm | 34-37"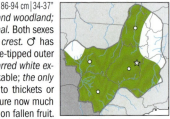
Rare to locally fairly common in *deciduous and gallery forest, and woodland; numbers reduced due to hunting, now most numerous in Pantanal*. Both sexes have *long recurved crown feathers forming expressive bushy crest*. ♂ has cere and base of bill bright yellow. All black with white belly, white-tipped outer tail feathers. ♀ has bicolored crest. Blackish above narrowly barred white except on neck; tail white-tipped; lower underparts buff. Unmistakable; *the only currasow here*. Mainly terrestrial, but disturbed birds retreat to thickets or tree limbs. Shy in most areas, but due to reduced hunting pressure now much tamer along Transpantaneira. Pairs or small groups feed mainly on fallen fruit. ♂'s song a deep low booming, "oom, oom, boo-boóp, boo-boo-boó," so low-pitched as to be "felt" than heard. Nervous birds give various squealing calls.

CHACO CHACHALACA
ARACUÃ-DO-PANTANAL

RUSTY-MARGINED GUAN
JACUPEMBA

CHESTNUT-BELLIED GUAN
JACU-DE-BARRIGA-CASTANHA

COMMON PIPING GUAN
CUJUBI

BARE-FACED CURASSOW
MUTUM-DE-PENACHO

HOATZIN (Opisthocomus) is a unique, almost prehistoric-looking bird found near water across Amazonia.

HOATZIN *Opisthocomus hoazin* 61-68.5 cm | 24-27"

Locally numerous in shrubbery and trees around lakes and along rivers in NW of our area (Ilha do Bananal). A *bizarre-looking*, superficially cracid-like bird. Unique and ungainly. The sole member of its family, the Hoatzin seems to be allied to the Cuckoos. Unmistakable. *Extensive bare facial skin bright blue. Iris red. Loose shaggy crest of long stiff feathers rufous.* Dark olive brown above streaked whitish on neck and back; tail dusky-brown *broadly tipped pale buff.* Bend of wing and conspicuous covert tipping white; *primaries mainly rufous. Below buff*, becoming rufous on flanks and lower belly. Hoatzins are found in sluggish, sedentary groups that rest in the open during the early morning but then retreat into shade during mid-day. Often tame, they allow a close approach as they peer about with an almost dazed expression, clumsily shuffling and often crashing into vegetation if flushed. They often hold their wings partially outstretched, as if for balance. Flight weak, never long sustained, just a few flaps followed by glide. Threatened nestlings drop out of the nest, plop into the water, and seem to disappear, eventually clambering out using vestigial claws on wings. Hoatzins feed mainly on young leaves, digested in its large crop (almost like a cow!). Often noisy, Hoatzins give voice to near-constant wheezy and nasal "whaaah" or "rruh" calls.

CORMORANTS & SHAGS (Phalacrocoracidae) and **DARTERS** (Anhingidae) represent two families of superficially duck-like aquatic birds that feed on fish; they differ in bill shape, etc.

NEOTROPIC CORMORANT *Phalacrocorax brasilianus* 63.5-68.5 cm | 23-27"

Widespread and locally common along rivers, on ponds and lakes; most numerous in Pantanal. Bill long, slender, and hooked at tip; gular pouch and facial skin dull yellow to orange (brighter when breeding), outlined narrowly by white. Breeding adult *black*, somewhat browner above, with a few white plumes on head and neck. Nonbreeders are duller black, lacking facial plumes, with less noticeable white border to facial skin. Immature grayish to dusky brown above, paler grayish brown below. Juvenile almost white below and on sides of head and neck. Not likely confused as it is *our only cormorant.* Anhinga has a longer and more pointed bill, longer and thinner neck and tail, and silvery white on wing. When swimming can look vaguely duck-like (cf. Muscovy). Notably gregarious, cormorants can occur almost anywhere fishes swim, and seem markedly nomadic. They swim low, sometimes with just head and neck protruding, bill tilted up jauntily; feeding birds dive frequently. As their plumage is not waterproof, cormorants are often seen perched on snags or trees with their wings outstretched to dry. Flight is strong and steady, often quite high in lines or a V-formation, neck held slightly kinked. Generally silent, but perched birds sometimes emit low-pitched grunting sounds, "urh-urh-urh," especially when nervous.

ANHINGA *Anhinga anhinga* 82.5-89 cm | 32.5-35"

Uncommon along rivers and around freshwater lakes and ponds. Distinctive shape: *small head, long slender neck,* and *long fan-shaped tail. Sharply pointed bill* yellow to orange, culmen duskier; facial skin bluish (brighter when breeding). ♂ *glossy black* with contrasting *upperwing-coverts silvery white,* some feathers edged black; tail narrowly tipped brown. Breeding birds have scattered white plumes on head and neck. ♀ similar but has *head, neck, and chest grayish buff.* Immature brown-backed with reduced white on wing. Though vaguely cormorant-like, the Anhinga should not be confused. Anhingas are much less gregarious than cormorants, often perching alone in trees or bushes along the water's edge, sometimes with wings outstretched to dry their feathers. Though its flight may seem more labored than a cormorant's, the Anhinga can also soar high, with wings outstretched and tail broadly fanned. They swim low in the water, moving ahead sinuously, often with only the head and neck protruding. Anhingas dive or simply submerge when foraging for food, skewering small fish with their bills, bringing prey to the surface before swallowing.

HOATZIN
CIGANA

NEOTROPIC CORMORANT
BIGUÁ

ANHINGA
BIGUATINGA

SCREAMERS (Anhimidae) are *very large, rather goose-like,* mainly terrestrial birds found near water. Our two species are not usually found together. In spite of their substantial weight, they can walk with long unwebbed toes on floating vegetation. A sharp spur protrudes from the bend of wing.

SOUTHERN SCREAMER *Chauna torquata*   84-91.5 cm | 33-36"

Fairly common and conspicuous in marshes and edges of lakes and ponds in Pantanal (where most numerous southward). Unmistakable. Facial skin, heavy legs coral pink. Mostly gray with a *curled crest and black-and-white neck "choker;"* mantle darker. In flight shows *large silvery white patch on wing-coverts* (not evident when perched). Pairs stand alertly, necks erect, in marsh vegetation or on low trees. Gathers in flocks when not breeding; family parties often seen Sep-Dec. Seems tamer than Horned Screamer. A noisy bird, giving a variety of loud, vaguely musical calls, many a variant of "chajá" (a local name), this call often given in flight. Birds at rest sometimes repeatedly give a vigorous "cheeree-ha, ha." Vocalizes both day and night.

HORNED SCREAMER *Anhima cornuta*   84-91.5 cm | 33-36"

Uncommon and local in marshy areas and around lakes; more numerous in N (Ilha do Bananal, Rio Vermelho E of Cuiabá); *scarce or absent in Pantanal.* Unmistakable. Beady iris yellow; heavy legs gray. *Mainly glossy black, feathers of crown and foreneck broadly edged silvery whitish;* belly white. In flight shows a *large silvery white patch on wing-coverts* (marginally evident on perched birds); wings long and broad, underwing-coverts white. A *long slender whitish "horn"* (actually a feather shaft) springs from forehead. Pairs stand around on grassy areas near water or in low trees; where numerous, can be more social. Often wary, they flush ponderously and struggle to get airborne, then alight atop a distant shrub or tree. Herbivorous, grazing on leaves and roots of aquatic plants. Sometimes soars high, then looking like a huge vulture. Vocal, with a powerful voice (not a "scream") audible for several kilometers. Call a deep, throaty "guu-uulp, güü" usually uttered by perched birds, sometimes given in series and interspersed with loud honking and gulping notes. The vernacular name "mahooka" is a good mnemonic.

DUCKS (Anatidae) are familiar swimming birds with webbed feet and dense waterproof plumage that range widely in marshes and around ponds and lakes.

MUSCOVY DUCK *Cairina moschata*   ♂ 76-84 cm | 30-33" ♀ 71-76 cm | 28-30"

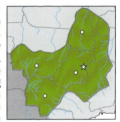

Widespread but generally uncommon around rivers and lakes; common in Pantanal. Bill pinkish with black mottling. ♂ has red caruncles on bill; blackish facial skin. *Mainly glossy greenish black,* with slight expressive bushy crest. In flight shows *large white patch on wing-coverts* (often hidden at rest). ♀ markedly smaller, lacking crest; they show less white. Older birds have more white. Juveniles dark brown to blackish with no gloss; white in wing reduced or absent. Nearly unmistakable, though juvenile might be mistaken for cormorant. Cf. also Comb Duck. Domesticated Muscovy Ducks (which sometimes range far from houses) almost always show scattered white feathers in their plumage, a pattern not seen in wild birds. Unlike their ungainly domestic brethren, wild Muscovies are elegant, alert, and wary, flying heavily but strongly. They graze on short grass, dabble in shallow water, and rarely associate with other waterfowl. Muscovies perch freely in trees, and also roost there. Nests are usually in tree cavities. Surprisingly quiet, they occasionally hiss and quack softly.

COMB DUCK *Sarkidiornis melanotos*   ♂ 68.5-71 cm | 27-28" ♀ 53.5-56 cm | 21-22"

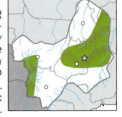

Rare to uncommon, seemingly erratic, in marshes, rice fields, and along rivers. In our area occurs mostly in Pantanal (where much less numerous than Muscovy), with smaller numbers in NE (wandering birds?). ♂ has *large fleshy knob over bill* (smaller after breeding). Head, neck, and underparts white with black head and neck speckling; sides broadly blackish, crissum yellow tinged, upperparts black glossed green and purple. In flight *wings all dark.* ♀ markedly smaller, *lacks bill knob.* Sides scalloped grayish, whitish crissum. Muscovy Duck lacks white on head, neck, underparts. Cf. also Neotropic Cormorant. Usually seen in small groups, at times with whistling ducks. Can be active at night; generally wary. Sometimes perches in trees; usually nests in tree hollows.

Coscoroba Swan (*Coscoroba coscoroba*) occurs as a vagrant in S Pantanal, N to around Porto Jofre (mainly Sep-Oct), on lakes and in open marshy areas. Unmistakable. A *white goose-like bird* with a *long slender often arched neck, black wing-tips, bright pink bill and legs.*

SOUTHERN SCREAMER
TACHÃ

HORNED SCREAMER
ANHUMA

MUSCOVY DUCK
PATO-DO-MATO

COMB DUCK
PATO-DE-CRISTA

WHISTLING DUCKS are handsomely patterned, long-necked, rather long-legged ducks found widely in open marshy areas. Noisy and gregarious, they are more active at night than other ducks.

WHITE-FACED WHISTLING DUCK *Dendrocygna viduata*   43-48 cm | 17-19"
Widespread and generally common in marshes and around ponds and lakes. Favors open terrain more than Black-bellied Whistling Duck. Bill and legs blackish. *Face and foreneck white (often stained brown from feeding) contrasting with rest of black head and neck*; brown above, feathers buff edged with streaked effect. Lower neck and breast rich rufous-chestnut; broad area on flanks narrowly barred blackish and white; midbelly black. In flight shows *dark wings and tail* (shoulders maroon-chestnut). The white face is obvious (discernable even when stained). No white in wings (contra Black-bellied) or on tail (contra Fulvous). Usually found in groups, looking about alertly, frequently standing at water's edge; sometimes joins other whistling ducks. Feeds in wet marshy areas and shallow water. Most frequent call a shrill "whee-see-see," often given in flight.

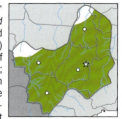

BLACK-BELLIED WHISTLING DUCK *Dendrocygna autumnalis* 46-51 cm | 18-20"
Locally common in freshwater marshes, ponds, rice fields, and flooded pastures. Most numerous in N of our area and Pantanal. Bill and legs pinkish red. Face and upper neck gray with crown and hind-neck stripe rufous brown; chest tawny-brown, breast grayish; *belly black*, crissum mottled white. Above rufous brown, wing-coverts whitish, tail black. In flight shows a *broad and conspicuous white wingstripe*, barely visible at rest. Juvenile has gray bill and legs and a pallid version of adult's pattern. The only whistling duck with reddish bill and legs and a white wingstripe. Favors open marshy areas; flocks often stand at water's edge. Sometimes calls overhead at night. Flocks can number in the hundreds, even thousands, and regularly consort with other whistling ducks. Perches freely in trees, especially on dead branches. Usually nests in tree holes. In flight has a distinctive "droopy" silhouette, with head held at a downward angle, legs dangling below tail. Call consists of high-pitched whistles: "wi-chi-tee" or "wit-chee, wit-chee-chee."

FULVOUS WHISTLING DUCK *Dendrocygna bicolor*   48-53.5 cm | 19-21"
Uncommon in marshes and around ponds and lakes in Pantanal; occasional wanderers elsewhere. The *least numerous whistling duck in our area*. Bill and legs bluish gray. *Head, neck, and underparts rich tawny-fulvous* with diagonal whitish furrows on sides of neck. *Creamy whitish tips to elongated flank feathers* (often gives effect of a broad side stripe); crissum creamy white. Above blackish, feathers tawny edged giving a barred effect. In flight wings look dark (chestnut and black). *Conspicuous white "ring" above tail*. Not likely confused; the most uniformly colored whistling duck. Behavior as in more numerous Black-bellied, but Fulvous rarely perches in trees and is particularly fond of rice fields. The "droopy" shape in flight is similar. Call a rather shrill whistle, "ki-wheeah," characteristically 2-noted, most often given in flight.

ORINOCO GOOSE *Neochen jubata*   56-63.5 cm | 22-25"
Now rare and local on sandbars along major rivers in N of our area (e.g., Ilha do Bananal), with wanderers further S (a rare vagrant even in Pantanal). A *large, goose-like* waterbird; bill red below; *legs reddish pink*. Head, heavy neck, and breast pale grayish buff, neck feathers often ruffled; *belly contrastingly rufous*. Back, rump, and tail black with purplish gloss, crissum white. *Wings mostly dark green*. ♀ smaller. Juvenile duller. Unmistakable. No other goose-like bird is found on river sandbars. Usually in well-dispersed pairs (sometimes small groups) that walk with an elegant upright carriage on sandbars and around associated lagoons. Unlike most waterfowl, Orinoco Geese engage in no courtship display, instead ♂♂ fight vigorously with others before pairs form up. Mainly herbivorous, grazing on short grass. Nests in tree cavities. Gives a distinctive nasal honking "unnhh?" and also a high-pitched whistled note.

WHITE-FACED WHISTLING DUCK
IRERÊ

BLACK-BELLIED WHISTLING DUCK
MARRECA-CABOCLA

FULVOUS WHISTLING DUCK
MARRECA-CANELEIRA

ORINOCO GOOSE
PATO-CORREDOR

BRAZILIAN TEAL *Amazonetta brasiliensis*   38-40.5 cm | 15-16"
Fairly common and widespread in shallow ponds and marshes. The most numerous duck in our area. Legs bright red (both sexes). ♂'s bill red, ♀'s gray. ♂ brown with contrasting *whitish rear cheeks and upper neck*; tawnier breast, dark spotting on flanks. ♀ has two *large white eye spots, whitish lower cheeks and sides of neck*. In flight both sexes have *large area of iridescent green on coverts and a white triangle on secondaries*. Pale brown patch above tail contrasts with blackish tail. This relatively long-necked duck is unlikely to be confused. It occurs widely in small groups of up to a dozen or so birds, among which there are invariably a few red-billed ♂♂. Its beautiful and flashy wing-pattern is unique. Sometimes loafs on open shorelines with whistling ducks, but usually maintains its own flock integrity. Feeds mostly by dabbling in shallow water. Flight fast, low, and direct. Oft-repeated call, given mainly in flight, a high-pitched whistle: "sweeu" or "swu-eét."

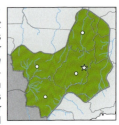

RINGED TEAL *Callonetta leucophrys*   36-37 cm | 14.25-14.5"
Vagrant to *ponds and marshes in our area*; a few records from Pantanal. Bill pale blue, legs pink. ♂ has *face and upper neck creamy buff outlined narrowly by dark crown and hindneck*; back reddish chestnut, *breast pinkish buff dotted black*, flanks finely vermiculated gray and white, crissum white. ♀ has *complex pattern of brown and white on face*, brown upperparts, breast and flanks irregularly barred brown and whitish. In flight both sexes have dark wings with an iridescent green speculum and a *unique white oval patch on central wing*. A lovely, delicate duck, unlikely to be confused (but flying Brazilian Teal also show a green-and-white wing pattern). Usually occurs in pairs, at most in groups of 6-8 birds. Swims in ponds and loafs at water's edge, sometimes perching in trees. Call a strange nasal "ko-weah."

WHITE-CHEEKED PINTAIL *Anas bahamensis*   44-47 cm | 17.25-18.5"
Vagrant to *ponds and lakes*. Bill blue gray with *conspicuous coral red base* (smaller and paler in ♀). ♂ has *face and foreneck snowy white contrasting with brown crown and hindneck*; blackish above with feathers edged tawny-buff; *pointed tail buffy whitish*. Below buff spotted black. ♀ similar but duller, especially on head; tail shorter. In flight wings show a green speculum bordered broadly by cinnamon. Should not be confused: no other duck combines red on bill with white face. In its usual range this elegant duck occurs in groups that swim and loaf on muddy or sandy shorelines, but here it as singletons with other ducks. They dabble and tip-up for food, mostly seeds.

SOUTHERN POCHARD *Netta erythrophthalma*   46-48 cm | 18-19"
Rare and very local on ponds and lakes. Recorded in our area from Brasília NP, where a flock of several hundred wary birds seemed to be resident on the reservoir from the 1970s to around 1990, presumably breeding there, but may not occur at present. ♂ iris red, ♀ brown. Both sexes have bill blue gray with black tip. Very dark ♂ has *head, neck, and breast glossy blackish chestnut*; otherwise dark brown, more chestnut on sides. ♀ mainly brown (not as dark), with *striking white facial pattern* (white foreface, throat, and stripe arching from cheeks to behind eye). Few other ducks are as dark as ♂; ♀'s face pattern unique. In flight both sexes show a *conspicuous white wingstripe*. At great distances the white wingstripe might recall a Black-bellied Whistling Duck but these species are utterly different in profile, the pochard being compact with a shorter neck and legs. The status of this rare duck remains poorly known in our part of Brazil. It feeds by diving and by dabbling on the surface of shallow water. Flight swift and direct, taking off after pattering along surface.

Rosy-billed Pochard (*Netta peposaca*) occurs as a vagrant to ponds and marshes (one record from our area at Serra das Araras). ♂ unmistakable, with *bright rosy pink bill* and knob at base of maxilla; head, neck, breast, and upperparts black; flanks vermiculated gray and white, crissum white. ♀ has blue-gray bill, is brown (darker above) with whitish eye-ring and postocular line, white throat. Both sexes have a bold white wingstripe.

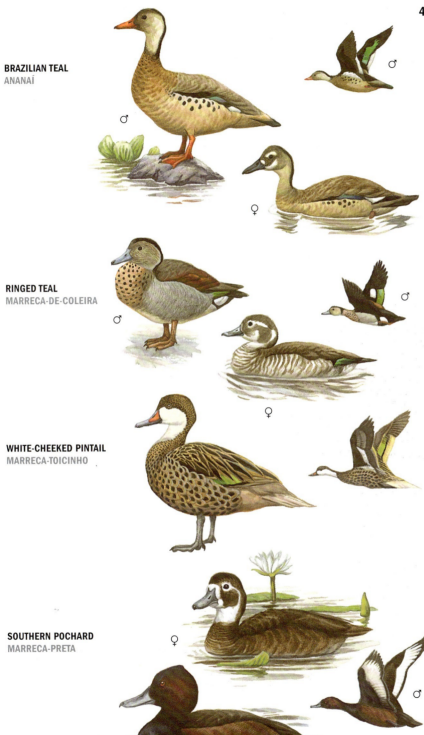

BRAZILIAN TEAL
ANANAÍ

RINGED TEAL
MARRECA-DE-COLEIRA

WHITE-CHEEKED PINTAIL
MARRECA-TOICINHO

SOUTHERN POCHARD
MARRECA-PRETA

BRAZILIAN MERGANSER *Mergus octosetaceus* 50-54 cm | 19.25-21.25"

CR *Extremely rare and local along clear streams and smaller rivers usually with at least a fringe of gallery forest.* Found only at few locations (Serra da Canastra and Chapada dos Veadeiros NPs; Jalapão area in Tocantins). Presumably once more widespread, but may never have been numerous. *Long slender serrated bill black;* legs pink. ♂ has *head, neck, and long wispy crest black with iridescent green sheen;* above blackish, below vermiculated dark gray and whitish. ♀ similar but browner above, crest shorter. In flight both sexes show *large "doubled" white patch on secondaries* (a black line separates the two white areas). South America's only merganser is hard to mistake. Neotropic Cormorant can occur on the same fast-flowing streams; it also has a long slender bill, similar silhouette. Now one of the world's rarest waterfowl, this merganser's decline is attributed to the development and siltation of watercourses across the vast majority of its former range. It persists in minuscule numbers at only a few sites, each with only a few pairs. Pairs or small family groups swim (easily, even in strong currents), rest on protruding rocks, and dive for small fish and invertebrate food. Flight swift and direct, rather strictly following watercourses. Usually nests in tree holes. Lack of enough suitable breeding sites may also be a critical limiting factor to population size.

MASKED DUCK *Nomonyx dominicus* 33-35 cm | 13-14"
Very local on marshy overgrown ponds and shallow lakes. A chunky, heavy-set duck with short thick neck and stiff tail held on surface of water (sometimes cocked up by ♂ in display). ♂ has *bill bright blue;* ♀'s bill dusky gray. Breeding ♂ *rufous-chestnut* with contrasting *black face; back spotted with black,* below pale buffy. ♀ and nonbreeding ♂ have *face buff crossed by two prominent dark horizontal stripes;* above blackish spotted with buff. Below tawny-buff, whiter midbelly, breast spotted blackish. Not often seen in flight, both sexes then showing a *square white patch on secondaries.* This rather secretive duck favors ponds (sometimes surprisingly small) with abundant emergent and floating vegetation and is adept at hiding; early in the day they are sometimes more in the open. Swims low in the water, not associating with other ducks, though sometimes with grebes. Feeds mainly by diving, also by simply sinking, grebe-like, beneath the surface.

GREBES (Podicipedidae) are small duck-like diving birds with lobed toes that are found on lakes and ponds. They rarely fly, and come to land only to nest.

LEAST GREBE *Tachybaptus dominicus* 21.5-23 cm | 8.5-9"
Uncommon and local on ponds and lakes. A small gray grebe with *a golden yellow iris* and *slender blackish bill* (looks slightly upturned). Adult has *blackish crown, slaty gray face and neck* (with black throat when breeding). Blacker above, grayer below tinged cinnamon-brown. When seen in flight (not often) shows a white wingstripe. Immature similar but eye less bright, bill paler; throat whitish; juveniles have whitish head striping. Pied-billed Grebe, often with this species, is larger and browner overall with a heavier pale bill. Pairs or small groups frequent quiet ponds and lakes, mostly near fringing woodland. Nesting pairs sometimes move to small and ephemeral ponds, departing as water levels drop. Nests are placed on floating rafts of vegetation attached to plants. Parents carry downy young on their backs. They dive for aquatic insects and small fish; they can slowly sink from sight without a ripple. Usually quiet, but breeding birds give soft purring calls and a descending churring.

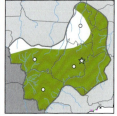

PIED-BILLED GREBE *Podilymbus podiceps* 28-33 cm | 11-13"
Rare and surprisingly local on ponds and lakes. Thick bill chalky grayish white *ringed conspicuously with black when breeding;* dark eye with eye-ring whitish. Adult *grayish brown,* darker and browner above, paler below with gray face and neck and *black throat;* midbreast, belly, and crissum white. In flight wings dark. Nonbreeding adult and immature have head and neck more cinnamon brown, throat whitish. Very young birds show whitish head striping. Least Grebe is smaller with finer bill, yellow eye. Cf. juvenile Common Gallinule and ♀ Masked Duck. Pied-billed Grebe is more sedentary than Least, and flies even less often; it favors deeper, more open water, though both species are often together. Breeding birds give a series of hollow notes, "cuk-cuk-cuk-cuk, cow, cow, cow," also various long-continued clucks and whinnies.

BRAZILIAN MERGANSER
PATO-MERGULHÃO

MASKED DUCK
MARRECA-DE-BICO-ROXO

LEAST GREBE
MERGULHÃO-PEQUENO

PIED-BILLED GREBE
MERGULHÃO

HERONS (Ardeidae) are *long-legged, long-necked wading* birds with long pointed bills. They feed mainly on fish and small vertebrates obtained in shallow water.

LEAST BITTERN *Ixobrychus exilis* 26.5-30 cm | 10.5-11.75"

Uncommon and local in marshes, mainly in Pantanal (but inconspicuous and overlooked). *Boldly patterned;* tiny. Iris yellow. ♂ has *black crown* with *ochraceous buff face, neck, and underparts; nape and upper back rich rufous.* Above *black with two thin buff stripes down back; wing coverts ochraceous-buff and rufous.* ♀ *dark brown* with grayer neck, below lightly streaked brown. In flight both sexes have *pale wing-coverts,* contrasting black primaries. Striated Heron is larger and stockier, has all-dark wings. A secretive skulker, Least Bittern climbs in reeds and feeds by stalking. When spotted in the open, it "freezes" with neck outstretched. Flushed birds fly low, and quickly pitch back into cover. Distinctive call a low-pitched, hollow "wohh" repeated steadily and slowly.

A very rare morph (in N America called **Cory's Least Bittern**) has been seen in the Pantanal. It is patterned like a normal bird but *chestnut replaces buff on underparts and wing-coverts.* Cory's also lacks the pale back stripes; bill blackish.

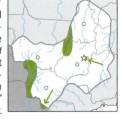

ZIGZAG HERON *Zebrilus undulatus* 30.5-33 cm | 12-13"

Uncommon and very local in undergrowth by sluggish streams and around edges of oxbow lakes in N (e.g., Ilha do Bananal) *and N Pantanal.* The Zigzag's "scarcity" likely results mainly from its inconspicuous behavior. Iris yellow. Adult *looks very dark, shaggy-crested; blackish above with wavy narrow buff barring;* primaries black. Below buffy whitish, densely patterned with blackish. Immature has *rufous-chestnut forecrown, face, and sides of neck;* below buff with sparse blackish streaking and vermiculations. Cf. larger, stockier Striated Heron. Solitary birds, Zigzags stay in heavy cover and usually forage from land near water. Rarely located except when calling, a far-carrying (but not loud), hollow "hhoow-oo" repeated at 4-6 second intervals for long periods; it can resemble call of Gray-fronted Dove. Vocalizes from low perches, mainly at dusk.

STRIATED HERON *Butorides striatus* 38-43 cm | 15-17"

Common and widespread near water and in marshes. A small, *chunky* heron *with short yellowish legs* (orangey when breeding.) Adult has *crown and shaggy crest black;* above greenish black with wing-coverts edged buff. *Sides of head, neck, and breast gray* with throat stripe white bordered brown; belly pale grayish. Immature has crown rufous-streaked, browner upperparts, brown-striped underparts. Familiar and easily recognized (but cf. Zigzag Heron). Essentially solitary, though where numerous (as in Pantanal) it may seem to be in groups. Generally hunts from a perch over water, not actually wading; takes prey with quick short bill jabs. Flicks tail sideways when nervous and crest often raised. Usually nests alone, at most in small groups, generally not with other herons. Flushed birds give a characteristic abrupt and complaining "kyow!"

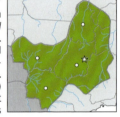

BOAT-BILLED HERON *Cochlearius cochlearius* 48-53.5 cm | 19-21"

Uncommon and local (mostly in N of our area and Pantanal). A stocky, *nocturnal,* inconspicuous heron with *very broad and heavy bill.* Adult has *crown and long wide crest black. Pearly gray above; face, forehead, and breast white;* belly rufous-buff and black. Immature has shorter crest and is browner above, buffy below. Black-crowned Night Heron lacks the shovel-like bill and looks less "white" at night. By day, roosts in thick foliage near water, often in groups, reluctant to flush. Disperses to feed alone at night, when wary. Rarely with other herons. Feeds much like other herons while standing in shallow water; also scoops with wide bill. Flushed birds give a low-pitched "qua" or "kwa" in flight (a bit duck-like).

BLACK-CROWNED NIGHT HERON *Nycticorax nycticorax* 60 cm | 23.5"

Locally common around lakes, ponds, and rivers, especially in Pantanal. *Stocky, hunched silhouette.* Stout black bill; iris red. Adult has *crown and back glossy black;* two long white occipital plumes; face, forehead, and underparts white; *wings pale gray.* Immature has yellowish iris, lower bill greenish yellow. *Above dark brown with prominent buffy white streaks, wing spots;* below streaked grayish brown. Brownish subadult echoes adult pattern (cf. Boat-billed Heron). The streaky young birds do not resemble other herons here (but cf. rare Pinnated Bittern). True to its name, this heron is most active at night, flying out from roosts at dusk to feed. When feeding young, sometimes forages by day. Call an abrupt "quok" or "wok," often given in flight.

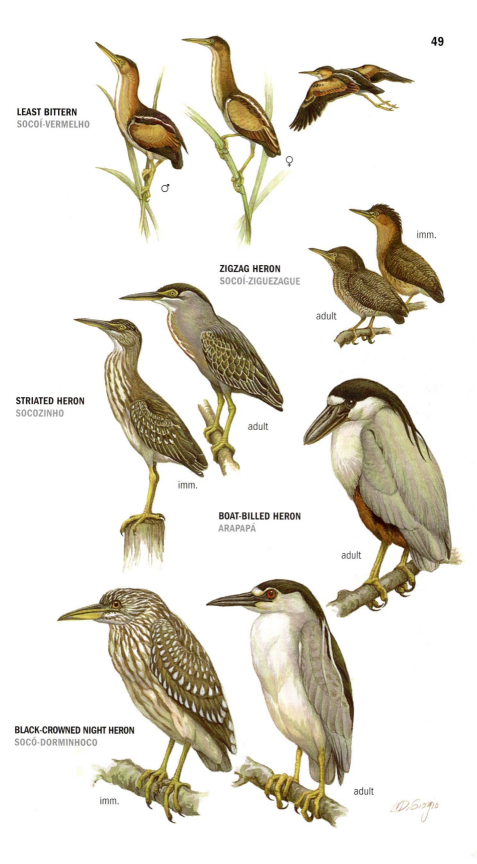

TIGER HERONS are solitary, heavy-bodied herons with subtly intricate plumage patterns. Young birds have bold "tiger" markings.

RUFESCENT TIGER HERON *Tigrisoma lineatum*   66-76 cm | 26-30"

Widespread and locally common in marshes, around ponds and lakes, and along forested rivers and streams; most numerous in Pantanal. Adult has *head and neck rich rufous-chestnut; above dusky finely vermiculated buff.* Bare skin around gape yellow; throat and foreneck white with dark median stripe; belly buff. Immature very different: *rufous-cinnamon with prominent black neck chevrons, bold banding across wings;* above boldly spotted buff and cinnamon; below paler with dark bars and chevrons. Tail has four narrow white bands. Attains full adult plumage over several years, gradually losing the barring and spotting of younger birds. Cf. much rarer Pinnated Bittern. Very similar immature Fasciated Tiger Heron is in different habitat. Stands still with neck hunched for long periods at water's edge or shallows, awaiting unwary prey. Allows a close approach, especially where used to people (as in Pantanal). Perches freely in trees. Mainly vocalizes at night giving a fast series of groaning notes, "honh-honh-honh-honh...," often ending in a more protracted "honnhhh-hh." Flushed birds often fly off, while giving a complaining "gwok."

FASCIATED TIGER HERON *Tigrisoma fasciatum*   61-66 cm | 24-26"

EN *Rare and very local along fast-flowing rocky rivers and streams in hilly regions* (but seems inexplicably absent from many seemingly suitable areas). Iris and bare skin around gape yellow. Adult has *black crown, gray face; neck slaty finely barred buff;* above dusky finely vermiculated buff. Foreneck white with dark median stripe; belly brownish buff. Immature essentially identical to much more common Rufescent Tiger Heron; in the hand, young Fasciated has three narrow white tail bands (four in Rufescent); bill slightly shorter and heavier with a more arched culmen. *The two tiger herons generally separate out by range and habitat;* some younger birds cannot be identified. With age, Rufescents develop increasingly rufous heads and necks. Often stands on boulders in turbulent water, also perching on large branches over streams. Generally wary.

PINNATED BITTERN *Botaurus pinnatus*   66-73 cm | 26-28"

Rare and local in extensive marshes, sometimes in adjacent wet pastures and rice fields. *Recorded primarily from Pantanal, but even there very rarely seen.* A large, *cryptically colored* heron. Crown dark, face grayish, *sides of neck buff narrowly barred blackish; above rich buff striped and barred blackish.* Throat white, *foreneck and underparts buffy whitish streaked pale brown.* Immature of more numerous, superficially similar Rufescent Tiger Heron has broadly banded upperparts (especially across wings); its markings are less intricate than the bittern's. Immature Black-crowned Night Heron is smaller and more grayish, with streaked upperparts. Usually remains in thick marshy cover but emerges to hunt in low-light conditions, stalking frogs and small rodents. Occasionally seen flying low over reeds. Generally solitary. Usually "freezes" when seen in the open or alarmed, crouching and retracting or stretching neck skyward. Also can "sink" down, sometimes disappearing into amazingly short grass. Usually silent, but breeding ♂ gives a low-pitched muffled "poo-doo" or booming "poo," mainly at dusk or during night.

AGAMI HERON *Agamia agami*   63.5-71 cm | 25-28"

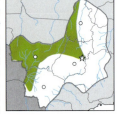

Rare and local along forested margins of oxbow lakes and streams in NW of our area and Pantanal. Multicolored with *exceptionally long and slender bill, long thin neck.* Iris orange, bare facial skin yellowish green; bill mostly blackish; rather short legs olive yellowish. Adult has crown and hindneck black with silvery plumes when breeding; *sides of neck chestnut* with short curved silvery gray plumes springing from lower neck; median neck stripe white and brown. *Above rich dark glossy green. Belly rich chestnut.* Immature browner above, coarsely brown-streaked below. The beautifully colored adult is unmistakable; even the duller immature is distinctive due to its thin, rapier-like bill. Rarely in the open, the gorgeous Agami is a solitary and shy heron, favoring thick, almost impenetrable vegetation in swampy areas. It apparently is quite active at night. Often wades in belly-deep water and hunts, waiting motionless to spear fish. When disturbed, sometimes flushes to land on an exposed branch, but usually sneaks back into cover and freezes, hoping to remain undetected. Usually quiet, but gives a low-pitched guttural croaking.

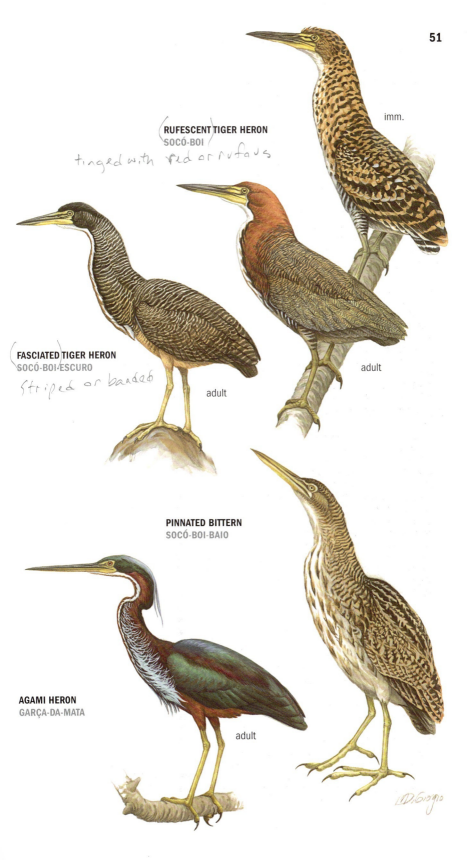

CATTLE EGRET *Bubulcus ibis*   47-52 cm | 18-20"
Very common and widespread in open agricultural terrain, especially near cattle; most numerous in Pantanal. Fairly small and *stocky*. Lores yellowish (bright green when breeding); *stout, fairly short* bill yellow (orange or reddish when breeding); legs olive yellowish to greenish (reddish when breeding). Adult white; when breeding acquires *buff plumes on crown, back, and chest* (a trace of buff color is always retained). Immature duller white. Cf. Snowy Egret and immature Little Blue Heron, both of which are larger and more slender, have different bill and leg colors. Cattle Egrets look heavy-jowled, rather different from the slim-faced appearance of other white herons. Highly gregarious, especially at their roosts from which flocks fan out to feed in open country. True to their name, Cattle Egrets often feed with cattle (sometimes also around Marsh Deer), pursuing insects that the animals flush. They also forage in other semiopen areas such as plowed fields and even roadsides where garbage has been dumped, but generally do not frequent marshy areas unless cattle are present. Despite its dominance and abundance, the Cattle Egret is a relatively recent (1960s) immigrant to Brazil, having colonized Guyana from Africa in the 19th century, from there spreading over all South America.

EGRETTA herons and egrets are long-billed, long-necked, and long-legged wading birds that range near water. The Great Egret has been considered congeneric. All of them have aigrettes (plumes) on head, neck, and back in breeding plumage.

LITTLE BLUE HERON *Egretta caerulea*   56-66 cm | 22-26"
Rare to occasionally uncommon in marshes and around lakes and ponds in Pantanal. Lores greenish (bluer when breeding); *fairly heavy bicolored bill bluish gray, tip black; legs greenish. The only all-dark heron in our area.* Adult unmistakable *dark blue-gray with reddish maroon head and neck*. When breeding, head and neck brighter and elongated plumes spring from crown, chest, and back. Immature *entirely white* but for inconspicuous dusky tips to outer primaries. Subadult birds molting into adult plumage look pied: white with *irregular slaty splotches*. Compare white immature Little Blue to the slimmer, more elegantly shaped, agile, and graceful Snowy Egret. Though a fairly active feeder, Little Blue will also often stand motionless in wait for unsuspecting prey. Unlike most other herons, it sometimes flies with its neck at least partially outstretched, especially soon after taking off.

SNOWY EGRET *Egretta thula*   54-63.5 cm | 21-25"
Fairly common and widespread in marshes and around lakes, ponds, and rivers; most numerous in Pantanal. Iris and *lores yellow* (almost orange when breeding); *slender bill black; legs black with bright yellow feet* ("golden slippers"), that color extending up rear of legs. Adult *entirely white*; when breeding, graceful filmy aigrettes spring from the crown, back, and chest. Immature similar, but bill can be yellowish; greenish yellow extends up rear of legs. Great Egret is much larger, with a proportionately longer neck and bright yellow bill. Immature Little Blue Heron has a bicolored, stouter bill and entirely greenish yellow legs. An elegant, lovely heron, the Snowy tends to feed more actively than many others, dancing and prancing about gracefully. Snowies also sometimes stir shallow water with their feet to startle hidden prey into view. When water levels are dropping, and food is temporarily abundant, they can gather in large numbers together with other wading birds.

GREAT EGRET *Ardea alba*   91.5-99 cm | 36-39"
Common and widespread in marshes, and around lakes and ponds; most numerous in Pantanal. Iris yellow, *bill bright yellow*, lores yellow-green (greener when breeding), *legs black*. White. When breeding, long filmy aigrettes spring from back and foreneck. Immature has yellow bill tipped dusky. This large, slender, long-necked heron can be confused with Snowy Egret (smaller; black bill and legs; yellow feet). Immature Little Blue Heron has a bicolored bill, greenish legs. Great Egret's feeding behavior is similar to the Cocoi Heron's but the egret is a more gregarious bird, gathering in groups of hundreds or more when and where feeding conditions are optimal. Nests in colonies, often with other herons. Usually quiet, but like other egrets it gives a low-pitched throaty "ahhrrr" when startled or flushed.

CAPPED HERON *Pilherodius pileatus* 56-58.5 cm | 22-23"

Uncommon but widespread in marshes, swampy areas, and along larger rivers; most numerous in Pantanal. Iris brown; *bill and facial skin cobalt blue* (more intense when breeding), sometimes tipped pale purplish; legs bluish gray. Adult has "*peaked*" *black crown* with long white head plumes (longer when breeding), white forecrown; *head, neck, and breast pale creamy buff* (deeper when breeding). Above very pale pearly gray; belly white. Immature lacks plumes, has grayish crown streaking and *no* buff. No egret has a black crown nor does any show bright blue on bill and face. In flight Black-crowned Night Heron's chunky shape somewhat recalls Capped; both fly with shallow wingbeats. Generally solitary, Capped Herons sometimes associate loosely as they feed on muddy or grassy shorelines where they forage mainly by standing quietly and spearing prey, less by stalking. Wary and quick to flush, Capped Herons are usually silent, only rarely giving short croaks.

COCOI HERON *Ardea cocoi* 112-122 cm | 44-48"

Common and widespread around ponds, rivers, and marshes; most numerous in Pantanal. Very large and long-necked. Iris yellow, facial skin greenish (bluer when breeding); heavy bill yellow (orange when breeding); legs greenish. Adult has *black crown, white neck and breast*. Black belly *contrasts with white thighs*; *wings and back bluish gray*. Breeding birds develop long plumes (black on head, white on breast and scapulars). Immature less patterned, more grayish throughout with blacker crown; lacks plumes; thighs already white. Flight slow and steady on deep wingbeats with much white showing on forepart of wing, the striking Cocoi is *the only large, boldly black-and-white heron*. It stands and waits for prey in shallows and at water's edge, sometimes remaining motionless for long periods; also stalks slowly. Gathers in groups where feeding conditions are favorable, typically at pools left isolated by receding water levels. Usually quiet, though disturbed birds utter a complaining "kwawk" or "kwaak" as they flush.

WHISTLING HERON *Syrigma sibilatrix* 56-58.5 cm | 22-23"

Fairly common and widespread in pastures and near ponds and marshes; most numerous in Pantanal. Rather thick bill pink with a black tip, bare facial skin pale blue; rather short legs greenish. Crown and occipital plumes slaty black; *face, neck, breast, and wing-coverts golden buff* (feathers of coverts edged darker); belly *white*. Above bluish gray with *white rump conspicuous in flight*. Immature has more streaked neck. Cf. Capped Heron, also with blue on face. Forages solitarily or in loose groups, often standing motionless for several minutes, striking out at large insects, worms, frogs, and small snakes. *More independent of water than most herons*, Whistling Herons rarely or never wade into water though they sometimes feed in muddy or damp grassy situations. Distinctive call (the "whistle"), given especially in flight and often with bill wide open, a piercing "weeee... weeee...," sometimes in series.

IBISES & SPOONBILLS (Threskiornithidae) are vaguely heron-like wading birds with, except in the spoonbill, *decurved bills*. They fly with their neck outstretched.

THERISTICUS are *very large*, relatively short-legged, *conspicuous* ibises of open terrain.

BUFF-NECKED IBIS *Theristicus caudatus* 71-76 cm | 28-30"

Fairly common and widespread in cerrado and pastures; most numerous in Pantanal. Large and strikingly patterned. Iris red; black bare face and stripe on sides of chin; long decurved bill blackish; *legs salmon pink*. Adult has *buff head, neck, and breast*; crown, hindneck, and lower breast rufous; paler throat and chest; black belly. In flight shows *conspicuous silvery white wing-coverts*; black flight feathers and tail. Immature duller with dusky streaks on head, neck, and underparts. An open country ibis, Buff-necked is not restricted to water and wades only rarely. Conspicuous, ranging in often noisy groups, it forages mainly on dry ground, favoring recently burned areas; walks in short grass, picking at and probing soil for insects, even frogs. Tends to be wary. Perches in trees, and sometimes roosts in groups in palm groves around ranch buildings, occasionally associating loosely with macaws (Hyacinths or Blue-and-yellows). Flight strong, often well above ground. Classic call a loud and far-carrying "tur-túrt," characteristically doubled, often heard long before birds are seen. Perched birds also vocalize frequently, sometimes giving lengthy choruses accompanied by bowing and other displays.

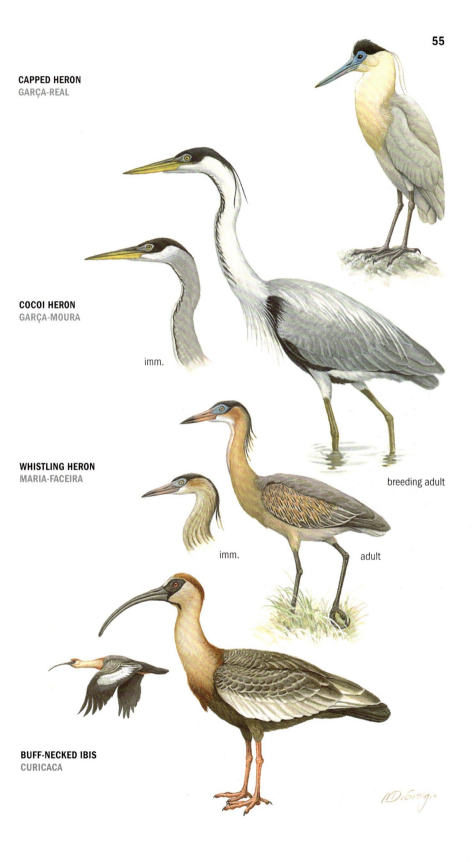

CAPPED HERON
GARÇA-REAL

COCOI HERON
GARÇA-MOURA

WHISTLING HERON
MARIA-FACEIRA

BUFF-NECKED IBIS
CURICACA

PLUMBEOUS IBIS *Theristicus caerulescens*　　　　　　71-76 cm | 28-30"
Fairly common in marshes and around ponds and lakes in Pantanal. Iris yellow-orange, long decurved bill dark gray, legs pink. *Uniform plumbeous gray;* long but often recumbent crest. No other ibis in our area is gray, and few are so large. Usually in pairs, occasionally small (family?) groups, that forage while walking in shallow water at the edge of small ponds and roadside ditches. Rarely associates with other wading birds, but pairs can join groups at drying pools where food is concentrated. Tame, especially compared to the wary Buff-necked. Call, a characteristic sound of the Pantanal, a loud and ringing "ko-ko-ko-ko-ko-ko-ko."

GREEN IBIS *Mesembrinibis cayennensis*　　　　　　56-58.5 cm | 22-23"
Uncommon to locally common around ponds, lakes, and streams. Most numerous in Pantanal. Iris gray, bare face dark green; long decurved bill; rather short legs green (sometimes bright, almost turquoise). Adult *dark bronzy green,* blacker below; short bushy crest usually not obvious, but sometimes ruffled; *shiny green feathers on nape and hindneck.* Immature duller and blacker overall, lacking the shiny green. Bare-faced Ibis is smaller and blacker, has reddish facial skin, bill, and legs. Usually solitary, at most in pairs or small dispersed groups. Walks slowly in shallows, probing the mud; perches freely in trees. Less conspicuous than many ibises, Green seems most active at dawn and dusk. Flight jerky and limpkin-like, with stiff quick wingstrokes. Loud, easily recognized call a mellow rolling "koro-koro-koro-koro" or "klu-klu-klu-klu," usually given in flight.

BARE-FACED IBIS *Phimosus infuscatus*　　　　　　48-51 cm | 19-20"
Fairly common in marshes and open areas around ponds and lakes; most numerous in Pantanal. Iris brown, *conspicuous bare face red; long and slender decurved bill reddish brown,* often dark-tipped; legs pinkish red. Adult *black,* glossed bronzy green above. Immature duller and less bronzy with a blackish bill. White-faced Ibis is larger and when breeding has more chestnut foreparts; it is more highly bronze-glossed on wings and also has white on face. In flight, unlike Bare-faced, White-faced's legs extend well beyond tail. Green Ibis has green bill, face, and legs; *no red.* Bare-faced is a gregarious ibis that usually is in often compact groups (up to several hundred individuals), sometimes with other wading birds. It feeds by probing damp ground and mud of marshes, pastures, lake edges. Flight strong, with steady fast wingbeats; glides less than White-faced. Like White-faced (but not Green), Bare-faced is often seen flying overhead, moving between its feeding and roosting sites.

WHITE-FACED IBIS *Plegadis chihi*　　　　　　56-61 cm | 22-24"
Seasonally fairly common in marshes and open areas around ponds and lakes, mainly in Pantanal. Seems to occur only as austral migrant from Argentina (Mar-Jul) and not known to breed here. *Iris red, facial skin reddish edged by white feathers;* long decurved bill dusky; legs grayish (reddish when breeding). Above dark metallic bronzy green. Breeding adult has *head, neck, and underparts rich chestnut;* has a bronzy iridescence above, especially on wings. Nonbreeding adult has *head and neck dark grayish brown, with narrow whitish streaking.* Immature even duller than nonbreeding adult. Bare-faced Ibis is smaller and stockier with a shorter neck, notably shorter legs, pinker bill, and blackish plumage (never showing streaking or chestnut). Cf. Limpkin. White-faced is a notably gregarious ibis that often wades in fairly deep water, probing mud. It flies with its long neck and legs outstretched, the flapping frequently interspersed with glides.

LIMPKIN *Aramus guarauna*　　　　　　66-71 cm | 26-28"
Locally common in and around marshes and ponds; conspicuous in Pantanal. Large and long-necked, with *long somewhat drooping bill* horn-colored tipped blackish; long legs dusky. Dark brown *streaked white on head, hindneck, and upper back.* Ibises have more slender and decurved bills, bare facial skin. Immature Black-crowned Night Heron is much chunkier with a short thick neck, heavier straight bill. Limpkins perch freely in shrubs and low trees. When nervous they often wing-flick; where not persecuted can be tame. Flight strong though stiff, with neck outstretched and legs often dangled; the wings jerk characteristically on the upstroke. Feeds mainly on apple snails (*Pomacea* spp.), occasionally frogs and larger insects. Often noisy (regularly vocalizing at night), with a variety of loud wailing calls, typically "carr-rr-rao" or "carr-reé-ow." The sole member of its family, Aramidae.

ROSEATE SPOONBILL *Ajaia ajaja*   71-79 cm | 28-31"
Local in marshes and along rivers; most numerous in Pantanal. Sometimes placed in genus *Platalea*. A spectacular, unmistakable *pink* wader with a *long flat spatulate bill*. Iris orange to red; bill gray to pinkish; legs mainly pink to red. Adult has *unfeathered head greenish gray with black nape*. Otherwise mostly *pink* except for white neck, upper back, and chest; *reddish stripe on coverts above and below wing* (brightest when breeding, and most evident in flight); tail orange. Buff chest-tuft when breeding. Immature has feathered head, is whitish overall (becoming pinker with age, over a three-year period), but already has pink tail, underwing coverts. Gregarious in small flocks, often joining other waders to feed at drying pools. Spoonbills wade in shallows, feeding by sweeping bill side to side. Especially lovely in flight when the pink and red on their wings can be seen to best advantage, spoonbills fly with their necks out-stretched and legs trailing behind, interspersing wingbeats with brief glides.

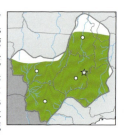

STORKS (Ciconiidae) are an ancient group of large, long-legged wading birds that fly with necks and legs extended and that can soar to great heights on their long flat wings.

WOOD STORK *Mycteria americana*   89-99 cm | 35-39"
Fairly common and widespread in marshes and around ponds, lakes, and rivers; most numerous in Pantanal, where it can be seasonally abundant. Iris brown; *blackish bill long and heavy, decurved at tip*; legs dusky with pinkish feet. Adult has *unfeathered head and neck dark gray, paler crown, black nape band* (locally called the *cabeça seca*). Otherwise *white* with *black flight feathers* (mainly hidden except in flight). Immature dingier, with head and neck partially feathered and browner. Maguari Stork has a straight bill and bare face, shows more black on closed wing, reddish legs. King Vulture's flight pattern is somewhat similar to Wood Stork's. Wood Storks congregate in flocks (sometimes large) at better feeding sites, there often with herons and egrets. Feeding birds shuffle in shallow water and mud, sweeping their bills to capture prey. Though somewhat ungainly on the ground or when perched, Wood Storks are powerful and graceful in flight and can soar very high. Their nesting colonies can be large, and though mainly quiet away from their nesting sites, there they often emit various grunting calls.

MAGUARI STORK *Ciconia maguari*   99-108 cm | 39-42.5"
Uncommon to locally fairly common in grasslands, marshes, and other open areas, usually near water; most numerous in Pantanal. Long straight bill grayish with reddish tip; *bare face orange-red*; iris yellow, legs orange-red. Mostly white (though head and neck may look "dirty," apparently from staining). *Black scapulars, greater wing-coverts, and outer primaries*, tail also black. Perched Maguaris show more black on closed wing than do Wood Storks, this difference being visible at a great distance. Jabiru is much larger and whiter. The Maguari is typically a solitary bird, most often seen standing in marshy or open terrain, or patiently stalking prey. Small groups may congregate around drying pools, occasionally with other wading birds though the Maguari is often more independent of water than they are. Nests solitarily on the ground in a secluded area of a marsh.

JABIRU *Jabiru mycteria*   127-140 cm | 50-55"
Locally fairly common in marshes and around ponds; most numerous, by far, in Pantanal. Unmistakable: a *huge* stork with *massive upswept bill*. Iris dark; bill and legs blackish. Adult has *bare black head and neck with a broad red band* around neck (which sometimes looks "swollen"). Less than one percent of population has the entire head and neck pinkish red. Otherwise *white*. Immature dingier (grayish brown; gradually whitens with age) with neck duller red. Wood and Maguari Storks are much smaller with black on wings, no red on necks, etc. Jabirus typically range in pairs, striding across grassy areas or in shallow water, striking at prey (fish, frogs, snakes, and even young caiman) with their powerful bills, gulping down prey with an upward head toss. When not breeding they can gather in flocks, occasionally numbering in the hundreds. The Pantanal is the "world center" for the spectacular Jabiru, even surpassing Venezuela's famed llanos. Occasionally one becomes tame around a fishing or tourist camp (but beware that enormous bill!). Though take-off is labored, once airborne Jabirus are strong fliers, often soaring effortlessly, sometimes very high. Nests are massive, conspicuous stick structures; pairs are apparently always solitary. Jabirus are usually silent, but bill-clacking sometimes occurs at the nest.

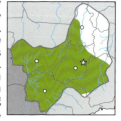

AMERICAN VULTURES (Cathartidae) are superficially eagle-like large birds whose small *heads and necks are featherless*, apparently to stay clean while feeding on carrion, their main food. Bills are hooked for ripping into flesh, but their *feet are relatively weak*, not suited for carrying food. Vultures regularly perch in the open, often on snags, usually looking "hunched." Eggs are laid on the ground in a secluded recess, often a hollow in a tree or bank. Their relationships are still debated, some considering them closest to diurnal raptors, others to storks.

TURKEY VULTURE *Cathartes aura* 66-76 cm | 26-30"

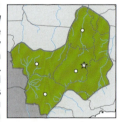

Very common and widespread in open lands and borders of forest and woodland. Bare head reddish and warty with a *whitish nape band*. Plumage brownish black, wing-coverts edged brown. In flight wings long and fairly broad, tail fairly long. Pale shafts on outer primaries show from above; from below *wings are two-toned, with silvery gray flight feathers and dark coverts*. Immature has grayish head and bill, browner upperparts. Lesser Yellow-headed Vulture is slightly smaller and blacker. It favors more open terrain and adults have head mainly yellow-orange. Whitish shafts of outer primaries form patch as seen from above. Turkey Vultures are familiar birds that are often seen in flight, typically *tilting* side to side with wings held in a *marked V-shaped dihedral*, flapping infrequently but deeply. Sometimes they soar very high. They locate carrion mainly by their keen sense of smell and can even find food beneath forest canopy. Black and King Vultures, lacking a sense of smell, watch for their descent, then follow.

LESSER YELLOW-HEADED VULTURE *Cathartes burrovianus* 57-64 cm | 23-25"

Locally fairly common in grassland, including campos and cerrado; often most numerous where marshy. Bare head and neck mostly orange-yellow with blue on crown. Plumage black. In flight wings as in Turkey Vulture but *shafts of outer primaries whitish*, from above forming a conspicuous patch. Immature has dusky head. Sometimes joins Turkey Vulture at carrion, but never as numerous. Lesser Yellow-headeds *rarely fly high*, more typically sailing low above the grass, almost harrier-like.

Greater Yellow-headed Vulture (*C. melambrotus*) occurs locally in forest along N edge of our area, e.g., Serra das Araras. Resembles Lesser Yellow-headed Vulture but substantially larger. In flight from below note blackish inner primaries contrasting with other silvery flight feathers. Wings are held flatter, so profile is less V-shaped than Turkey Vulture's.

BLACK VULTURE *Coragyps atratus* 56-63.5 cm | 22-25"

Very common, widespread, and familiar *in open terrain; most numerous in settled areas* (especially around garbage dumps), fewer where forest is extensive. Bill blackish with pale tip. *Bare head and upper neck dark gray*. Plumage dull black. In flight has broad, fairly long wings, *short tail; outer primaries have whitish base conspicuous from above and below*. Immature King Vulture is much larger, shows white mottling on underwing-coverts. Great Black Hawk has similar broad wings and short tail, but its larger, feathered head protrudes more and it shows prominent white in tail, none in primaries. Black Vultures are often tame around towns, roosting in large groups in trees; they mainly feed on refuse and carrion, but also capture live prey. They hop on the ground with some agility (Turkey Vultures just shuffle). Unlike *Cathartes* vultures, which it dominates at food sources, the Black Vulture has no sense of smell. In flight its shallow, stiff, fast wingbeats typically alternate with short bouts of sailing, but it can also soar to great heights.

KING VULTURE *Sarcoramphus papa* 71-81 cm | 28-32"

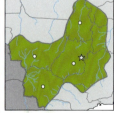

Uncommon and now rather local in wilder, mainly forested, areas; rare in more open areas. *Large; unmistakable. Bill caruncles orange red; bill tip red*. Spectacular adult *mostly white. Head and neck bare and multicolored*; wrinkled head mostly purplish gray; *lower neck ruff dusky. Flight feathers, rump, and short tail black*. In flight long broad wings are held flat. Immature sooty gray with *whitish mottling on underwing-coverts, grayish head*. Adult features are acquired over four-plus years. Adult only to be confused, and only at great distances, with Wood Stork (which has a similar flight pattern). Immature King can be confused, in distant flight (when size difference can be unclear), with much smaller Black Vulture (but Black has pale patch on outer primaries, no mottling on underwing-coverts). Solitary or in pairs, circling high in the sky on broad flat wings, generally not associating with "lesser" vultures, Kings dominate other vultures at carcasses. Unlike the others, only infrequently is it seen perched in open.

HAWKS, EAGLES, & KITES (Accipitridae) are the primary family of diurnal birds of prey, and are characterized by their strongly *hooked bills* and *sharp talons*, adapted for killing and carrying prey. Some species soar regularly, but many others do not. *Identification can be difficult*, with many plumage variations. ♂♂ are smaller than ♀♀.

✓ **OSPREY** *Pandion haliaetus* 54-58.5 cm | 21-23"
Uncommon boreal migrant to areas near larger bodies of water. A unique, fish-eating hawk whose rough soles and reversible hind toe aid in grasping its slippery prey. Slight bushy crest. Iris yellow. *Head and underparts white;* some dark streaking on crown and chest. *Dark brown eyestripe* extends to nape and *dark brown upperparts;* tail barred grayish and dark brown, paler and grayer from below. Immature has brown-streaked crown, feathers above with buffy whitish edging. In flight long narrowish wings are *held distinctively "crooked" with black patch on bend of wing.* Ospreys have a distinctive silhouette, habitat, and behavior. Generally solitary, they typically perch on bare branches near water and hunt while hovering, then plunge feet-first into water after fish. In normal flight usually glides on set wings; flapping is deep and deliberate. Has a variety of yelping or piping calls, the most frequent an upward-inflected "cleeyp." Breeds in North America (and Old World); pre-breeders stay in our area year-round, but do not breed.

SWALLOW-TAILED KITE *Elanoides forficatus* 56-61 cm | 22-24"
Widespread and conspicuous in and above canopy and borders of forest and woodland. A beautiful, graceful raptor with long pointed wings and unmistakable very long, deeply forked tail. Head, neck, and underparts white contrasting with *black upperparts, wings, and tail.* At close range back and upperwing-coverts can be seen to be glossed blue or green. In flight underwing-coverts white, contrasting with blackish primaries. Preeminently aerial, with resting birds huddling inconspicuously in forest canopy or on snags. Flight graceful and buoyant, with deep slow wingstrokes and long periods of easy gliding and soaring. Rather gregarious, with flocks of 5-10 or even more birds frequent. Feeds on insects captured in the air and snatched from foliage; occasionally swoops down to drink water on the wing. Also captures frogs, lizards, and snakes. Usually quiet, but gives shrill piping calls and whistles, mainly in flight. Our local breeding population is augmented by North American migrants that spend the boreal winter in our area.

WHITE-TAILED KITE *Elanus leucurus* 38-40.5 cm | 15-16"
Uncommon in pastures and open agricultural areas, also locally in natural grasslands. This conspicuous open-land raptor is most numerous in cultivated regions, and is increasing due to the expansion of agriculture. Iris red. Adult has *head, neck, and underparts white,* black in front of eye; back and wings pale pearly gray with *black on wing-coverts; long white notched tail.* In flight wings long and pointed, underwing white with blackish primaries, *black wing patch.* Juvenile has a more orange eye, dark brown streaking on crown and nape, mainly brown back, and pale gray tail. This lovely kite should not be confused, Plumbeous and Mississippi Kites are much grayer and favor more wooded terrain; they do not hover like White-taileds so characteristically do. Perches in the open, foraging mostly in the early morning and late afternoon. Flight easy and graceful with deep wing strokes and wings held in a shallow dihedral while gliding. Hunts mainly while *hovering* with body angled upward at ca. 45°, slowly dropping to the ground after small rodents, its primary prey.

PEARL KITE *Gampsonyx swainsonii* 23-25 cm | 9-10"
Uncommon but probably increasing in open and agricultural areas, even around towns. Small and quite falcon-like. Iris carmine. Adult *slaty blackish above* with forehead and cheeks creamy buff and narrow white nuchal collar; tail gray above, paler below. *White below* with blackish patch on sides of chest and rufous-buff thighs. Juvenile slightly browner above. In flight wings pointed, underwing pale with *narrow white trailing edge.* Not likely confused; no other falcon-like raptor is *so pale and small.* American Kestrel has a similar shape, behavior, and even open habitat, but is longer-tailed, barred above, streaked below, etc. Perches in open on wires or atop low trees, scanning for prey, capturing lizards and insects after a short dive to the ground. Often rather tame. Not very vocal, but gives an occasional high-pitched scolding.

OSPREY
ÁGUIA-PESCADORA

SWALLOW-TAILED KITE
GAVIÃO-TESOURA

adult   adult

WHITE-TAILED KITE
GAVIÃO-PENEIRA

adult   adult

juv.

PEARL KITE
GAVIÃOZINHO

adult

ICTINIA kites are attractive, predominantly gray, unusually aerial raptors with long pointed wings.

PLUMBEOUS KITE *Ictinia plumbea* 34.5-37 cm | 13.5-14.5"

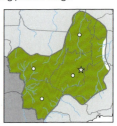

Fairly common in and above canopy of forest and woodland. Iris red; short legs yellow-orange. Adult *leaden gray*, slightly paler on head and darker on mantle; wings blackish, *inner webs of outer primaries rufous* (usually hidden when perched); tail blackish with two white bands (not apparent from above). In flight *wings long and pointed* (though primaries are spread when soaring), *rufous in primaries usually obvious*. Immature has whitish head streaked blackish; above slaty, feathers edged paler; below *coarsely streaked dark gray*; primaries have little rufous. Readily known from its overall gray coloration and *long wings, with wings extending well beyond tail tip on perched birds*. Cf. similar but rarer Mississippi Kite. Wings of immatures are *notably shorter* so when perched they can be confused with other young raptors with streaked underparts. Soars and glides gracefully in open air, often high; rather gregarious, especially when flying about, sometimes joining Swallow-tailed Kites. Subsists primarily on insects captured in flight, sometimes also snatched from canopy; often eats prey on the wing. Perched birds regularly rest on high limbs or snags. Notably quiet.

MISSISSIPPI KITE *Ictinia mississippiensis* 35.5-38 cm | 14-15"

Rare to uncommon boreal migrant to woodland and borders in Pantanal; only recently has it been realized that it occurs in Brazil (mostly Oct-Nov). Iris red. Adult has *whitish head* contrasting with gray upperparts; *pale gray secondaries show as band on closed wing*; primaries and fairly long tail black. In flight wings long and pointed, with *contrasting pale secondaries* visible from above. Immature *has reddish brown streaking below*. Much more numerous Plumbeous Kite has rufous primaries when adult; its head color is more uniform with back, not whiter. Plumbeous immature is streaked blackish, not rusty. On perched birds, Mississippi's proportionately longer tail and shorter wings are evident (wings barely extend to tip, unlike Plumbeous). Habits much as in Plumbeous Kite. In our area Mississippis can occur in monospecific flocks of up to several hundred birds. Breeds in North America.

HARPAGUS kites are relatively sluggish forest-based kites with a *dark throat stripe and two notches in the maxilla.*

DOUBLE-TOOTHED KITE *Harpagus bidentatus* 31.5-35.5 cm | 12.5-14"

Uncommon in canopy and borders of forest in NW of our area. Iris amber; legs yellow. Adult ♂ has *head bluish gray*, brownish gray upperparts, scapulars sometimes mottled whitish; rather long tail blackish with three narrow pale grayish bands and tip. *Throat with dusky median stripe; below mostly rufous*, crissum white, belly with a little grayish barring. ♀ more uniform rufous below, with little barring. In flight wings fairly long and rounded, with *underside mainly white* (contrasting with dark body); *fluffy white crissum feathers* are often conspicuous. Immature dark brown above, scapulars often mottled whitish; tail as in adult; whitish buff below with coarse dark brown streaking (variable); *median throat stripe* as in adult. In flight recalls an *Accipiter*, though the kite's wings are longer and more bowed, and white crissum more conspicuous. Cf. immature of larger Gray Hawk, which lacks throat stripe and has a prominent pale brow and malar. Perches quietly in forest midlevels and subcanopy, regularly at borders, often allowing a close approach. Hunts lizards and large insects; soars regularly, sometimes high, especially in mid-morning hours. Elsewhere this kite follows monkeys, capturing prey disturbed by their movements through trees (this behavior not yet recorded here). Quiet, but perched birds occasionally give a high-pitched, thin, whistled call, "wheeey-whit, pii, wheeey-whit!"

RUFOUS-THIGHED KITE *Harpagus diodon* 31.5-35.5 cm | 12.5-14"

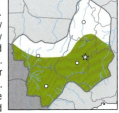

Rare and local in canopy and borders of more humid forest and woodland. Status not well known in our area, where there are very few records (may not breed here; may be transient?). Iris orange; legs yellow. Adult slaty gray above; rather long tail blackish with three narrow pale grayish bands and tip. *Throat stripe dusky; below pale gray* with white crissum; *thighs rufous.* In flight wings long and rounded, *underwing-coverts rufous*. Immature whiter below with throat stripe and coarse brown streaking; thighs already rufous. Bicolored Hawk adult has similar overall coloration, but lacks throat stripe and has a "capped" fierce look rather than the kite's rounder head and benign visage. Behavior much as in Double-toothed Kite.

PLUMBEOUS KITE
SOVI

MISSISSIPPI KITE
SAUVEIRO-DO-NORTE

DOUBLE-TOOTHED KITE
RIPINA

RUFOUS-THIGHED KITE
GAVIÃO-BOMBACHINHA

ACCIPITER hawks are forest-based, bird-eating raptors whose *rather short and rounded wings* and *squared-off tails* assist them in maneuvering through dense cover in pursuit of their prey. Though the sexes are alike, ♂♂ are *markedly* smaller than ♀♀.

TINY HAWK *Accipiter superciliosus*   ♂ 20.5-22 cm | 8-9" ♀ 26-28 cm | 10-11"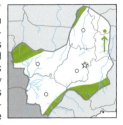
Rare and local in forest and borders. Iris red. Adult slaty gray to blackish above, blackest on crown; white below *narrowly barred with dark gray* except on throat; tail blackish with three or four dark gray bands. Immatures dimorphic: normal morph grayish brown above with darker crown, tail banded brown and grayish brown, *buff below barred rufous.* Rufous morph *rufous brown above* with darker crown, tail banded rufous and blackish; *buff below barred rufous.* This fierce little *Accipiter* can generally be identified by its small size alone. Barred Forest Falcon is larger with a dark chest. Infrequently seen, the Tiny Hawk ambushes birds from hidden perches in the subcanopy or midlevels; it is known to capture hummingbirds on a regular basis. Occasionally, in the early morning, one will perch in the open, sunning itself. It rarely or never soars. Quiet, only infrequently giving voice to a weak "kree-kree-kree" call.

RUFOUS-THIGHED HAWK *Accipiter erythronemius* ♂ 25.5-27 cm | 10-11" ♀ 28-30.5cm | 11-12"
Uncommon and apparently local in forest, woodland, and borders, mainly in S and E of our area. Status in our area unclear, and may occur at least partially as an austral migrant from breeding areas further south. Sometimes considered conspecific with Sharp-shinned Hawk (*A. striatus*) of North America. Iris yellow. Adult has cheeks and *thighs solid rufous*, slaty gray upperparts; tail blackish with three or four dark gray bands. Below white with a *variable amount of rufous barring*, white crissum. Immature brown above, feathers with some rufous edging; below buffy whitish with a variable amount of brown streaking and barring. An aggressive predator on birds that dashes swiftly through thick cover; it can be quite bold. Soars occasionally. Usually quiet; but ♂ sometimes gives a soft "ku-ku-ku-ku-ku" call, especially when breeding.

BICOLORED HAWK *Accipiter bicolor*   ♂ 33-36 cm | 13-14" ♀ 43-46 cm | 16-18"
Rare and local in forest, woodland, and borders. Iris yellow. In southern half of our area (**A**), adult ♂ has blackish crown, slaty gray upperparts; *below variable, usually paler gray becoming more rufescent on belly* with indistinct paler barring; thighs rufous and crissum white; tail blackish with two or three gray bands. ♀ slightly browner above, and rufescence below extends up onto breast. Immature dark brown above, except tail much as in adult. Below buffy whitish *with extensive dark brown streaking*; thighs rufous with some dark barring. Northward (**B**), differs in its *uniform pale gray underparts*. Immature dark brown above with blacker crown; *partial nuchal collar and underparts buff to white* (rarely even rufous). Rufous-thighed Kite is superficially similar, though structurally and behaviorally it is very different; note especially its median throat stripe, lack of a blackish crown. Immature is most likely mistaken for a forest falcon, in particular a small Collared, though the latter shows a conspicuous dark crescent on ear-coverts, has longer legs, and a more graduated tail. Sneaky, despite its size, and inconspicuous, but can be bold, indeed at times seeming almost fearless. Perches quietly on semiconcealed branches, from there swiftly dashing off in pursuit of small or medium-sized birds. Rarely perches in the open and almost never soars. Usually quiet, but territorial birds give a loud cackling "keh-keh-keh-keh...."

HOOK-BILLED KITE *Chondrohierax uncinatus* 38–43 cm | 15–17"
Rare and local in more humid forest and woodland in N and W of our area. Iris whitish; bare crescent above eye greenish yellow; heavy and strongly hooked bill with greenish cere and facial skin. Complex plumage variation. Light-morph adult ♂ *slaty gray to slaty black above; below barred gray and whitish or pale buff;* tail black with two broad whitish bands. In flight distinctive *broad, rounded wings are held somewhat forward and narrowing at base,* mainly grayish with primaries banded blackish. Light-morph adult ♀ dark brown above with gray face; *prominent rufous nuchal collar; below coarsely barred rufous brown and creamy whitish;* in flight shows rufous-barred underwing-coverts, banded flight feathers. Scarce dark-morph adult *more or less uniform brownish black,* including underwing; tail bands as in light morph. Light-morph immature brown above with *blackish crown; creamy white nuchal collar, sides of neck, and underparts,* the latter with variable coarse dusky barring (sometimes extensive); tail with three gray bands, narrower than adult's. Dark-morph immature as in adult but tail bands narrower. This infrequently seen kite is often best recognized by characters such as bill shape, facial skin, and the unique patch above eye (the latter, in concert with the lack of a ridge over eye, imparts an odd visage), rather than by plumage. The "paddle-shaped" wings on flying birds are diagnostic (cf. especially Roadside and Gray Hawks). A sluggish raptor, often perching for long periods as it searches for its primary prey, land snails; also eats small lizards, frogs. Soars regularly, though usually not for long or very high. Not very vocal, but perched birds occasionally give a fast chattered "weh-keh-eh-eh-eheheheh."

SNAIL KITE *Rostrhamus sociabilis* 40.5–45 cm | 16–17.75"
Locally common in marshes, ponds, and wet pastures; most numerous in Pantanal. Iris red; *bill slender and very sharply hooked,* with cere, lores, and orbital ring yellow-orange (duller in juveniles); legs orange-red. Adult ♂ *slaty black; uppertail-coverts, crissum, basal half of tail, and tail-tip white.* Adult ♀ blackish brown above with buffy whitish forehead and superciliary; throat buffy whitish; *below heavily mottled and streaked creamy buff and dark brown; tail as in ♂.* Immature like ♀ but browner above, more streaked below. In flight wings broad and rounded; ♂ underside blackish; ♀ and immature's browner; flight feathers mottled whitish (whitest in primaries). Immature Great Black Hawk has prominent tail barring. Dark-morph Hook-billed Kite has very different soft-part coloration, tail pattern. The Snail Kite's behavior is normally diagnostic. This conspicuous, marsh-loving kite is notably gregarious, gathering in groups where feeding conditions are good, and roosting and nesting communally. Usually perches in the open, often on fence posts or the ground, even on wires. Flight surprisingly agile, quartering slowly low over water and marsh on somewhat bowed wings, searching for the *Pomacea* snails that comprise the majority of its diet; it eats crabs and turtles when snails are in short supply. The bill is used to pry out snails at regularly-used feeding perches. The Snail Kite wanders (perhaps migrates?), shifting around in response to fluctuating water levels. Usually quiet, but perched birds occasionally give a raspy "kahhrrrr."

LONG-WINGED HARRIER *Circus buffoni* ♂ 48–50 cm | 19–20" ♀ 55–57 cm | 20–22.5"
Rare to uncommon and local in campos, cerrado, and agricultural areas, apparently preferring dry habitats in our area (elsewhere often in marshes). Plumage variable, but *always shows contrasting silvery gray flight feathers with black barring.* Light-morph ♂ black above with *white forehead and eyestripe,* white below with *black band across upper chest;* rump white, tail with several gray bands. In flight wings long and narrow, upper wing-coverts black, underwing uniformly barred. Light-morph ♀ browner above, facial markings buff, breast and belly pale buff lightly streaked brown. Rare dark morph (both sexes) *mostly sooty black* but wings and tail as in light morph. Flying birds are striking and not likely confused, but perched dark-morph birds can be confused with Zone-tailed Hawk. Rarely seen perched. Most often seen in flight as it glides over open areas, wings held in a dihedral and tilting from side to side; it usually remains close to the ground. When prey is spotted, it pulls up, circles, and plunges feet-first to the ground. Generally silent, but nesting birds occasionally emit a rather *Milvago* caracara-like high-pitched scream "kreeeyr, kreeeyr, kreeeyr."

WHITE HAWK *Leucopternis albicollis*   43-49.5 cm | 17-19.5"
Uncommon in forest and borders in NW of our area. Iris brown; cere gray; legs yellow. Adult has *head, neck, upper back, and underparts white*, upper back spotted black. Above black, wing-coverts and scapulars fringed white; tail white with broad black subterminal band. In flight wings broad and rounded, *underwing mainly white*, trailing edge black. Juvenile has some dusky streaking on head, especially crown. Beautiful but lethargic, the White Hawk regularly perches at forest edge, quietly surveying its surroundings. It soars frequently, sometimes circling high with its broad wings held flat and tail fanned wide. Feeds mainly on reptiles, but also on small mammals, large insects. Gives a loud, husky scream, "shrreeeyr," both while soaring and when perched.

GRAY-HEADED KITE *Leptodon cayanensis*   46-53.5 cm | 18-21"
Uncommon in forest, woodland, and borders. Iris dark brown; cere, orbital area, legs bluish gray. Adult has *pearly gray head, slaty black upperparts, white underparts*; tail black with two pale grayish bands. In flight shows fairly long, broad, and rounded wings, long tail. Seen from below has *black underwing-coverts*, flight feathers barred whitish and blackish, the *dark wings contrasting with white underparts*. Immatures dimorphic; cere, orbital area, and legs always yellow. Light morph has *head, neck, and underparts white* with small black crown patch; upperparts dark brown. Dark morph has blackish brown head, variably blackish-streaked underparts (some birds almost solidly dark across chest, others with throat white except for a dark median stripe, others with rusty on neck). In flight from below immatures of both morphs have underwing-coverts white. Adults nearly unmistakable, and *often look strangely small-headed*. Black-and-white Hawk-Eagle resembles light-morph immature but has orange (not yellow) cere, black (not pale) lores, and blacker upperparts; kite lacks the hawk-eagle's white leading edge to wing. Cf. also immature Gray Hawk and immature Double-toothed Kite. Ranges in forest canopy, perching on open branches in early morning, otherwise staying mostly within cover. Soars frequently, often quite high, especially during territorial display. Wide diet includes insects, snakes, lizards, and birds, all procured primarily in the canopy of trees. Gives a loud nasal cackling "keh-keh-keh-keh-keh-keh," usually from a perch but sometimes in flight.

CRANE HAWK *Geranospiza caerulescens*   46.5-51 cm | 18.25-20"
Uncommon in forest and woodland. Lanky and long-legged; small-headed. Local race's iris yellowish white; cere gray; *legs reddish*. Adult *gray with whitish barring on underparts* and wings; long tail black with two white bands. In flight wings rather long and rounded; underwing-coverts gray with whitish vermiculations, *conspicuous white band across black primaries*. Immature similar but has more whitish barring below. Some individuals show tawny-buff on lower belly and crissum; the pale tail bands can be buff as well. Distinctive shape, reddish legs, and dark face of this unusual raptor should be enough to identify perched birds, while flying examples can be recognized by their unique white primary band. Encountered singly, sometimes while foraging clumsily on branches and tree trunks, probing into crevices, epiphytes, and even bird nests with its long double-jointed legs, flapping wings to maintain its balance. Soars regularly, though not too high or for very long. Infrequently gives a whistled "sweeeoo."

BUTEO hawks are "typical" midsized raptors, with broad rounded wings and broad rounded tails that are adapted for *frequent soaring*. Buteo favor semiopen areas and forest edges, preying mainly on small mammals, also birds.

ROADSIDE HAWK *Buteo magnirostris*   33-38 cm | 13-15"
Widespread, common, and conspicuous in clearings, agricultural areas, and forest and woodland borders. Sometimes placed in genus *Rupornis*. Iris pale yellow; *cere orange-yellow*. In most of our area (**A**) adult brownish *gray above*, throat and chest somewhat paler, often with some rufous on chest; *breast and belly barred whitish and dull grayish rufous*. Tail banded dull gray and blackish. Immature similar but breast streaked and browner. In Mato Grosso do Sul (**B**) *dark brown above* with *blackish head and chin*; below cinnamon-rufous irregularly barred whitish; tail banded blackish and rufescent. In flight throughout underwing barred grayish with *extensive rufous in primaries* (evident from below, but especially conspicuous from above). *In most areas this is the most frequently seen raptor.* No other similar hawk shares the prominent rufous in the wings, though note that this color is usually hidden on perched birds. Cf. immature Gray Hawk. Usually perches in the open; can be rather sluggish, allowing a close approach. Perched birds often wiggle tails sideways. Usually flies with shallow wingbeats interspersed with short glides and except in display does not soar high. One of our more vocal raptors. Perched birds give a high-pitched nasal, descending scream "kreeeeah," while in flight (less often from perch) gives a fast series of nasal "keh" notes.

GRAY HAWK *Buteo nitidus*   40.5-45 cm | 16-17.75"
Uncommon in woodland, forest borders, and clearings. Iris brown; cere and legs yellow. Adult gray above, *paler pearly gray on head and neck*, with wavy darker gray barring throughout; white below with narrow dense dark gray barring, tail black with two or three white bands. In flight *wings whitish with faint gray barring*. Immature dark brown above, *face and eyestripe contrastingly pale buff* with a *dark brown malar streak*. Below whitish to creamy buff with dark brown spotting and streaking, often heavy, *thighs often barred*. Often some whitish shows at base of tail, which is blackish with three or four gray bands. Attractive adult easy to recognize. Cf. the browner Roadside Hawk. Immature best known by its combination of relatively small size and striking facial pattern. Gray Hawk favors fragmented forest areas, where it perches at varying heights; hunts a wide variety of prey, usually attacking from a perch. Soars regularly, but generally not for long or very high. Gives a series of high-pitched, clear, whistled phrases, "wuu-yeéuw, wuu-yeéuw, wuu-yeéuw...," usually from a perch. Call an abrupt clear "keeeuw."

SHORT-TAILED HAWK *Buteo brachyurus*   39.5-43 cm | 15.5-17"
Uncommon in semiopen areas and forest and woodland borders. Iris brown. Light-morph (predominant) adult *blackish brown above*, this extending onto *sides of head and neck, imparting a hooded effect*; forehead whitish; *below white*; tail grayish brown with several indistinct pale bars. In flight has *white underwing-coverts*, flight feathers whitish with blackish tips and barring. Dark-morph adult *entirely sooty black* except whitish forehead; tail as in pale morph. In flight *blackish underwing-coverts* contrast with whitish flight feathers. Light-morph juvenile has head streaked whitish buff, sides with dark streaking. Dark-morph juvenile as in adult but underparts and underwing-coverts variably marked with buffy whitish. Light morph readily known from its dark "hood." Light-morph White-tailed Hawk shares this character but it has a more barred underwing and obvious black band on white tail. Rarer dark morph can be more difficult. Cf. especially Zone-tailed Hawk, which has comparatively longer and narrower wings and tail with more obvious white banding; it flies in a dihedral. Dark-morph White-tailed Hawk is larger, with mainly white tail. Dark-morph Swainson's Hawk has an all-dark underwing. Short-tailed Hawk—whose tail is not particularly short—is *almost always seen in flight*, soaring high with other raptors and vultures. Perched birds generally remain hidden in leafy tree canopy. Hunting, it glides on flat wings, peering intently below. Upon sighting prey (usually a bird), it folds its wings and plummets after it. Usually silent, but occasionally gives a high-pitched whistle, "kleeeeu."

ROADSIDE HAWK
GAVIÃO-CARIJÓ

GRAY HAWK
GAVIÃO-PEDRÊS

SHORT-TAILED HAWK
GAVIÃO-DE-RABO-CURTO

ZONE-TAILED HAWK *Buteo albonotatus* 47-56 cm | 18.5-22"
Rare and local in forest and woodland borders and semiopen areas. *In flight bears a superficial but striking resemblance to Turkey Vulture.* Cere and legs yellow. Adult *uniform blackish; rather long, narrow tail* blackish with one wide and 1-3 narrow *white bands* (one in ♂; *2-3 in ♀). In flight wings relatively long and narrow, two-toned with black underwing-coverts and pale grayish flight feathers.* Immature has a variable amount of white flecking; tail *lacks* bold bands (instead is narrowly barred pale gray and blackish). Turkey Vulture flies with a similar tilting dihedral and shows similar two-toned underwing; but the larger vulture has a small naked head and lacks tail-banding. Dark-morph Short-tailed Hawk has more typical *Buteo* proportions (wings shorter and broader), shows much less distinct tail-banding. Zone-taileds are not often seen perched; when they are, separation from other dark raptors can be difficult. Cf. especially the larger, heavier Great Black Hawk, and dark-morph Long-winged Harrier. Usually seen in flight, tilting with wings held in a marked V-shaped dihedral. Generally flies low, sometimes accompanying Turkey Vultures. Appears to mimic vultures, presumably to render potential prey complacent and incapable of fleeing in time. Eats birds, small mammals, and lizards. Not very vocal, perched birds occasionally giving a rough scream, "reeeeah."

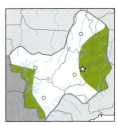

SWAINSON'S HAWK *Buteo swainsoni* 48-56 cm | 19-22"
Rare boreal migrant to agricultural and other open terrain, occurring mainly as a transient on southward passage (Oct-Nov; only a few records from our area). Cere and legs yellow. *Variable.* Light-morph adult (predominant) dark brown above with *whitish on uppertail-coverts; throat white* contrasting with *brown chest;* below mostly whitish; tail gray with blackish bars, tail underside pale grayish with faint dusky bars and wider subterminal band. In flight wings rather long, *underwing-coverts creamy white and flight feathers contrastingly dark grayish.* Dark-morph adult uniform dark brown to blackish with some buff barring on lower belly, crissum, and *whitish undertail-coverts.* Intermediates occur, also a rufous morph (these being browner overall, more rufous on lower underparts). Immature even more variable: buff face with dark eyeline and malar streak; otherwise dark brown above, buff below with *variable amount of dark spotting, often heavy,* and usually a *dark blotch on either side of chest; whitish on uppertail-coverts;* tail quite uniform. Light-morph adults are relatively easy to identify, but the more confusing immatures are seen at least as often in Brazil. On them, note especially the *absence* of any obvious tail pattern (except for whitish at base of uppertail) and the two-toned wing pattern from below (pale coverts, dark flight feathers). Swainson's often flies with a slight dihedral, a helpful field mark. Cf. immature of broader-winged White-tailed Hawk. Favors open country; migrants often rest on the ground. Breeds in North America, passing nonbreeding season on the Argentine pampas.

WHITE-TAILED HAWK *Buteo albicaudatus* 51-61 cm | 20-24"
Uncommon to locally fairly common in open terrain (cerrado, fields, etc.); avoids forested areas. Cere and legs yellow. Light-morph adult *slaty gray above and on sides of head* (the "hood"), with *prominent rufous on shoulders.* Below white with fine dark barring on sides and thighs; *rump and tail white* with *wide subterminal black band; wings extend beyond tail.* In flight wings long and broad; underwing-coverts white and flight feathers grayish. Dark-morph adult (less numerous) *uniform slaty gray,* sometimes with *a little rufous on shoulders;* tail as in light morph. Immature variable, blackish brown above with pale brow, blackish malar, and variably mottled whitish below; *tail pale grayish, palest basally and on uppertail-coverts.* Light-morph Short-tailed Hawk is smaller with a whiter underwing, less pattern on tail. Immatures generally show enough rufous on shoulders and/or whitish at base of tail as to be recognizable. The White-tailed Hawk is a magnificent, powerful raptor that tends to perch low, sometimes on fence posts or even on the ground, but also sometimes soars to great heights. It often hunts while hovering or volplaning into the wind, not too high, stooping down to ground after prey. Call repeated "klee," but usually silent.

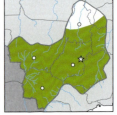

HARRIS'S HAWK *Parabuteo unicinctus*   48-53.5 cm | 19-21"

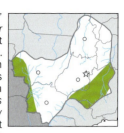

Rare and local in agricultural terrain and scrubby woodland in S of our area, where there are surprisingly few records. Cere, lores, legs yellow. Adult *blackish to sooty brown* with *rufous wing-coverts and thighs*; crissum white. *Uppertail-coverts, base and tip of rather long tail white*; rest of tail black. In flight wings rather long and narrow, *underwing-coverts rufous*, flight feathers blackish (*but in most lighting conditions looking all dark*). Immature brown with buff to whitish streaking and mottling, especially below; *already shows rufous on wing-coverts*, thighs more barred. *Underwing-coverts also already show some rufous*, flight feathers paler than in adult. Tail pattern as in adult but often less clear-cut. Handsome adult's rufous is distinctive, but can look all dark except in good light; the rufous is more obscure in younger birds, but generally is still apparent. Wings are longer and narrower than in any *Buteo*. Cf. larger and more rufescent Savanna Hawk and ♀/immature Snail Kite (with more hooked bill, etc.). Often perches in the open, even on the ground, sometimes several together. Eats small mammals as well as birds, captured in fast low flight. At least elsewhere known to hunt cooperatively, several birds together. Soars regularly, sometimes quite high. Quiet, but occasionally gives a descending "krreeaarrh."

SAVANNA HAWK *Buteogallus meridionalis*   53.5-61 cm | 21-24"

Widespread and relatively numerous in semiopen and agricultural areas. A *large, long-legged*, open-country raptor. Legs yellow. Adult *mostly dull cinnamon-rufous*. Upperparts and wings variably mixed with gray; underparts with inconspicuous fine black barring; rather short tail black with white median band and tip. In flight wings very long and broad, *underside mainly rufous*, flight feathers tipped black. Immature sooty brown above with buffy whitish forehead and superciliary and *some buff to rufous on wing-coverts and scapulars*. Below deep buff heavily streaked and mottled dusky; tail blackish with narrow pale buff banding. The handsome rufescent adults are unlikely to be mistaken. Immature Great Black Hawk resembles a young Savanna but lacks Savanna's rufous on wings and is shorter-legged. Cf. also young Harris's Hawk. A conspicuous raptor, Savanna is often seen perched on low posts or trees, and sometimes rests on ground. It feeds on a variety of prey including small mammals, birds, frogs, reptiles, and even large insects. Soars freely on broad flat wings. Usually silent, but has a drawn-out and fairly high-pitched "keeeeuu," given especially in flight.

BLACK-COLLARED HAWK *Busarellus nigricollis*   46-51 cm | 18-20"

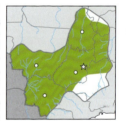

Common in marshy areas and near ponds in Pantanal; much less numerous (or absent) elsewhere. Unmistakable adult *bright rufous above* with *contrasting creamy whitish head and throat*, nape and back sparsely streaked black. *Below bright cinnamon-rufous with conspicuous broad black crescent across chest*. Short tail blackish with narrow rufous bands near base, white tip. In flight *wings very broad* and rounded, *mostly cinnamon-rufous with contrasting black primaries and trailing edge to secondaries*. Immature much duller, though it *already shows a contrasting pale head and dark brown collar*; above dark brown mottled and barred with rufous, below pale buff mottled and streaked with dark brown. Usually perches low, pouncing on fish in shallow water or aquatic vegetation; its spiny soles aid in grasping prey. Often allows a close approach. Though often sluggish, Black-collareds are powerful flyers, capable of soaring to great heights on their broad wings. Perched birds regularly give a low, raspy "reh-h-h-h."

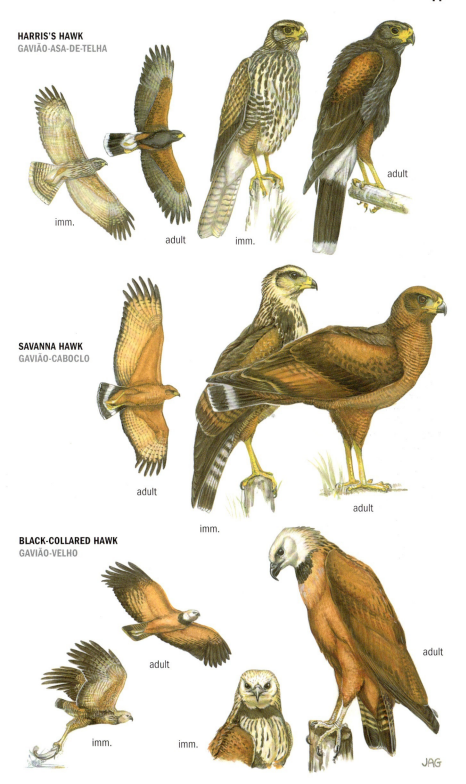

GREAT BLACK HAWK *Buteogallus urubitinga*   53.5-61 cm | 21-24"

Locally common in forest and woodland borders, often near water, in Pantanal; less numerous elsewhere, but widespread. Iris brown; *longish legs* yellow. Adult *slaty black; basal half of tail white, terminal half mostly black*. In flight wings broad and rounded. U*nderwing blackish, showing little pattern* (sometimes a little white at base of primaries). Immature has a buff face with *dark brown eyeline and malar streak;* dark brown above with buff and rufous streaking, mottling; below buff coarsely streaked dark brown, thighs densely barred; tail longer than adult's, grayish with 5-8 blackish bars, below paler. Perched immatures resemble immature Savanna Hawk, though even younger individuals of that species have usually begun to show some rufous, especially on their wing-coverts. This powerful raptor eats a variety of prey (birds, small mammals, amphibians, and even fish). It soars regularly, sometimes ascending very high. Most frequent call a drawn-out, high-pitched scream "kueeeeeeeeeeeeu!" given both while perched and in flight; also a fast "kukukukukukukukukú."

CROWNED SOLITARY EAGLE *Harpyhaliaetus coronatus*   71-84 cm | 28-33"

VU *Rare and local in woodland and cerrado, mainly in areas with some topographic relief*. A *huge eagle*, now seriously reduced in numbers. Adult *brownish gray*, darker on bushy crest and mantle; *short tail black with one white band*, pale tip. In flight adult has *gray wings with black wing-tips and trailing edge*. Immature brown above with buffy whitish face and brow. Below buffy whitish, irregularly streaked dusky, often darker on sides of chest; thighs dark brown. In flight immature has *rather pale underwing*, feathers with black tips and trailing edge. This tremendous eagle can usually be known by its immense size alone. Adult Great Black Hawk is blacker, lacks crest; immature also lacks crest, has barred (not dark) thighs. Immature Black-chested Buzzard-Eagle, also a large bird (though not so extremely so), is darker and more heavily marked below, including its underwing. Rarely encountered, the spectacular Crowned Solitary Eagle always occurs at low densities and now is confined to wilder cerrado regions; it likely is more at risk than current Brazilian lists would indicate. It takes prominent perches, most often in trees but also sometimes on the ground or even termitaries. Preys on mainly terrestrial mammals and birds, but habits still relatively poorly known.

BLACK-CHESTED BUZZARD-EAGLE *Geranoaetus melanoleucus*   62-68 cm | 25-27"

Local and very uncommon in semiopen, hilly terrain in S and E parts of our area. A *large* and impressive raptor of open lands. Sometimes placed in the genus *Buteo*. Iris brown. Adult nearly unmistakable, with a *short wedge-shaped tail* that barely protrudes past wings in flight. *Upperparts as well as throat gray, becoming black on breast; belly white with a little fine dusky barring*. Wing-coverts pale gray vermiculated black; tail dark gray. In flight wings long and broad, especially at base (combined with short tail, giving an unusual *delta-wing effect*); underwing-coverts whitish, flight feathers grayer. Immature dusky brown above with *pale superciliary but without gray shoulders;* whitish to buff below, heavily streaked brown; *breast often extensively dark; tail longer than in adult*, thus imparting a more "normal" silhouette. Immature Crowned Solitary Eagle is even larger and is markedly paler below, with less dark blotchy streaking on underparts and a whiter underwing. A powerful bird of prey that mostly takes small mammals. Regularly seen in pairs, often gliding effortlessly along cliffs and ridges on flat or slightly raised wings. Often perches on rocks or the ground. Not very vocal, but sometimes gives a surprisingly high-pitched, broken "kee-ee-u" or "kee-kee-kee," both while perched and in flight.

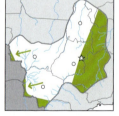

GREAT BLACK HAWK
GAVIÃO-PRETO

CROWNED SOLITARY EAGLE
ÁGUIA-CINZENTA

BLACK-CHESTED BUZZARD-EAGLE
ÁGUIA-CHILENA

HAWK-EAGLES are magnificent *large, forest-based* raptors. Very uncommon in our area, they sometimes are seen soaring at mid-day on outstretched wings.

BLACK-AND-WHITE HAWK-EAGLE *Spizaetus melanoleucus* 53.5-61 cm | 21-24"

Rare and local in forest and borders; few records from our area. Formerly placed in the genus *Spizastur*. Iris yellow; *cere bright orange*, contrasting *black lores and bill*. Adult has *short black crest; snowy white head, neck, and underparts. Above otherwise black.* Tail blackish with three gray bands, its underside whitish with 3-4 dark bars (the terminal band widest). In flight wings broad and relatively long, *underside mainly white* with faint dark barring on flight feathers. Also shows a prominent white leading edge ("headlights") to inner wing. Immature slightly browner above. Immature Ornate Hawk-Eagle has a long pointed crest, only dusky lores (Black-and-white's *black* lores are remarkably conspicuous). Ornate shows bold black barring on flanks and thighs; in flight its wings are proportionately shorter, its tail longer, and Ornate lacks Black-and-white's white leading edge to wing. Immature light-morph Gray-headed Kite lacks orange and black on face. This stunning raptor is most often seen in graceful soaring or gliding flight. It hunts mainly from the wing, stooping swiftly onto unsuspecting prey (primarily birds), but also waits in ambush from a partially hidden perch. Unlike the other hawk-eagles, the Black-and-white seems quiet.

ORNATE HAWK-EAGLE *Spizaetus ornatus*   58.5-68 cm | 23-26.75"

Rare and local in forest and borders. A *beautifully patterned* hawk-eagle whose *long erectile pointed crest is usually held straight up*. Iris yellow. Adult has *black crown and crest; face, neck, upper back, and sides of chest rich rufous* bordered by long and bold black malar stripe; above brownish black. Below white with *breast, belly, and thighs prominently barred black*. Long tail boldly banded black and pale gray. In flight wings broad and rounded, *mostly whitish* with flight feathers banded blackish. Immature has *head, neck, and underparts white*, and already has crest and *black barring at least on flanks and thighs*; above browner. The smaller Black-and-white Hawk-Eagle has conspicuous black lores and an orange cere; it lacks the long pointed crest and never shows black barring below. Cf. also younger Black Hawk-Eagle, Gray-headed Kite. Behavior similar to Black Hawk-Eagle though Ornate seems not to soar as often or as high, more often just circling fairly low above the forest canopy; it seems less likely to soar over adjacent semiopen areas than the Black. Flying birds vocalize frequently, with quality of their call similar to Black's though its pattern differs with *long note first*, "wheeeeer, whip, whip, whip, whip."

BLACK HAWK-EAGLE *Spizaetus tyrannus*   58.5-66 cm | 23-26"

Rare and local in forest, borders, and adjacent clearings, mostly in NW of our area. Iris yellow. Adult *black; erectile bushy* crest basally white; *flanks and thighs barred white. Long tail* banded black and dusky-gray, *tail underside more boldly banded black and paler gray*. In flight wings long, broad, and rounded held slightly forward with notably indented rear margin ("butterfly wings"), with *underwing-coverts black, flight feathers boldly banded black and white*. Immature quite variable, at first with a creamy white superciliary, crown and sides of head mottled black and white; above blackish brown, below mixed buff and blackish (streaked on chest, barred lower down, especially on thighs); underwing-coverts more marked with white than adult's. Cf. Ornate Hawk-Eagle. Some Hook-billed Kites have somewhat similar flight patterns. No other blackish raptor has such conspicuous banding on the underside of its wings. A powerful raptor that hunts birds (up to the size of macaws) and certain small to midsized mammals, stooping on them from a perch. More often seen soaring (especially during sunny midmornings) than when perched. The frequent calling of soaring birds often draws attention to them, a loud and far-carrying whistle "wheep, wheep, wheeteeeeeer" with a distinctive long final drawn-out note (which can be given on its own).

BLACK-AND-WHITE HAWK-EAGLE
GAVIÃO-PATO

ORNATE HAWK-EAGLE
GAVIÃO-DE-PENACHO

imm. adult imm. adult

BLACK HAWK-EAGLE
GAVIÃO-PEGA-MACACO

adult imm. adult

imm.

CRESTED EAGLE *Morphnus guianensis* 71-84 cm | 28-33"

Very rare in forest in NW of our area (Guaporé); very few records. Very large. Almost the length of a Harpy Eagle but more slightly built, with *broad tail proportionately longer. Long, erectile, single-pointed crest*. Iris brown; heavy bill blackish. ♀ much larger than ♂. Light-morph adult has *head and chest pale ashy to brownish gray* with crest black; above blackish with wing-coverts fringed whitish; below white with some fine cinnamon barring, especially on thighs; tail boldly banded black and pale gray. In flight wings broad and rounded with *underwing-coverts white*, flight feathers white barred blackish. Rarer dark-morph ("banded") adult darker above, chest blackish, *below boldly banded black and white*. Intermediates occur. Younger birds somewhat variable. *Head, neck, and underparts white* (foreparts gradually becoming grayer), wing-coverts (especially) and back with whitish marbling, tail at first with narrow banding. The even larger Harpy Eagle is more heavily built, with broader wings, shorter tail, and much thicker legs. Adult Harpies are obvious enough, but younger individuals of the two species can be easily confused (especially ♀ Cresteds vs. ♂ Harpies). Note especially the Harpy's bifurcated crest, and its black on underwing-coverts. Cf. also the smaller *Spizaetus* hawk-eagles. Surprisingly secretive for such a large bird, the Crested Eagle usually remains within the canopy, but occasionally one will rest in the open at the forest's edge. Like Harpies, Crested Eagles basically ignore people and thus are all too often shot. Occasionally one circles above the canopy or sails across a valley, but it does not soar high. Feeds on small to midsized mammals (even small monkeys), snakes, and some birds. Call a loud, hawk-eagle-like scream, "wheyr-wheyr-wheyr-wheyr-wheyr-whéyr-br."

HARPY EAGLE *Harpia harpyja* 89-99 cm | 35-39"

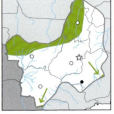

Rare and local in forest in NW of our area, e.g., a well-known eyrie at Serra das Araras. Doubtless declining, but was never a numerous bird here. A *huge, powerful* eagle; very heavily built, with *massive bill, legs (especially), and talons. Long, erectile, double-pointed crest*. Iris amber. ♀ much larger than ♂. Adult has *head and neck pale gray*, with crest blackish; above blackish; *broad chest band black*; below white with *thighs black-barred*; broad, fairly long tail boldly banded black and pale gray. In flight wings very broad and rounded, *underwing-coverts white with a conspicuous black bar, black axillars*, flight feathers white with black barring. Juvenile and immature variable, needing four years to attain adult plumage: *head (even crest), neck, and underparts white, coverts and back very pale gray* (gradually darkening); tail at first with more narrow banding; underwing-coverts at first lacking the black. The Harpy is so huge that confusion is improbable except with the somewhat smaller and slighter Crested Eagle (cf. under that species). It is helpful that traces of the Harpy adult's black chest band appear early on. Though considered the most powerful bird of prey in the world, given its size the Harpy is a remarkably inconspicuous bird. Usually remaining inside forest, one will occasionally perch at the forest's edge or launch across an open area such as a river. Harpies are unafraid of people, doubtless due to their great size, and rarely flee. Perched birds thus present tempting targets and all too often end up getting shot. The Harpy never soars and almost never is seen above canopy level in forest. It feeds on various midsized mammals (including monkeys and sloths) and larger birds. If the prey it has killed is too heavy, it may dismember it on the spot. A whistled "wheeeee" or "wheeee-wheeea" is given infrequently, especially around the nest.

FALCONS & CARACARAS (Falconidae) form another group of diurnal birds of prey, differing from Accipitridae in their notched upper mandibles and other anatomical characters. Recent evidence indicates the two families may not be that closely related. ♂♂ are substantially smaller than ♀♀.

CARACARAS are classified in no less than four genera. The first two species are noisy conspicuous forest birds; the last two are open-country species that mainly eat carrion.

BLACK CARACARA *Daptrius ater*   40.5-43 cm | 16-17"

Fairly common and conspicuous at forest edge, clearings, and along rivers in *NW of our area*. *Extensive bare face bright yellow orange* (immatures yellower); legs orange-yellow. *Glossy black* with *base of tail white*. Immature has a variable amount of buff spotting or barring below. Red-throated Caracara is considerably larger and more heavily built with white on belly, not tail; it is more a true forest bird. Black Caracara ranges in pairs or small groups of birds, often flying quite high as they patrol riverbanks and other open areas in search of food. It is nearly omnivorous. Often seen resting in the open on snags or sandbars, and also perches freely in trees. Regularly gives voice to a harsh, loud scream, but nowhere near as raucous as in Red-throated.

RED-THROATED CARACARA *Ibycter americanus*   51-56 cm | 20-22"

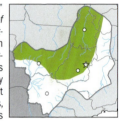

Uncommon and local in canopy and borders of forest and woodland in *N of our area*. Declining. *Bare face and throat red*. Adult *glossy black* with *contrasting white belly*. Immature's face more yellowish, bare throat black. Occurs in small, noisy groups of five to six birds that fly with deep, slow wingbeats, generally through or just above the canopy, never soaring. Foraging birds sometimes descend lower, and at such times can be notably unsuspicious. Feeds mostly on bee and wasp eggs and larvae obtained by ripping apart nests, without being stung; also eats some fruit. No other bird, not even the great macaws, makes as much noise as a group of this species. Most frequent is a raucous "kra-kra-kra-kra-kow" or just a "kra-kow;" also a more mellow "kowh" repeated several times; in flight often a far-carrying "krraah."

YELLOW-HEADED CARACARA *Milvago chimachima*   40-45 cm | 15.75-17.25"

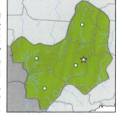

Common and widespread in semiopen and agricultural areas. A *small caracara with weak bill and talons*. Adult *creamy to yellowish buff* with *narrow dark postocular streak*; back and wings blackish brown. *Uppertail-coverts buff*; rather long and rounded tail yellowish buff with dusky bars and broad dusky tip. In flight wings dark brown with *buff patch on primaries*. Immature *much darker* with head and neck streaked brown, throat and breast mottled and streaked. Southern Crested Caracara is much larger, blackish and white, etc. Cf. Laughing Falcon. Seen singly or in pairs, conspicuous in open country and towns; patrols highways for roadkill, and also walks on fields. Eats mainly carrion, but also takes some live food. Frequently accompanies cattle, sometimes capybaras, even perching on their backs (apparently to pick off ticks). Flight consists of several shallow flaps followed by a glide on downbent wings; does not soar. Has a rather weak, nasal scream, "wreeeah."

Chimango Caracara (*M. chimango*) occurs as a vagrant from further south along S edge of our area. Resembles immature Yellow-headed but much darker and more uniform, with less streaking. Behavior similar, with equally catholic habitat preferences.

SOUTHERN CRESTED CARACARA *Caracara plancus*   51-58.5 cm | 20-23"

Common and widespread in semiopen and agricultural areas. Cere, base of heavy bill, and lores reddish orange to yellow (yellower in younger birds?); bill pale blue, long legs yellow. Adult has *black crown with slight bushy crest* contrasting with *buffy whitish sides of head, throat, and chest*; neck, back, and breast with blackish barring; belly and thighs black, crissum white; uppertail-coverts and basal half of tail white with dark barring. Tail tip black. In flight long wings black with *prominent patch of white in primaries*. Immature has *blackish crown*, is browner overall; adult's white areas (including wing patches) are buffier; upper back and underparts vaguely streaked. Adults are unmistakable and even younger birds should show enough head pattern and bill color as to be recognizable. A conspicuous bird, usually perching in the open, often low on fences or on the ground, where it strides about authoritatively. Flies strongly with steady, deep wingbeats; sometimes ascends very high, but soars only briefly. An opportunistic feeder, mainly eating carrion (regularly with vultures) but also taking some live prey. Regularly seen along highways, where it feeds on roadkills. Usually quiet, though gives various low guttural calls.

BLACK CARACARA
GAVIÃO-DE-ANTA

RED-THROATED CARACARA
GRALHÃO

YELLOW-HEADED CARACARA
CARRAPATEIRO

SOUTHERN CRESTED CARACARA
CARCARÁ

LAUGHING FALCON *Herpetotheres cachinnans* 46-51 cm | 18-20"
Uncommon in canopy, borders of forest and woodland, and adjacent open areas with scattered trees. A *distinctive* "puffy-headed" raptor, with *head and underparts pale buff, contrasting broad black mask from lores to nape,* blackish streaking on crown. Above dark brown, tail blackish with 3-4 buff bands. In flight blackish wings show a prominent buff patch at base of primaries. Much slighter Yellow-headed Caracara has similar buff coloration, but no broad mask (only a thin eyestripe). Usually seen singly, perched upright on a prominent branch with good visibility; often remains motionless for long periods, quietly watching for snakes, its primary prey. Flies with fast, stiff, shallow wingbeats (recalling an *Amazona* parrot); does not soar. Well known from its distinctive, loud, "laughing" vocalizations, given mainly around dawn and dusk, but also at other times (even at night). Most frequent are "guá-co, guá-co, guá-co..." or simply "guá, guá, guá...," often repeated for several minutes. At times the calls are less measured, breaking into what sounds like maniacal laughter. Occasionally pairs vocalize together, calling in syncopation.

MICRASTUR forest falcons are *inconspicuous* forest-dwelling raptors with short rounded wings, *long graduated tails*, long legs, and *fairly distinct facial ruffs*. Our two species have *loud vocalizations*, and are heard far more often than seen.

BARRED FOREST FALCON *Micrastur ruficollis* 33-38 cm | 13-15"
Uncommon *in forest and woodland*, mainly in *humid habitats. Facial skin and wide orbital ring bright yellow*; long legs yellow. Many color forms. Gray-morph adult *slaty above*, wings somewhat browner (especially ♀); throat pale; *below white evenly barred black*. Graduated tail blackish with 3-4 narrow pale bars and tip. Rufous-morph adult browner above and with *rufous foreneck*. Immature *above dark brown*, head blacker, *usually with a partially broken whitish to pale buff nuchal collar*; *buff to whitish below* with *coarse dusky barring* (some birds are all barred below, others almost unmarked). Adult Tiny Hawk is much smaller (especially ♂), lacks extensive yellow facial skin (yellow only on cere), and has wider gray tail banding (not white). Immature Bicolored Hawk also has wider grayish tail banding, and never shows barring below (the vast majority of Barred Forest Falcons show at least some). Cf. also immature of larger Collared Forest Falcon. Perches quietly inside forest, mainly low and at midlevels; sometimes confiding, but more often flushes and disappears. Attends army antswarms, then often on the ground. Eats some birds, though birds at antswarms don't seem to be especially afraid of it. Like other forest falcons, Barred never soars or flies above the forest canopy. Its usual vocalization is a repeated sharp stacatto "our!" or "ahnk!" given at several-second intervals (speeds up when excited). Calls primarily around dawn and dusk, sometimes even in full darkness. Agitated birds also give a much faster series of notes, rising and falling in pitch and loudest in the middle: "koh-koh-koh-ko-kó-kó-ke-ke-ke-keh-keh-kuh."

COLLARED FOREST FALCON *Micrastur semitorquatus* 51-58.5 cm | 20-23"
Uncommon *in forest, woodland, and borders* in deciduous as well as humid habitats. *Large* and *lanky*, with a *very long graduated tail*. Cere, base of bill, and orbital ring yellowish olive. Many color forms. Adult blackish above, black of crown *arching across cheeks as a crescent; white to buff (rarely tawny) below, the color extending up onto nuchal collar*. Rare dark morph *all sooty black to blackish brown*, some birds with pale spotting or barring below. Long tail blackish with 3-4 narrow whitish bands. Immature like adult above though browner, some with buff spotting or barring; nuchal collar may be less distinct; below whitish to deep buff, *coarsely barred or chevroned dark brown*. The facial crescent and nuchal collar are conspicuous in all plumages; along with its large size and lanky shape, these should be sufficient to identify it. Bold and rapacious but usually furtive, remaining in dense leafy or viny cover. Feeds primarily on birds. Vocalizes mainly at dawn and dusk, with most frequent call a series of hollow resonant "oow" notes repeated steadily, sometimes long-continued. Also gives a rising series of 8-11 "ko" notes that ends with two lower-pitched "oow" notes, and occasionally a faster series with more of a laughing quality. Calling birds almost always remain hidden.

FALCO falcons vary in size from the small kestrel to the large and very powerful Peregrine. ♂♂ are much smaller than ♀♀. They fly very fast on *long pointed wings.*

BAT FALCON *Falco rufigularis* 24-28 cm | 9.5-11"
Uncommon at forest and woodland borders, also often perching in isolated trees in clearings. Head and upperparts dark slaty; throat, sides of neck, chest whitish buff; *broad black "vest" on breast, flanks narrowly barred whitish/buff;* lower belly, crissum rufous. In flight wings pointed, narrow; underwing blackish, white patterned. Cf. the very rare Orange-breasted Falcon. White-collared Swift might be taken for a high-flying Bat Falcon. Singletons or pairs may perch for long time on favored snag. Flight fast, dashing. Captures prey on wing (birds, large insects, and bats). Both sexes give a loud shrill "kee-kee-kee-kee..." or "kiy-kiy-kiy-kiy...."

ORANGE-BREASTED FALCON *Falco deiroleucus* 35.5-40.5 cm | 14-16"
Rare and very local at forest edge and around gorges with extensive cliffs. No local breeding sites are known (recent sightings include Chapada dos Guimarães). Robust. *Very heavy talons* notable when perched. Like Bat Falcon, but blacker above; *white throat; orange-rufous chest and sides of neck; dark "vest" on breast with coarse and more prominent buff to whitish barring.* Immature paler below. In flight broad-based wings are more like Peregrine than Bat Falcon. Much more numerous Bat Falcon can be hard to distinguish, its ♀ being almost as big as ♂ Orange-breasted. Takes commanding perches on snags and cliffs (where it nests). Hunts birds up to parrot size. Stoops less than Peregrine, instead overtaking prey by surprise and speed. Infrequently heard call a loud "kyah-kyah-kyah-kyah...," less shrill than Bat Falcon's, given mostly around mating sites.

PEREGRINE FALCON *Falco peregrinus* 38-48 cm | 15-19"
Uncommon in open areas, generally near water. Boreal and austral migrants occur (status unclear); not known to breed here, but could. *Large and heavily built with long broad-based wings.* Boreal migrant adult (illustrated) *dark bluish gray above,* blacker on head; *wide black moustache, whitish sides of neck.* Below white, tinged pinkish buff; *breast and belly barred black.* Austral migrant adult slightly darker above, blacker sides of head; below averages buffier. In flight has buffy whitish underwing barred blackish. Immature *browner above;* head buffier (forehead and superciliary often whitish); *extensive dark streaking below.* Seen singly, perching in the open. Hunts during fast, strong flight ("the fastest bird on earth"). Feeds mostly on larger birds (sometimes bats) taken in flight after spectacular stoops or direct pursuits. Usually quiet here. Nearly cosmopolitan, nesting locally on cliffs and buildings.

APLOMADO FALCON *Falco femoralis* 37-43 cm | 14.5-17"
Uncommon in open terrain, cerrado, agricultural areas. Slender, with long wings and tail. Adult bluish gray above. *Pale buff superciliary encircles gray crown;* black moustache, whitish cheeks; *throat white; chest and sides of neck buff; sides and breast band blackish narrowly barred white; lower belly and thighs rufous;* tail blackish with narrow whitish bars. In flight wings pointed, underside dark with white bars and spots, and *narrow white trailing edge.* Immature browner above, chest more streaked, lower underparts paler. Larger, heavier Peregrine lacks rufous in plumage, has broader wings and a shorter tail. Smaller Bat Falcon lacks Aplomado's buff brow, long tail. Seen singly or in pairs; usually perches in the open, often on a rock or promontory, sometimes on ground. Hunts a variety of prey, capturing birds in fast level flight and also hawking insects. Often attracted to grass fires. Other than occasional cackles, usually silent.

AMERICAN KESTREL *Falco sparverius* 25.5-29 cm | 10-11"
Fairly common, widespread, and conspicuous in open, semiopen agricultural and built-up areas; tolerates disturbed habitat if agricultural chemical use is not too intense. *Small.* Long tail, long narrow pointed wings. ♂ has *rufous back,* blue-gray crown, white face with *black moustache and ear stripe;* wings mostly blue-gray. Below pale cinnamon, deepest on breast, *with black spots; long tail rufous with black subterminal band,* white on outer feathers. ♀ *above rufous barred blackish,* brown streaked below; brown tail barred. Similarly sized Pearl Kite has all gray upperparts and a shorter tail that lacks barring. Flying Pearl Kite is whiter below; it never hovers. Kestrels perch on wires and roadside fences and hunt while hovering, plunging to ground after insects (less often small mammals, lizards). Call a shrill "killy-killy-killy..." or "kree-kree-kree...."

BAT FALCON
CAURÉ

ORANGE-BREASTED FALCON
FALCÃO-DE-PEITO-LARANJA

PEREGRINE FALCON
FALCÃO-PEREGRINO

APLOMADO FALCON
FALCÃO-DE-COLEIRA

AMERICAN KESTREL
QUIRIQUIRI

RAILS, GALLINULES, & COOTS (Rallidae) typically are skulking chicken-like birds that occur near water; many are best known from their calls. Some gallinules often swim.

OCELLATED CRAKE *Micropygia schomburgkii* 14-15 cm | 5.5-6"

Locally fairly common in tall-grass campos, now found mainly in protected areas, e.g., Emas and Brasília NPs. Tiny; secretive. Legs coral red. Above buffy brown *with black-edged white spots*. Face and underparts bright yellowish buff. Cf. Ash-throated Crake (much larger, etc.). Almost invisible in tall grass, emerging mainly after tape playback or if forced into the open by fire. Far-carrying ventriloquial song, often given as a duet, a combination of almost musical tinkling, "titititi," often long-continued, and high-pitched growls: "tzreeer, tzreeer...."

LATERALLUS crakes are small, *inconspicuous* rails with *short stout bills* and short cocked tails. Heard more often than seen, they range in marshy areas.

RUFOUS-SIDED CRAKE *Laterallus melanophaius* 15.5-16 cm | 6-6.25"

Uncommon and somewhat local in marshy edges and damp areas with luxuriant grass (usually near standing water); strangely scarce in Pantanal. Bill pale greenish; legs yellowish brown. Olive brown above. *Face, sides of neck, and breast rufous*; white median underparts; *flanks white banded black*. Never easy to see, but sometimes responds to tape playback; soon after dawn may emerge to feed on floating vegetation at water's edge. Presence usually revealed from its abrupt descending churrings, "tr-r-r-r-r-r-r-r-r," that last several seconds and are sometimes answered by another bird, presumably its mate; usually longer and stronger than Gray-breasted Crake's. Also gives high-pitched tinkling calls like Gray-breasted and a short "treeeeeng" call.

RUFOUS-FACED CRAKE *Laterallus xenopterus* 14-15 cm | 5.5-6"

Poorly known. Apparently rare and very local (overlooked?) in *marshy or seasonally flooded campos*. Recorded only from Brasília NP and W Minas Gerais. Bill and legs bluish gray. Head and neck rufous, tail black; *bold black-and-white wing-barring*. Foreneck ochraceous buff, throat and lower underparts white, flanks banded black. Rufous-sided Crake lacks obvious wing-barring, has rufous sides, greenish bill, brownish legs. Voice a descending churring similar to Rufous-sided's.

GRAY-BREASTED CRAKE *Laterallus exilis* 13-14 cm | 5-5.5"

Locally fairly common in wet areas with tall grass, shrubs, and marshy vegetation around lakes, mainly in Pantanal. Bill blackish, pale green at base of mandible; legs grayish brown. *Head, neck, and breast gray; chestnut nape and upper back*; white throat. Olive brown above, often some white barring on wing-coverts; belly white, flanks and crissum banded black. Smaller than other *Laterallus*, this species' chestnut nape patch is unique. Furtive and usually nearly invisible, scurrying mouse-like in dense grass, sometimes creeping up stems. Rarely emerges from cover, never straying far. Even with tape playback, hard to see. Call a short descending churring, often introduced by scratchy notes; typically harsher and briefer than Rufous-sided's call. Also gives high-pitched tinkling calls: "ti, tee-tee-tee-tee."

RUSSET-CROWNED CRAKE *Laterallus viridis* 15.5-16 cm | 6-6.25"

Rare to locally uncommon in dense undergrowth of woodland borders and regenerating clearings; *not tied to vicinity of water*. Bill blackish; iris bright red; legs reddish. *Crown rufous; gray sides of head*; above olive brown. *Below cinnamon-rufous*, paler on throat. The larger Uniform Crake (rare) has a greenish yellow bill, no rufous on cap or gray on face. Furtive and hard to see, but sometimes more in the open in early morning; occasionally climbs up into thickets. Call a long descending churring, more sputtering and hesitant toward end; more staccato than Rufous-sided Crake's call.

ASH-THROATED CRAKE *Porzana albicollis* 22 cm | 8.5"

Fairly common and widespread in damp grassy areas and marsh edges. Not a typical marsh rail; *regularly far from open water*. Bill olive yellowish; legs dusky. Above blackish brown, *feathers edged buffy appearing streaky*. *Face and underparts gray*; white midthroat, flanks banded black and white. Gray-breasted Crake is *much* smaller. Ash-throateds sneak into the open more than most rails and flush more readily, flying a short distance before dropping back into the grass. Distinctive call a loud far-carrying "dr-dr-dr-dr-drow," repeated rapidly up to 15-20 times, subsiding toward end.

PARDIRALLUS are midsized *marsh-dwelling* rails with *fairly long, brightly colored bills*.

SPOTTED RAIL *Pardirallus maculatus*   27 cm | 10.5"

An unmistakable, boldly patterned rail found very locally in marshes in S Pantanal. Bill greenish yellow with red spot at base; legs coral pink. *Above blackish boldly spotted and streaked white*. Throat and foreneck spotted and streaked black and white; below coarsely barred black and white. Immature has duller softparts with a variable amount of barring below. Usually stays in thick marshy vegetation; at dawn and dusk occasionally emerges to feed on mud or floating vegetation. Gives a variety of loud grunting or squealing calls: "skeey, skeeuh-skeeuh."

BLACKISH RAIL *Pardirallus nigricans*   28.5 cm | 11"

Uncommon and local around marshy ponds and adjacent wet pastures. Bill lime green; legs coral pink. Head and underparts *slaty gray*; above dark olive brown. *Long bill plus overall uniform dark plumage distinctive*. Favors dense cover but can be less reclusive than other rallids, sometimes even poking about in the semiopen though never very far from cover; can even scoot across openings and roads. Distinctive call a loud duet of a shrill and sharply rising "drüü-ee-ee-eet?" (presumably given by ♂) with various softer, more guttural clucks and growls (presumably given by ♀).

UNIFORM CRAKE *Amaurolimnas concolor*   20.5-21.5 cm | 8-8.5"

Rare and seemingly very local (overlooked?) in *swampy forest, woodland, and near streams*. Has been recorded only from Emas NP and far W. Bill greenish yellow; iris red; legs reddish. *More or less uniform rufous brown*; brighter and paler below with whitish throat. Cf. Russet-crowned Crake (with dark bill, gray face, different voice). In shape Uniform Crake is reminiscent of a wood rail. Hardly known here, elsewhere found singly or in pairs. Rarely leaves dense cover, and thus very hard to see. Walks erectly, often flicking or pumping its cocked tail; when alarmed, crouches and usually scurries off. Distinctive song a series of loud upslurred whistled notes that first rise and become louder, then fade away, basically a repetition of a "tooeee" note. Seems mainly to sing in the calm at dusk.

WOOD RAILS are *large, colorful, long-billed* birds of marshes and swampy woods. Easier to see than many members of their family, they walk with head and neck held upright, often flicking their usually cocked tails. They have notably loud vocalizations.

GRAY-NECKED WOOD RAIL *Aramides cajanea*   33-38 cm | 13-15"

Fairly common and widespread around streams, ponds, swampy and marshy areas; *especially numerous in Pantanal*. Greenish yellow bill, orange-yellow near base; legs coral red. *Head, neck, and chest gray*, browner on crown, whiter on throat, *cinnamon-rufous below*. Above brownish olive; rump, tail, and lower belly black. Flight feathers rufous (seen mostly in flight). Not likely confused, but cf. Giant Wood Rail (local). Usually wary, remaining in wooded or shrubby cover, though in Pantanal this species is much tamer and easier to observe, frequently emerging to feed at muddy margins and often crossing roads. Probes mud or leaf litter, flicking aside leaves with its bill. Well known from its loud cackling calls, often given as an antiphonal duet or chorus; heard mostly at dawn or dusk, sometimes even at night. Phrasing varies, but a repeated "kok" or "ko-kee" is often followed by a "kow-kow-kow-kow-kow;" this can vary to a "kok-a-lok" or "ko-wey-hee." During vigorous singing bouts the birds can sound truly maniacal, with the chorus continuing for several minutes.

GIANT WOOD RAIL *Aramides ypecaha*   44-47cm | 17-18.5"

Uncommon but conspicuous in marshes, adjacent wet grassy areas, and pastures in N of our area (upper Rio Araguaia drainage, N Minas Gerais); may also occur in S Pantanal, but seemingly not yet recorded. *Very large*. Bill greenish yellow; long legs coral pink. *Hindcrown and nape cinnamon-rufous* blending into pinkish brown on most of underparts and *contrasting with pale bluish gray face, throat, and chest*. Above brownish olive; rump, tail, and lower belly black. Flight feathers rufous (seen mostly in flight). A handsome if gangly bird, not likely confused. Gray-necked Wood Rail is much smaller, has mainly gray head and neck (no rufous). Behavior similar to Gray-necked, though Giant seems more willing to walk onto dry land farther from water. Calls vary, and consist of a series of loud squeals, squawks, and more complex phrases that sometimes sound almost deranged.

PURPLE GALLINULE *Porphyrula martinica*   28-32 cm | 11-12.5"

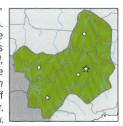

Uncommon to locally fairly common in marshes, around ponds and lakes. Sometimes placed in the genus *Porphyrio*. *Beautiful* adult has *pale blue frontal shield, red bill tipped greenish yellow*, and *very long legs and toes yellow*; immature's bill and legs yellowish. Adult unmistakable with *head, neck, and underparts deep violet-blue*; mid-lower belly blacker, *crissum white* (conspicuous on flushed birds). *Above bronzy bluish green*. Immature *much duller: brown above* with a *bluish tinge on wings*; *whitish below*, tinged buff across breast, with *white crissum*. Azure Gallinule is considerably smaller. Immature Common Moorhen is much grayer generally, especially below. Walks about on floating vegetation, using its long toes to good advantage; tail usually held cocked, exposing the fluffy white crissum. Unlike Common Moorhen, the Purple rarely swims, though it does often clamber about in shrubby vegetation. Flight usually weak and slow, but birds do disperse, and they sometimes turn up in unexpected places. Gives various clucking and cackling calls, usually in a fast series: "kuhkuhkuhkuh-kurrah-kuh."

AZURE GALLINULE *Porphyrula flavirostris*   23-25 cm | 9-10"

Uncommon and decidedly local on floating vegetation mats and in grassy areas around lakes and ponds; more numerous in the rainy season (Dec-Mar). A *small* gallinule with *"faded" plumage*. Sometimes placed in the genus *Porphyrio*. Bill and frontal shield yellowish green; legs yellow to orange-yellow. Mottled olive brownish above, back with some dusky streaking; *wings strongly tinged cerulean blue* and tail dusky. *Face and sides of neck and chest azure blue*, breast tinged azure blue; *otherwise white below*. Immature browner above and buffier on face and sides of neck, but with *wings already tinged blue*. So much smaller and more delicate than the Purple Gallinule that confusion is unlikely. Immature Purple lacks immature Azure's mottling on back. Cf. also Ash-throated Crake (no blue, legs brownish). Shyer and more skulking than Purple Gallinule, only infrequently in the open, Azure Gallinules tend to remain hidden in grassy vegetation and rarely or never swim. They especially favor floating grass mats. When pressed flushes a short distance with legs dangling, then crashes back into vegetation; just after landing sometimes briefly holds up wings. Much quieter than other gallinules.

SPOT-FLANKED GALLINULE *Porphyriops melanops*   26-28 cm | 10.25-11"

Very rare and local on small, usually marsh-fringed ponds in generally open terrain; spreading and increasing in many regions, though so far not in our area. Sometimes placed in the genus *Gallinula*. *Small frontal shield and stout bill lime green*. Face and crown blackish, becoming slaty gray on sides of head, neck, and breast. Above olive brown with chestnut on wing-coverts; short tail blackish. *Flanks brownish boldly spotted white, midbelly and crissum white*. This diminutive gallinule is well-named, for its flank spotting is conspicuous even on swimming birds; no other rallid shows anything like it. Common Moorhen is considerably larger and has a red bill. Spot-flankeds are almost entirely aquatic, and unlike Common Moorhens are rarely or never seen walking on the ground. Usually in small groups, they sometimes swim on open water but more often are seen as they putter about on floating vegetation, picking at water's surface or vegetation, often nodding head. When alarmed, swimming birds usually retreat into dense vegetation. Has a variety of clucks and cackles, sometimes given while chasing each other on the water.

COMMON MOORHEN *Gallinula chloropus*   33-35.5 cm | 13-14"

Uncommon and local on and around *ponds and marshes*. Though a widespread and numerous bird elsewhere, in our region quite scarce. *Frontal shield and stout bill red, bill yellow-tipped*. Adult *mostly slaty gray* (browner on back and wings, blacker on head and neck) with a *conspicuous white stripe down sides and flanks* and white sides of crissum. Immature drabber and (especially below) paler, though *already showing the white on flanks and crissum*. Purple Gallinule adults are very different, while immatures are much browner. Common often swims (unlike Purple Gallinule), usually with head nodding back and forth; also walks on marshy vegetation or damp ground, jerking its tail with each step. Flight weak, usually low over water. Gives a variety of cackled calls, often long-continued, "kruunh... kruunh...," a harsh "kreeenh," and a more complex "krehdehdehdeh, kreh, kreh, kreh, kreh."

PURPLE GALLINULE
FRANGO-D'ÁGUA-AZUL

AZURE GALLINULE
FRANGO-D'ÁGUA-PEQUENO

SPOT-FLANKED GALLINULE
FRANGO-D'ÁGUA-CARIJÓ

COMMON MOORHEN
FRANGO-D'ÁGUA-COMUM

SUNBITTERN *Eurypyga helias* 43-48 cm | 17-19"

Fairly common along edge of shady ponds, streams, and rivers; most numerous in Pantanal. Bill long and straight, mandible yellow to orange; iris red; short legs yellow to orange. Unmistakable, with *small head, slender neck, and complex plumage pattern*. Head black with *white superciliary and another long white stripe below cheeks*; throat white, neck and breast rufous barred and vermiculated black. Above dusky brown *with buff banding*; belly buff with some black barring, midbelly whiter. Wing-coverts with large white spots. In flight wings *reveal a large "sunburst" of buff and orange-rufous in primaries*; tail crossed by two wide black and chestnut bands. The beautiful Sunbittern is a solitary bird that walks deliberately along water's edge, often with an odd mincing gait, extending head and neck back and forth; it's willing to get its feet wet, but only rarely does it wade very deeply. Catches its insect or vertebrate prey with lightning-fast jabs. When threatened, leans forward and swivels body from side to side, spreading and flaring wings, raising and fanning tail. Stiff-winged, jerky flight is interspersed with glides on bowed wings. Gives a long high-pitched, penetrating whistled note with tinamou-like quality; alarmed birds give a hissing and a loud "kak-kak-kak-kak" call.

SUNGREBE *Heliornis fulica* 28-31 cm | 11-12"

Uncommon and local on sluggish forest-fringed rivers and streams, marshy lakes, and ponds. An *inconspicuous* aquatic bird with a small head and slender neck; *usually seen swimming*. Legs striped black and yellow, *lobed feet boldly banded black and yellow* (prominent when out of water). Short straight bicolored bill (red above in breeding ♀, which also has a red eye-ring). *Crown and hindneck black* with *white superciliary and another white stripe down neck*; ♀ has cinnamon-buff cheeks. Otherwise olive brown above, whitish below. Long wide tail black narrowly tipped white. Superficially grebe- or duck-like, the Sungrebe has a rather different, more rounded shape; no duck has a black-and-white-striped head and neck. Cf. immature Wattled Jacana. Generally seen swimming alone, low in the water, sometimes only its neck protruding; pumps head and neck back and forth. Remains close to cover, favoring areas where shoreline vegetation actually overhangs water; when pressed, swimming birds may clamber back into thick low vegetation, sometimes disappearing without a trace. Flight weak and usually not long-sustained, at first pattering along surface. ♂ has unique folds of skin under wings where young can shelter or even be carried in flight. Usually quiet, but occasionally gives loud and nasal cackling calls, "kah, ka-ka!" or "kra, kah-kah!" or a deeper "kro-kro, ko-kwa!" vaguely recalling a wood rail.

WATTLED JACANA *Jacana jacana* 23-24 cm | 9-9.5"

Common and widespread in *marshes and around ponds* but, like so many of our waterbirds, *most numerous in Pantanal*. Unique, with *extremely long toes and nails—readily apparent in the field—that enable them to walk with ease on floating vegetation*. Bill yellow with *bi-lobed red frontal shield and lappets*. Adult has *head, neck, and underparts black; above rufous-chestnut*. In flight shows *unmistakable pale greenish yellow flight feathers*. Both sexes have a carpal spur on wing, often visible while the wings are held aloft briefly after alighting. Immature very different, lacking red on bill though with the *same greenish yellow flight feathers*: brown above with blackish crown and hindneck and *conspicuous white superciliary*; below white. Intermediate-plumaged birds mottled with black are frequently seen. Even immatures should not be confused: no rail has such long toes, any yellow on the wing, or the eyestripe. Cf. Sungrebe. Attractive and conspicuous, jacanas parade around on marsh vegetation, grass, or mud, sometimes gathering in small groups (especially when water levels have receded). Flight usually low on stiff, often bowed wings, their necks outstretched and their long legs and toes trailing behind and often held somewhat elevated. Noisy birds, jacanas give a variety of loud cackling and chattering calls. Flushed individuals may retreat a short distance, voicing a protest but—unlike so many marsh birds—they usually remain in view. Jacana breeding systems are unusual in that ♂♂ are responsible for incubation and care of the young; typically several ♂♂ reside within the territory of a slightly larger ♀.

SUNBITTERN
PAVÃOZINHO-DO-PARÁ

SUNGREBE
PICAPARRA

WATTLED JACANA
JAÇANÃ

imm. adult

PLOVERS & LAPWINGS (Charadriidae) are shorebirds with stout short bills. Typically they wade little; some are found in open dry areas. Though many are migratory, three of our four species are resident.

SOUTHERN LAPWING *Vanellus chilensis* 33-35 cm | 13-14"

Common, widespread, and conspicuous in open areas with short grass, often near water and around houses. Unmistakable. *Fairly long wispy crest. Bill reddish pink tipped black*; legs pink. *Head and neck brownish gray*, whitish on foreface, black forehead and midthroat; *breast glossy black*. Above brownish gray with bronze and green scapulars, white rump; mainly black tail. In flight shows *broad, rounded wings* with carpal spur at the bend, *bold white band on wing coverts* contrasting from below with black flight feathers. Usually in pairs or small groups that attract attention through their bold demeanor and loud calls. Often active at night, sometimes even feeding then. Nesting birds are aggressive in the defense of their small patches of "turf," flying at the intruder, protesting loudly, and sometimes landing in front of you with wings outstretched. Calls include a strident "keh-keh-keh-keh-keh...," given at the merest provocation and quickly becoming tiresome.

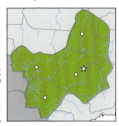

PIED PLOVER *Hoploxypterus cayanus* 23 cm | 9"

An *elegant plover of sandbars and muddy areas along rivers and around ponds*. Sometimes placed in the genus *Vanellus*. Bill dusky; narrow but often conspicuous *narrow eye-ring coral red; rather long legs coral red*. Face black, extending down hindneck to upper back and to a wide pectoral band. A *conspicuous white diadem* encircles the grayish midcrown. Grayish brown above with *black scapulars bordered by white stripe down sides of back*; rump white, tail black. Mostly white below. In flight shows a *bold, three-part pattern* with grayish brown forewing, white inner flight feathers and coverts, and black primaries; a carpal spur shows at bend of wing. Nearly unmistakable, but cf. Southern Lapwing. An uncommon bird, here ranging in well-dispersed pairs, rarely in small groups. Feeds by running forward in short bursts, pausing to pick at prey; wades only infrequently. Though often quiet, gives a soft "whoyt" or "whee-whoyt," also a querulous "wheeyp?" call, often given at night.

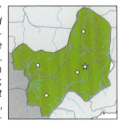

COLLARED PLOVER *Charadrius collaris* 16 cm | 6.25"

Uncommon on sand or gravel bars and short grass near water. A dainty plover with slender black bill, legs yellowish. Adult sandy brown above with white on forecrown, *black loral stripe, and broad black frontal band*; hindcrown, nape, and sides of neck tinged cinnamon, brighter in ♂. Below white with a *narrow black chest band*. Immature has duller facial pattern, browner chest band only on sides. In flight shows a faint white wingstripe and white sides to rump and tail. This species is unique in our area, far from the coast (where there are other similar species). When nesting occurs in pairs, at other times gathering in small groups that sometimes range away from rivers onto wet grassy areas and even airstrips. Though quiet, breeding birds give a simple sharp "chip" or "krip" call.

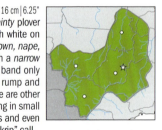

STILTS are unmistakable, elegant, *black-and-white* shorebirds with *extraordinarily long legs*. The two species replace each other geographically, and were formerly considered conspecific.

WHITE-BACKED STILT *Himantopus melanurus* 36-39 cm | 14-15.5"

A *slender, very long-legged* shorebird found around shallow ponds and marshes in open areas; in our area *most numerous in Pantanal, but even here quite local*. Needle-like black bill; legs an *almost garish reddish pink. Black above, white below*; crown and band across upper back white. ♀ somewhat browner on mantle, young birds even more so. In flight wings all black with white rump extending up lower back as a wedge. *In flight legs trail far behind tail and are often dangled loosely*. Other than the Black-necked Stilt, unmistakable. Very conspicuous birds, stilts are often found in groups that feed mainly while wading. They nest in loose colonies on muddy ground or in very shallow water, generally fully out in the open. Noisy, excitable, and alert, stilts quickly become aware of one's approach and will often circle giving short strident barking calls that are repeated endlessly if young birds are about.

BLACK-NECKED STILT *Himantopus mexicanus* 36-39 cm | 14-15.5"

Replaces White-backed Stilt in NE of our area. Differs in *smaller area of white on forehead, white patch above eye, and lack of white across upper back*. Habitat and behavior similar, as is voice.

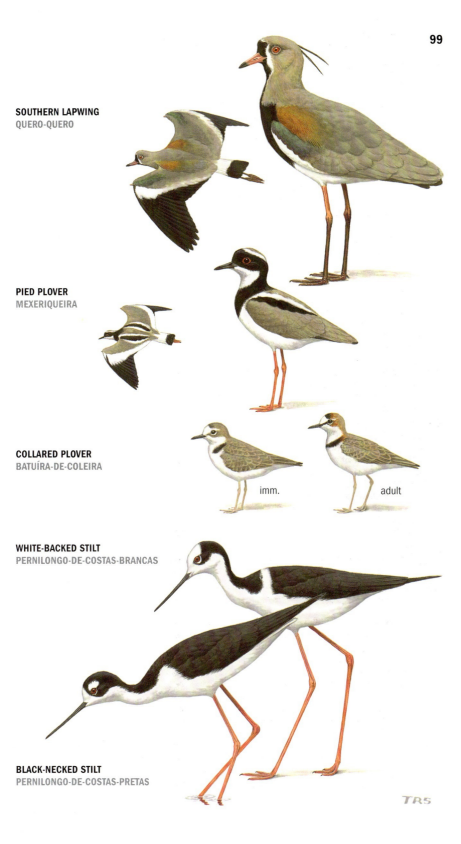

SOUTHERN LAPWING
QUERO-QUERO

PIED PLOVER
MEXERIQUEIRA

COLLARED PLOVER
BATUÍRA-DE-COLEIRA

imm.   adult

WHITE-BACKED STILT
PERNILONGO-DE-COSTAS-BRANCAS

BLACK-NECKED STILT
PERNILONGO-DE-COSTAS-PRETAS

AMERICAN GOLDEN PLOVER *Pluvialis dominica* 25 cm | 10"
Uncommon boreal migrant to open grassy or sandy areas, usually near water though sometimes far from it. Small numbers of this plover can drop in just about anywhere, but they are most numerous westward (especially in Pantanal?) in Sep-Oct. Short stout bill. Nonbreeding birds grayish brown above mottled whitish or golden, with *fairly prominent whitish superciliary, dusky crown and ear-patch*. Whitish below, breast and flanks mottled grayish. Juvenile more heavily mottled with golden above, grayer below. Breeding plumage (rarely seen in Brazil) blackish brown above *spangled with golden yellow*; forehead white extending back as stripe around face and down sides of breast; *face and underparts black*. In flight shows *dark wings and rump* and a smoky gray underwing. Considerably stockier than the other shorebirds with which it is likely to occur (e.g., Upland and Buff-breasted Sandpipers). Feeding birds tend to scatter on short grass, mud, or sand, but in flight they usually are in compact flocks. Common flight call a distinctive "ch-weet" or "qweed-leet." Breeds in North America.

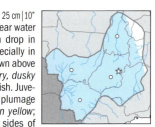

SANDPIPERS, SNIPES, & PHALAROPES (Scolopacidae) are a widespread and variable group of shorebirds, *the vast majority of our species (all but the two snipes) breeding in North America*, passing the boreal winter here. Most are found *near water*, though a few forage on dry ground. They can be *hard to identify*. Though locally numerous on the Brazilian coastline, in our area shorebirds are scarce and infrequently seen.

BUFF-BREASTED SANDPIPER *Tryngites subruficollis* 20 cm | 8"
Rare boreal migrant (mainly during southward passage, Sep-Nov) to short-grass fields and sandbars along rivers, So far only in Pantanal. Short bill black; *legs yellow*. Adult has crown and upperparts blackish brown, *feathers margined with buff imparting a scaly appearance*; face and underparts buff. A whitish eye-ring accentuates the dark eye. Juvenile's whiter feather-edging above *imparts an even more marked scaly effect*. In flight dark flight feathers contrast somewhat with brown coverts; *underwings contrastingly white*. An attractive and often tame shorebird, not likely confused (but cf. Upland Sandpiper). Like Uplands, Buff-breasteds often are found far from water, favoring *short grass*; they usually occur in small monospecific groups, walking about quietly with their heads bobbing, a charming sight. When flushed, "Buffies" often fly a considerable distance, then circle and return to the same area.

UPLAND SANDPIPER *Bartramia longicauda* 29 cm | 11.5"
Rare to uncommon boreal migrant (mainly during southward passage, Aug-Oct) to open grassy areas; shows no affinity for water. Profile characteristic, with *small dove-like head, long slim neck*, and *long tail* (at rest extending beyond wing-tips). Rather short slender bill yellowish; legs yellow; *large dark eye imparts a distinctive "wide-eyed" expression*. Brown above mottled blackish and with buff feather edging. Buffy whitish below with brown streaking, chevrons, and barring. In flight blackish primaries contrast with mottled brown upperparts; underwing barred. Buff-breasted Sandpiper is smaller and much buffier overall; American Golden Plover is chunkier and thicker-billed. "Uppies" stride—graceful, alert, and upright—through grass (if tall, can be hard to see). Often flies with shallow, stiff wingbeats; upon alighting the wings may be held up briefly. Call a characteristic mellow "huu-huuit," most often given in flight, frequently as they fly high overhead.

HUDSONIAN GODWIT *Limosa haemastica* 38-40.5 cm | 15-16"
Rare boreal migrant to muddy or sandy margins of lakes, rivers, and marshes; few records, all from Pantanal and during southward passage, Sep.-Nov. Bill long (8-10 cm/3-4") and slightly upturned, pinkish with terminal half blackish. ♀ larger and longer-billed than ♂. Nonbreeding plumage brownish gray above with whitish superciliary; *foreneck grayish*, throat and lower underparts whitish. Juveniles similar but browner. Breeding plumage birds handsome: face whitish with fine gray streaking, dark brown upperparts with buff and whitish spotting, and *underparts chestnut* (♂) or *buff scaled with chestnut* (♀). Flight pattern striking with *white upper tail-coverts contrasting with black tail*, a *bold white wingstripe*, and *black underwing-coverts*. This large, long-billed shorebird should be easily recognized. It often associates loosely with other shorebirds, when feeding wading into deep water (up to its belly) and probing deeply.

AMERICAN GOLDEN PLOVER
BATUIRUÇU

breeding adult

nonbreeding adult

BUFF-BREASTED SANDPIPER
MAÇARICO-ACANELADO

UPLAND SANDPIPER
MAÇARICO-DO-CAMPO

HUDSONIAN GODWIT
MAÇARICO-DE-BICO-VIRADO

nonbreeding adults

TRINGA are elegant, long-legged shorebirds with slender bills.

GREATER YELLOWLEGS *Tringa melanoleuca*   30.5-33 cm | 12-13"
Widespread boreal migrant (Sep-Apr) to muddy or sandy margins of ponds, rivers, marshes; most numerous in Pantanal. Long straight bill (*often perceptibly upswept*) blackish with blue-gray base; *legs bright yellow*. Nonbreeding plumage brownish gray above spotted with white; head, neck, chest streaked grayish and white, white below. Breeding plumage darker above, more coarsely streaked blackish on neck, barred on flanks. In flight shows *dark wings, white rump*. Lesser Yellowlegs is smaller with shorter, finer, blacker bill. Greater is tall and rangy, occurring in small groups and often associating with Lesser and other shorebirds. Sometimes feeds by sweeping bill sideways in shallow water. Frequent call a loud and ringing "tew-tew-tew."

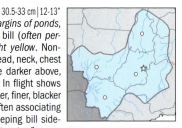

LESSER YELLOWLEGS *Tringa flavipes*   25.5-26.5 cm | 10-10.5"
Widespread but uncommon boreal migrant (Sep-Apr) to muddy margins of ponds, marshes; most numerous in Pantanal. Long straight bill black; *legs bright yellow*. Plumages similar to Greater Yellowlegs. Lesser is smaller (not always evident without direct comparison) and more delicate in shape, has shorter *all-black bill lacking upturn* and different calls. Solitary Sandpiper is smaller, more compact, darker above; has shorter olive legs. Stilt Sandpiper has a slightly decurved bill and whitish superciliary. Lesser's behavior much as in Greater but it tends to feed by picking insects from surfaces. Calls much less strident than Greater's, only rarely trebled, most often a single "kyew."

SOLITARY SANDPIPER *Tringa solitaria*   20.5-21.5 cm | 8-8.5"
Widespread but uncommon boreal migrant (Sep-Apr); true to its name often occurs alone around isolated shallow pools or ponds. Long slender bill mostly black; *legs olive*. Dark olive brown above with *prominent white eyering*; lower underparts white. Breeding plumage more white-spotted above; head and neck white-streaked. In flight shows *dark wings; sides of tail white barred dark brown*. Spotted Sandpiper teeters more, nods less; it has a less upright posture, paler legs, and flies with stiff shallow wingbeats (not the Solitary's deep quick wingstrokes). Lesser Yellowlegs has yellow legs, paler more spotted upperparts, and a less obvious eye-ring. Solitary often nods head, especially when nervous. When flushed, sometimes briefly towers upward; swooping flight can recall a swallow. Call a sharp clear "pt-weet," given especially in flight.

SNIPES are dumpy, *cryptically patterned* shorebirds with *long straight bills* whose flexible tip aids in their probing mud for food.

SOUTH AMERICAN SNIPE *Gallinago paraguaiae*   26.5-29 cm | 10.5-11.5"
Fairly common and widespread in *marshes and wet grassy areas*; inconspicuous except when displaying. *Very long bill (6.5-7.5 cm/2.5-3")*; short olive legs. Dark brown above with buff-and-white streaking (forming "V's" on back), head striped. Foreneck streaked, mottled dusky; belly white with dark barring on flanks, crissum. In flight shows dark wings with *white trailing edge on secondaries, short tail orange-rufous* with blackish subterminal band, *white sides and tip*. Except for Giant Snipe (q.v.), not likely confused. A close sitting bird that favors dense cover, only infrequently emerging to feed in the open. It typically flushes almost at your feet, giving a raspy call in protest, usually flying off a long way. Displaying birds fly overhead, emitting loud harsh vibrating sounds produced by air rushing through tail feathers during "power dives." Also gives a repeated "kak" call when on ground or a fence post.

GIANT SNIPE *Gallinago undulata*   43-45 cm | 17-17.75"
Rare and very local in wet grassy areas (*campos alagados*); recorded only from Brasília, Emas NP, and Chapada dos Guimarães (more widespread?). *Extremely long bill (10-11.5 cm/4-4.5")* thicker at base, flattened on forehead. Similar to South American Snipe, but considerably larger and "bulkier." Differs in its *bolder cinnamon edging above, bolder dark bars and chevrons below*. In flight shows *barred flight feathers*, no white in tail. Very hard to see on the ground (birds preferring to squat and creep away). Apparently feeds primarily at night. Displaying birds fly overhead mainly in full darkness (less often at dawn) giving a loud accelerating "whoa kor-cho, whoa kor-cho, whoa kor-cho" simultaneous with a low muffled sound from the rush of air through tail feathers. On the ground, often from a grass hummock, gives a fast nasal "wer-jr, wer-jr, wer-jr...."

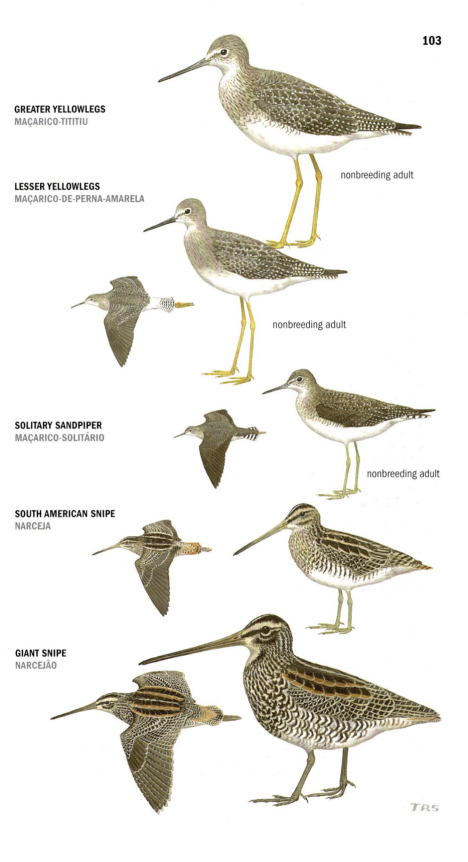

GREATER YELLOWLEGS
MAÇARICO-TITITIU

nonbreeding adult

LESSER YELLOWLEGS
MAÇARICO-DE-PERNA-AMARELA

nonbreeding adult

SOLITARY SANDPIPER
MAÇARICO-SOLITÁRIO

nonbreeding adult

SOUTH AMERICAN SNIPE
NARCEJA

GIANT SNIPE
NARCEJÃO

SPOTTED SANDPIPER *Actitis macularius* 19 cm | 7.5"
Uncommon boreal migrant (Oct-Mar) to muddy or sandy shorelines; most numerous in Pantanal. Bill dusky; legs yellowish. Nonbreeding plumage olive brown above with *short white superciliary, vague eye-ring, some wing-covert barring*; *white shoulder mark*. Breeding plumage *spotted black below*. In flight shows white wingstripe; flies with *shallow stiff wingbeats below the horizontal*. Usually feeds alone, with frequent teetering distinctive. Typical posture horizontal, with head held low. Call a loud shrill "peet-weet" or "peet-weet-weet," often given in flight.

CALIDRIS are small to midsized sandpipers.

LEAST SANDPIPER *Calidris minutilla* 15 cm | 6"
Rare boreal migrant (Oct-Mar) to shallow pools and marshes. Legs olive yellowish. Nonbreeding plumage brownish gray above, indistinct whitish superciliary; white below, chest streaked dusky. Juvenile's upperparts edged rufous, buff, whitish. Breeding plumage *more rufescent above* with rufous or buff feather edging. In flight shows faint white wingstripe, dark center to white rump. Best recognized by small size, leg color, and *hunched, short-necked posture*. Cf. Pectoral Sandpiper. Found singly here, sometimes with other shorebirds, often allowing a close approach. In flight gives a shrill, reedy "kree-eep."

WHITE-RUMPED SANDPIPER *Calidris fuscicollis* 18 cm | 7"
Uncommon boreal migrant (Oct-Mar) to shallow ponds and their grassy or muddy margins. *Wings longer than tail at rest*; legs black. Nonbreeding plumage *grayish above*, whitish superciliary; white below, with *grayish streaking across chest, sparsely down flanks*. Juvenile has rufous on crown and scapulars, sometimes a buff tinge on breast. Breeding plumage is extensively streaked on foreneck. In flight shows faint white wingstripe, *fully white rump* (no black center). Feeds on mud and in shallow water. Frequent flight call an almost squeaky "jeeyt."

Baird's Sandpiper (*C. bairdii*) occurs as vagrant in S Pantanal. Similar to White-rumped but browner above with a dark-centered rump.

PECTORAL SANDPIPER *Calidris melanotos* 20.5-23 cm | 8-9"
Uncommon boreal migrant (Sep-Mar) to grassy areas, marshes, mudflats, sandbars. Legs yellowish. Adult dark brown above with buffyish streaking and feather edging. *Face and foreneck pale buff streaked dusky-brown, contrasting with white belly*. Juvenile has sharper edging above, a more prominent superciliary, buffier foreneck. In flight shows virtually no wingstripe, dark center to white rump. Compared to other *Calidris*, longer-necked and more upright; *the sharp contrast below is unique*. Cf. Upland Sandpiper. Occurs in small groups that scatter as they feed, joining as a flock if flushed, then sometimes giving a throaty "krrik" or "krurk."

STILT SANDPIPER *Calidris himantopus* 22 cm | 8.5"
Rare boreal migrant (mainly Oct-Nov) to shallow ponds, adjacent muddy areas, mainly in Pantanal. *Long bill droops toward tip*. Long legs olive yellowish. Nonbreeding plumage brownish gray above, *whitish superciliary*; white below, fine gray streaking on foreneck. Juvenile has buff-edged feathers above (may be lost by the time birds reach Brazil). Breeding plumage (rarely seen here) with *rufous ear-coverts* and sides of crown, *dark barring on underparts*, streaking on throat. In flight shows *plain wings, white rump*. Lesser Yellowlegs behaves differently, is daintier with a slender straight bill, brighter yellow legs. Often wades into water up to its belly when feeding, submerging head with repeated up-and-down thrusts.

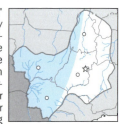

WILSON'S PHALAROPE *Phalaropus tricolor* 23 cm | 9"
Rare boreal migrant (mainly Oct-Nov) to shallow ponds; so far only *in or near Pantanal*. Needle-like black bill; *short legs olive to yellow*, black when breeding. Nonbreeding plumage *pale gray above*; face and underparts white; *dusky ear-patch*. Breeding plumage ♀ colorful: crown blue-gray, black mask continuing down sides of neck, chestnut across chest, back blue-gray with broad chestnut stripes; ♂ duller. In flight shows *unpatterned wings* with *no wingstripe, white rump*. Smaller and whiter overall than Lesser Yellowlegs; behavior different. Sometimes alights on water, spinning in circles, daintily picking tiny insects off surface; also forages on mud, then often dashing about with head held low, rearparts high.

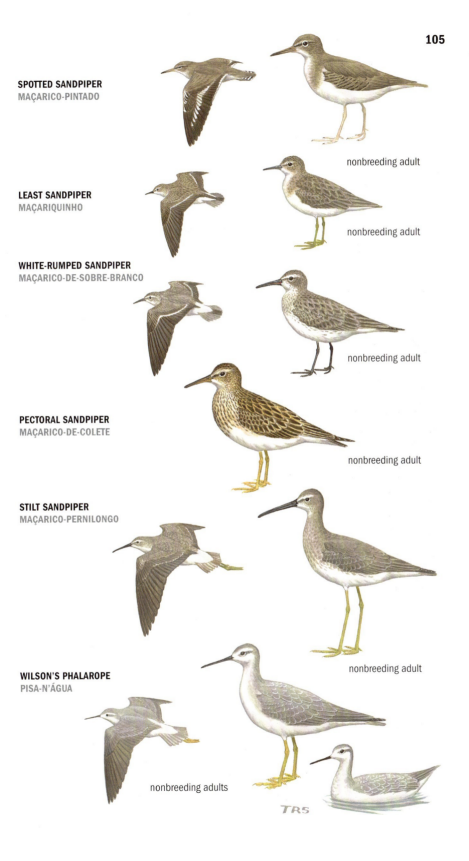

GULLS, TERNS, & SKIMMERS (Laridae) are cosmopolitan aquatic birds; note that no gulls are found in our area. Terns are mainly gray and white, and have forked or notched tails; they feed by diving. The skimmer is a flashy black-and-white bird that skims the water's surface for small fish.

YELLOW-BILLED TERN *Sternula superciliaris*   23-25.5 cm | 9-10"

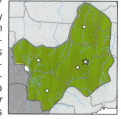

Fairly common and widespread *along larger rivers,* also feeding over nearby lakes and ponds. Formerly placed in the genus *Sterna.* Bill yellow with dusky tip in immatures; legs greenish yellow. Tail quite deeply forked. Breeding adult has black crown and nape with *white forehead extending back as narrow stripe over eye*; otherwise pale gray above, white below. Nonbreeding adult and immature similar but crown mottled gray and black (remaining blacker on nape). In flight upperwing mainly pale gray with outer two primaries black and middle primaries dusky. So *much smaller than other terns found in our area that confusion is unlikely.* Usually seen in pairs or small groups, resting on sandbars or winging purposefully over the water. Forages by hovering above water with fast wingbeats, plunging straight down after small fish near the surface. Nests in small colonies on sandbars. Has various sharp calls: a repeated "kik" or "keek," also a "kirrik" or "ki-ri-rik," all given especially in flight.

COMMON TERN *Sterna hirundo*   33-38 cm | 13-15"

Vagrant to *sandbars along rivers and around ponds; only a few records.* Bill blackish with some red near base (yellowish in first-year birds); legs reddish. Tail deeply forked. Nonbreeding adult mostly white with *black hindcrown and nape*; mantle pale gray with *dark bar along leading edge of inner wing* (the carpal bar). Breeding adult (unlikely in our area) similar but with solid black crown, *coral red bill with black tip, no carpal bar,* gray tinge below. In flight shows blackish tipping to primaries and dusky outer web on outer pair of tail feathers. Yellow-billed Tern is much smaller. Graceful, with a light, buoyant flight and deep wingstrokes. Generally silent on its wintering grounds, but occasionally gives a sharp excited-sounding "kip" call, sometimes in series. Breeds in North America, and in Old World.

LARGE-BILLED TERN *Phaetusa simplex*   38-40.5 cm | 15-16"

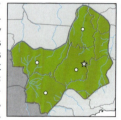

Locally fairly common, *mainly along rivers,* sometimes feeding around nearby lakes and ponds. A *large tern with a bold wing pattern and short, slightly forked tail.* Heavy bill yellow; legs dull yellowish green. Breeding adult has crown and ear-coverts black; *back, rump, and tail gray*; lores and underparts white. Nonbreeding adult similar but with some white on forecrown. In flight shows a very contrasty pattern with *large area of white on greater coverts and secondaries,* black primaries. Large size, stout yellow bill, and flashing wing-pattern should preclude confusion. Often found in small groups, regularly resting on sandbars but also sometimes perching on low branches. Nests in colonies with Yellow-billed Terns and skimmers on sandbars. Most often silent, but this species can be quite vocal around its nesting colonies, giving loud squealing calls: "kreeah."

BLACK SKIMMER *Rynchops niger*   40.5-46.5 cm | 16-18.5"

Fairly common locally along larger rivers, mainly in N of our area and in Pantanal. Striking and unmistakable. *Bill stout but compressed laterally, with lower mandible longer than the upper; basal half bright red and terminal half black.* Legs also red. Breeding adult *black above with forehead and underparts white.* Wings long, narrow, and pointed, black above with narrow white trailing edge to inner flight feathers; underwing smoky gray. Rather short tail somewhat forked, dusky. Nonbreeding adult somewhat browner above with a vague whitish collar; bill colors duller. Juvenile even browner and more scaled above, usually with dusky on crown and through eye; bill and legs even duller. Usually seen in small groups, often loafing on sandbars with Large-billed and Yellow-billed Terns; nests colonially, locally in large numbers, when water levels are low. Flight languid with slow graceful wingbeats. Skimmer feeding is unique: they fly back and forth over a stretch of water, the bill open with lower mandible "plowing" the water's surface and snapping shut upon contacting a small fish or crustacean. Nocturnal feeding is frequent. Has a distinctive sharp, nasal, barking call, "aow," given mainly in flight and around colonies.

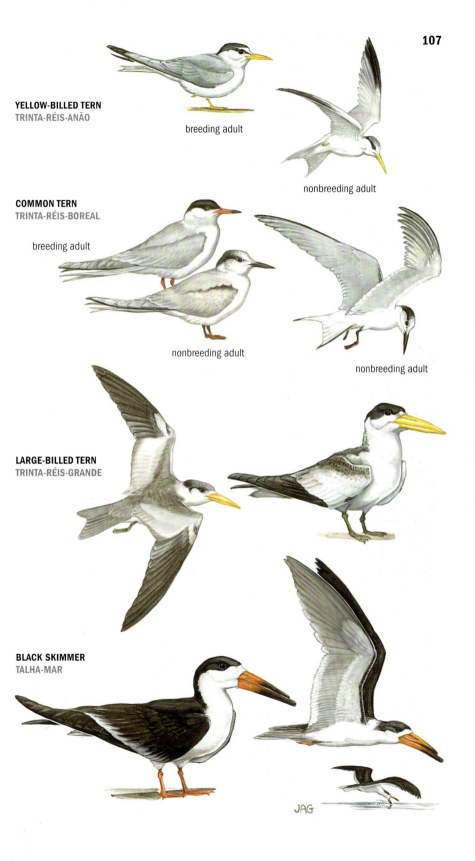

PIGEONS & DOVES (Columbidae) form a cosmopolitan family of attractive but usually subtly colored birds of varying sizes and habits. The term "pigeon" generally refers to larger species, "dove" to smaller ones, but usage can be inconsistent. They are *proportionately large-bodied and small-headed*; a few species have pointed tails. Some are relatively inconspicuous forest birds, but many others are numerous and conspicuous in open country. Note that the Common or Domestic Pigeon (*Columba livia*) occurs widely around most towns; introduced from the Old World, there are no truly wild populations in Brazil.

PATAGIOENAS pigeons are *large* and mainly *arboreal*, with broad slightly rounded (not pointed) tails. Most have dark vinaceous plumage. Except for the Picazuro, they mainly eat fruit. The genus *Patagioenas* was recently separated from *Columba*.

SCALED PIGEON *Patagioenas speciosa* 33-35 cm | 13-13.75"

Uncommon and local in canopy and borders of forest and woodland, *mainly in NW of our area*. A *unique, boldly scaled pigeon. Bill red with yellow tip, narrow reddish eye-ring. Head and upperparts rich purplish chestnut; neck boldly scaled white and blackish, upper back boldly scaled rufous and blackish*. Breast and belly vinaceous with dusky scalloping, fainter on whiter lower belly and crissum. ♀ duller. In poor light this species' scalloped pattern can be hard to discern, but its mainly red bill always stands out. Most apt to be confused with Pale-vented Pigeon, which likewise often perches in the open; cf. also Picazuro Pigeon. Usually noted singly or in pairs, often resting on exposed limbs in canopy; seems most numerous (perhaps just more easily seen?) in partially deforested country. Usual song a low-pitched, lazy, rhythmic "whooo, wh-wh-whooo, wh-wh-whooo."

PALE-VENTED PIGEON *Patagioenas cayennensis* 31 cm | 12.5"

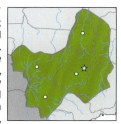

Fairly common and widespread in borders of forest, woodland, and clearings; often most numerous along rivers. Iris reddish with narrow red eye-ring. *Head bluish gray with more vinaceous forecrown and green iridescence on nape*. Above rich vinaceous *becoming gray on rump and uppertail-coverts; tail pale grayish brown with darker terminal half*. Throat whitish; breast vinaceous, *becoming pale gray on belly and crissum*. Plumbeous Pigeon is more uniformly dark and lacks the paler belly; in flight Pale-vented's gray rump and bicolored tail are usually obvious; Plumbeous is much more a forest bird. Pale-vented is an arboreal but essentially non-forest pigeon that often perches in the open and, unlike Plumbeous, regularly flies high. In display flight it climbs up with strong deep wingstrokes, then glides back down in an exaggerated dihedral. Song a frequently heard "wooooh; wok, wuh-woooh; wuk, wuh-woooh" with phraseology similar to Scaled Pigeon but faster and higher-pitched.

PLUMBEOUS PIGEON *Patagioenas plumbea* 30 cm | 12"

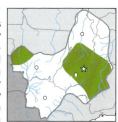

Fairly common in canopy and borders of forest in SE of our area (S Goiás and Minas Gerais) and in far NW (Serra das Araras). *Iris pale grayish. Mostly uniform vinaceous*, somewhat paler and grayer on head, neck, and underparts; wings and tail darker and bronzier. Pale-vented Pigeon is larger and less uniform, with a notably paler belly and bicolored tail. Found singly or in pairs, remaining in the canopy and not often perching in the open so usually (unlike Pale-vented) hard to see; several may gather at fruiting trees. Typically flies through the canopy, rarely at any height above it. Heard much more often than seen. Song in SE a distinctive five-noted "wuk, wuk-wuk, wuk-whoh," in NW an equally distinctive three-noted rhythmic "whuk-whuk-whuoó;" two species are probably involved. All birds also give a throaty growl, "rrrow."

PICAZURO PIGEON *Patagioenas picazuro* 36 cm | 14"

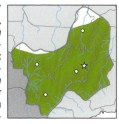

Fairly common to common and widespread in settled and cultivated areas, woodland borders, parks, and gardens. The most numerous large pigeon in our region, having benefitted substantially from the expansion of agriculture. Iris reddish orange. Mostly grayish vinaceous, *hindneck and upper back with narrow but prominent white scaling*, belly grayer; wings duskier with *conspicuous white edging on greater coverts*, forming a *conspicuous band in flight*; tail bluish gray with darker terminal half. Pale-vented Pigeon is smaller and lacks the white in wing and scaling on neck. A *notably gregarious* pigeon that often flies long distances between its roosting and feeding sites, sometimes high above the ground. Feeds mostly on the ground, regularly on open fields and lawns, but perches and nests in trees. The hooting song is a slow "whooo-ooo, whoó-whoo-whoo, whoó-whoo-whoo" that starts with a distinctive long slurred note and is sometimes interspersed with low growling calls (the latter can also be given independently).

SCALED PIGEON
POMBA-TROCAL

PALE-VENTED PIGEON
POMBA-GALEGA

PLUMBEOUS PIGEON
POMBA-AMARGOSA

PICAZURO PIGEON
ASA-BRANCA

LONG-TAILED GROUND DOVE *Uropelia campestris* 18 cm | 7"
Local and uncommon in campos and pastures. Long-tailed. Iris bluish gray; yellow eye-ring and legs. Forecrown bluish gray, brown above; wings with *two purple, black, and white bands; outer tail feathers black with white tips* (most evident in flight). Unlikely to be confused; *other ground doves are nowhere near as long-tailed.* Scaled Dove is shorter-tailed with a scaly pattern; when it flushes the rufous in its flight feathers is easily visible (this absent in Long-tailed). Generally found in pairs, not with other doves. Feeds while pottering about on the ground and often seen as it flushes from roadsides. Song a series of high-pitched notes, often sounding somewhat doubled (vaguely trogon-like), "kow-o, kow-o, kow-o...."

SCALED DOVE *Scardafella squammata* 20.5 cm | 8"
Fairly common and widespread in semiopen and agricultural areas, especially near buildings; favors places with extensive bare ground. Sometimes placed in genus *Columbina*. Long-tailed. Sandy brown above, head grayer; below buffy whitish; *scaled heavily throughout with black*. Wing-coverts show white; *extensive rufous patch in flight feathers* (visible only in flight); outer tail feathers black with white tips. *No other ground dove shows this species' scaly pattern.* Found in pairs or small groups, shuffling on ground, the tail usually held elevated; regularly accompanies Picui Ground Doves. Disturbed birds flush with a wing-whirr. Oft-heard song a monotonously repeated "ko, k-ko; ko, k-ko..." with a distinctive cadence, routinely given in heat of the day.

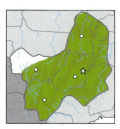

COLUMBINA are *small* ground doves with *short, squared tails* and pinkish legs.

PICUI GROUND DOVE *Columbina picui* 18 cm | 7"
Common and widespread in semiopen and agricultural areas in SW of our area and in far NE; very numerous in Pantanal. In SW (illustrated) above pale brownish gray, *whitish on foreface and grayer on crown;* below creamy whitish. Wings have an iridescent bluish stripe on lesser coverts, white edge on greater coverts forming stripe even on closed wing; *outer tail feathers white* (conspicuous when flushed). ♀ duller and browner. In NE more silvery-crowned, with even more white in tail. Shows more white in tail than other *Columbina*. Usually in pairs or small groups, pottering about on the ground. Oft-heard hollow-sounding song a repeated "k-woo, k-woo, k-woo..." (usually 6-9 notes in a sequence).

PLAIN-BREASTED GROUND DOVE *Columbina minuta* 16 cm | 6.25"
Fairly common but somewhat local in campos and cerrado; less often in agricultural areas than other *Columbina*. ♂ olive brown above, gray on crown and nape; *below pale grayish vinaceous.* Wings with shiny violet spots on coverts, *rufous in primaries* (conspicuous as they flush); underwing-coverts rufous; outer tail feathers *narrowly tipped white*. ♀ lacks gray; below pale drab, throat and midbelly whiter. ♀ Ruddy Ground Dove is larger, more rufescent on rump, has buff tail tipping. Found singly or in pairs, less often in small groups (sometimes larger when not breeding). Song a steady "who-oop, who-oop, who-oop...."

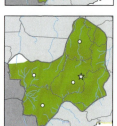

RUDDY GROUND DOVE *Columbina talpacoti* 17 cm | 6.75"
Common and widespread in semiopen and agricultural areas and around houses. ♂ ruddy vinaceous with paler face and throat; *bluish-gray crown.* Wing-coverts with black spots and bars, primaries rufous (visible mainly in flight); underwing-coverts black; tail black, outer feathers tipped rufous. ♀ paler and duller: olive brown above, grayish buff below; outer tail feathers tipped buff. Cf. the rare Blue-eyed Ground Dove. Ruddy ♂ distinctive; ♀ can be confused with smaller, shorter-tailed Plain-breasted. Occurs in pairs or small groups that feed mainly on seeds; confiding, when pressed flushing with a whirr, usually landing nearby. In nonbreeding season can occur in large flocks. Song a steady repeated "k-whoo, k-whoo, k-whoo...."

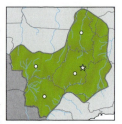

BLUE-EYED GROUND DOVE *Columbina cyanopis* 16.5 cm | 6.5"
CR *Exceedingly rare.* Apparently occurs in campos and cerrado, but so rare that even this species' preferred habitat is uncertain. Has been seen, once, at *Serra das Araras*. Iris apparently pale bluish. Above rufous-chestnut, duller on back and rump; *wing-coverts with iridescent bluish-black spots;* tail mostly black. Below paler buffy, belly and crissum whiter. ♀ paler and duller. Essentially unknown in life, and one can only hope for its rediscovery.

LONG-TAILED GROUND DOVE
ROLINHA-VAQUEIRA

SCALED DOVE
FOGO-APAGOU

PICUI GROUND DOVE
ROLINHA-PICUÍ

PLAIN-BREASTED GROUND DOVE
ROLINHA-DE-ASA-CANELA

RUDDY GROUND DOVE
ROLINHA-CALDO-DE-FEIJÃO

BLUE-EYED GROUND DOVE
ROLINHA-DO-PLANALTO

BLUE GROUND DOVE *Claravis pretiosa* 21 cm | 8.25"

Uncommon but widespread in lower growth and borders of gallery and deciduous woodland and forest. Legs pink. Unmistakable ♂ *bluish gray*, paler on face and underparts. Wing-coverts with black spots, tertials with two black bars; tail mainly black. ♀ brown above, *rump and most of tail rufous* (conspicuous in flight), lateral feathers black. *Wing-coverts with shiny purplish spots and two bars on tertials*. Breast pale grayish brown, belly whitish. ♀ of smaller Ruddy Ground Dove lacks the purplish wing markings, rufous rump, and uppertail. Less conspicuous than *Columbina* ground doves; generally in pairs that feed on the ground, rarely emerging too far from cover. Far-carrying song a distinctive slow-paced series of abrupt "boop" or "whoop" notes given from a hidden perch.

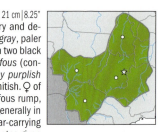

EARED DOVE *Zenaida auriculata* 26 cm | 10.25"

Locally common in semiopen and agricultural areas. Legs coral pink. Above brown with bluish gray crown, *two dark spots on ear-coverts*, and a purple sheen on neck. Wings with *large black spots on tertials; tail graduated, outer feathers with broad white tips and edging* (conspicuous in flight). Below pale vinaceous, lower belly pale buff. All the ground doves are markedly smaller. The larger and heavier White-tipped Dove lacks Eared's obvious head and wing spots. Conspicuous in settled areas, and sometimes quite tame. Potters about on the ground, head nodding, most often in small groups; larger flocks sometimes feed on open fields (and are locally considered an agricultural pest). Flight strong and fast. Not especially vocal, but has a subdued, low-pitched cooing, "whoo, whoo-whoo, whooh."

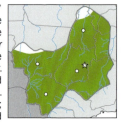

LEPTOTILA are large, plump terrestrial doves of *wooded habitats*. They have coral pink legs, rufous underwing-coverts, and *distinctive white tail-tipping*.

WHITE-TIPPED DOVE *Leptotila verreauxi* 26.5-28 cm | 10.5-11"

Common and widespread on or near ground in deciduous forest and woodland, borders, and clearings. Red orbital ring (blue in far SE). Grayish brown above, whitish on forehead; tail blackish, *outer tail feathers white-tipped* (in flight looks like a narrow terminal band). Throat white; *below pale vinaceous*, whitish lower belly. Gray-fronted Dove has a blue-gray forecrown, distinctive buff tinge on face and sides of neck, and is a more forest-based bird. Cf. also Eared Dove. Usually found singly, walking on ground, head bobbing, most often just inside cover but coming into the open much more often than Gray-fronted, especially soon after dawn. Flushed birds sometimes land on low perch where they may nod, dip their tail, and pace nervously. Flight fast and strong, almost always low. Song a soft, hollow "wh-whooó" with distinct 2-noted effect (Gray-fronted's is *single*-noted).

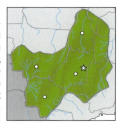

GRAY-FRONTED DOVE *Leptotila rufaxilla* 26-27.5 cm | 10.25-10.75"

Fairly common on or near ground in forest and woodland, *especially near water*. Red orbital ring. Above olive brown with *forecrown pale grayish and midcrown blue gray*; tail blackish, *outer tail feathers white-tipped. Face and sides of neck buff*; throat white, underparts pale vinaceous, lower belly whiter. White-tipped Dove is plainer, lacks gray on crown, buff on face; despite its name, the extent of its white tail-tipping is similar. Behavior similar to White-tipped but seems shyer, favoring such dense habitats that it is much less often seen; only occasionally does it emerge into the semiopen. Heard much more often than seen. Song an abrupt and mournful single-noted cooing, "whooh."

RUDDY QUAIL-DOVE *Geotrygon montana* 23 cm | 9"

Uncommon on or near ground in forest and woodland, mainly in N of our area. *Bill, loral line, orbital ring, and legs purplish red*. ♂ above rufous-chestnut glossed with purple; *prominent pinkish buff stripe across lower face bordered below by a reddish brown malar stripe*; breast pale vinaceous, belly dull buff; often shows a whitish bar in front of wing. ♀ much duller with *facial stripe pale cinnamon, malar stripe brown*; above dark olive brown. Breast brownish to grayish, belly dull buff. Juvenile like ♀ but feathers of upperparts have cinnamon edging. ♂♂ can be known from their overall ruddy coloration, unique among our pigeons, while ♀♀ show enough facial pattern to be recognizable. Usually found singly inside forest, foraging for seeds and fallen fruit; shy and not often seen. Unlike *Leptotila* doves, Ruddy Quail-Doves flush quietly without noisy wing-flapping. Song a soft, descending "oooo" given from a hidden perch.

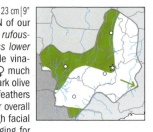

BLUE GROUND DOVE
POMBA-DE-ESPELHO

EARED DOVE
AVOANTE, POMBA-DE-BANDO

WHITE-TIPPED DOVE
JURITI-PUPU

GRAY-FRONTED DOVE
JURITI-GEMEDEIRA

RUDDY QUAIL-DOVE
PARIRI

PARROTS (Psittacidae) are familiar, colorful birds with short hooked bills and dextrous feet, used for handling food. They vary greatly in size, and some are long-tailed. Notably noisy and gregarious, the vast majority nest in tree holes; all eat fruits and seeds.

HYACINTH MACAW *Anodorhynchus hyacinthinus*　　　　96-101 cm | 38-40"

VU

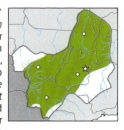

Formerly widespread in gallery forest and woodland, *now most numerous in Pantanal*; elsewhere very rare or absent. Agriculture's spread across interior Brazil has caused this macaw to decrease markedly in most areas. Though Hyacinths declined in recent decades due to trapping for the cagebird market, the Pantanal population may have stabilized and even be increasing thanks to the protection and good will of many landowners. A *spectacular macaw, the largest parrot in the world. Massive bill black, lower mandible outlined bright yellow. Plumage violet-blue, underside of wings and tail black.* This splendid parrot—the Pantanal's iconic bird—is virtually unmistakable, though in poor light other macaws can also look surprisingly dark. Hyacinth's profile with its outsized head is distinctive. Unlike *Ara* macaws, Hyacinths rarely fly very high. They occur in pairs or small groups and when feeding, mainly on *Scheelea* palm nuts, regularly descend to the ground, waddling with "pigeon-toed" gait. Flocks of up to 30-40 birds congregate where food is abundant. In our area nests are placed in large tree holes, usually in *Sterculia*. Vocalizations are incredibly loud, deeper in pitch and throatier than other macaws, "wraaanh!" often repeated and sometimes more drawn out, sometimes more a fast gabbling.

ARA macaws are *very large parrots with very long pointed tails, colorful plumage*, and bare facial skin. They are well known for their loud, far-carrying calls.

BLUE-AND-YELLOW MACAW *Ara ararauna*　　　　81-86 cm | 32-34"

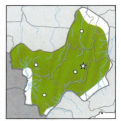

Locally fairly common in *gallery forest and woodland*, but declining or extirpated in many areas; absent from Pantanal. Facial skin white with lines of small black feathers. *Bright rich blue above*, flight feathers darker. Small throat patch black; *below rich golden yellow.* From below wings and tail yellow. Though seemingly hard to confuse, flying birds in low-angled light can look deceptively red. Most often seen in pairs or small groups, but gathers in large flocks (up to 30-50 birds) in some areas, e.g., Emas NP. Can commute long distances between roosting and feeding sites, flying overhead with flight steady and deceptively fast, long tails streaming behind—in good light, a lovely sight indeed. Perched birds can be hard to see, and are often quiet, sometimes best located by listening for falling fruit and debris. In our area often feeds on palm nuts, on or near ground. Voice less harsh than in other macaws, but still prodigiously noisy. Most characteristic is a loud throaty "rraaah!" given especially in flight (perched birds give it mainly in alarm or just before taking off); other calls include a querulous "kurreeorek."

RED-AND-GREEN MACAW *Ara chloropterus*　　　　89-96 cm | 35-38"

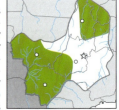

Uncommon and now local in *semihumid forest and woodland*, sometimes feeding out in cerrado; has declined in most areas due to trapping and development. Bill bicolored; facial skin with lines of small red feathers. *Mostly deep red; lower back, rump, and crissum pale blue. Wing-coverts mostly green*, flight feathers blue. From below wings and tail red. Not likely confused; in good light, the deep red plumage is obvious. Habits similar to Blue-and-yellow Macaw, but Red-and-green seems to occur mainly in areas away from water and never is in large flocks. Nest sites vary from tree hollows to cliff recesses. Calls similar to other large macaws', typically deeper and less varied than Blue-and-yellow's.

RED-BELLIED MACAW *Orthopsittaca manilata*　　　　46-48 cm | 18-19"

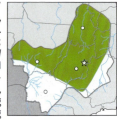

Fairly common in canopy and borders of várzea forest and swampy areas, *mainly in N of our area*. Formerly placed in genus *Ara*. *Facial skin yellowish.* Mostly green, bluer on crown, with mealy suffusion on throat and breast; *small belly patch red.* Flight feathers bluish. *Underside of wings and tail pale greenish yellow.* Blue-winged Macaw has a whitish face, brighter and deeper green plumage with red frontal band and red on lower back, no yellow under wings. Cf. the smaller Red-shouldered Macaw. Usually in groups, sometimes flocks of 25 or more. Often seen commuting between its roosting and feeding areas, typically in buriti (*Mauritia*) palms, sometimes roosting inside their curled-up fronds. In flight looks slender, with slim pointed wings and narrow, often upswept tail; wingbeats shallow. Flight call higher-pitched than other macaws', a repeated "kree-ee-ee" or "kree-ee-ak;" it carries far.

HYACINTH MACAW
ARARA-AZUL-GRANDE

BLUE-AND-YELLOW MACAW
ARARA-CANINDÉ

RED-AND-GREEN MACAW
ARARA-VERMELHA-GRANDE

RED-BELLIED MACAW
MARACANÃ-DE-CARA-AMARELA

PRIMOLIUS macaws are *considerably smaller* than other macaws (Red-shouldered is even smaller). Like the others, they have *bare facial skin* and long pointed tails. All were formerly in the genus *Ara*.

GOLDEN-COLLARED MACAW Primolius auricollis 36-40 cm | 14.25-15.75"
Locally fairly common in deciduous and gallery woodland, adjacent clearings in Pantanal; isolated population around Ilha do Bananal. Bare facial skin white. Bright green with *blackish forecrown, golden yellow collar on hindneck*. Primaries blue; tail reddish, blue toward tip; underside of wings and tail olive yellow. Blue-winged Macaw lacks yellow collar and black on foreface; has some red on forehead, lower back, and belly. Golden-collared ranges singly and in small groups, and often perches conspicuously on snags; it seems to persist well in man-modified areas. Noisy, though its vocalizations lack the raucous quality of larger macaws, e.g., a repeated "reenh."

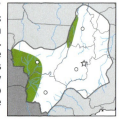

BLUE-WINGED MACAW Primolius maracana 38-42 cm | 15-16.5"
Rare to uncommon and local in semihumid forest and woodland, locally ranging out into cerrado. Though small numbers occur in Chapada dos Guimarães area, Blue-winged appears absent from Pantanal. Bare facial skin yellowish white. Bright green with *dark red forehead, bluish crown, red patch on lower back* (visible mainly in flight), and reddish patch on lower belly. Primaries blue; tail reddish, blue toward tip; underside of wings and tail olive-yellow. Golden-collared Macaw has an obvious yellow collar. Red-bellied Macaw has yellower facial skin, no red on forehead; its habitat preferences differ. Occurs in pairs or small groups, more often flying high overhead than Golden-collared. Vocalizations similar to Golden-collared.

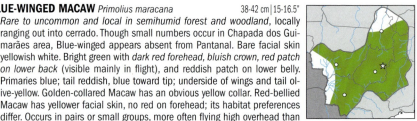

RED-SHOULDERED MACAW Diopsittaca nobilis 33-35 cm | 13-13.75"
Uncommon to locally fairly common in cerrado, deciduous and gallery woodland, clearings; *often numerous around Mauritia palms*. Seems to persist well in settled areas. *Small*, no bigger than a large *Aratinga* parakeet. Small area of bare *facial skin white; bill bicolored*. Bright green with bluish forecrown, *red at bend of wing* (the "shoulders"). Underside of wings and tail olive-yellow; unlike other macaws shows *no blue on primaries*. Blue-crowned Parakeet has more blue on head, a whitish orbital ring (not facial skin), no red on shoulders, reddish tail underside. Both *Primolius* macaws are larger and *lack* red on wing. Red-bellied Macaw is larger still; has facial skin yellow, no red on wing. Usually found in small groups that often perch in palms and can be quite tame. Calls are *Aratinga*-like, somewhat nasal, typically a "nyaah," often repeated rapidly.

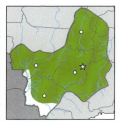

ARATINGA parakeets are midsized parrots with *long pointed tails* and *mainly green plumage*. They are gregarious and feed on a variety of fruits and seeds, sometimes flowers.

WHITE-EYED PARAKEET Aratinga leucophthalma 33-35 cm | 13-14"
Widespread, locally fairly common in wooded and forest edge habitats. Bill horn-colored; bare orbital ring yellowish white (the "white eye"). Bright green with scattered red feathers on face and foreneck, often extensive. Edge of forewing red; *lesser underwing-coverts red, greater coverts bright yellow*; underside of flight feathers and tail yellowish olive. Blue-crowned Parakeet has blue on head, lacks red feathers on foreparts and the red and yellow under wing. Most often seen in flocks (sometimes large) flying overhead (sometimes high) between roosting and feeding sites. Not a true forest bird, favoring borders and clearings; sometimes roosts and nests on cliffs. Noisy, with a variety of raspy screeching calls given in flight and while perched: "scree-screéah" or "scrah-scrah-scra-scra."

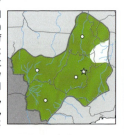

Dusky-headed Parakeet (*A. weddellii*) occurs in forest borders and adjacent clearings in N of our area, e.g., Serra das Araras. Smaller than White-eyed Parakeet with blackish bill, grayish head, yellower belly.

BLUE-CROWNED PARAKEET Aratinga acuticaudata 33-35 cm | 13-14"
Uncommon and local in deciduous and gallery woodland in Pantanal and far NE. In Pantanal (**A**), bill bicolored; bare orbital ring whitish. *Head mostly dull blue*, otherwise bright green, breast sometimes tinged blue. Underside of tail basally reddish. In NE (**B**), bill entirely horn-colored, and *blue restricted to crown*. Red-shouldered Macaw has bare facial skin, red on shoulders. White-eyed Parakeet lacks blue and shows red and yellow on underwing in flight. Behavior as in White-eyed Parakeet but seems more sedentary, much less apt to fly long distances high overhead; group site is never as large. Vocalizations as in White-eyed Parakeet and Red-shouldered Macaw but can be somewhat screechier.

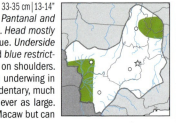

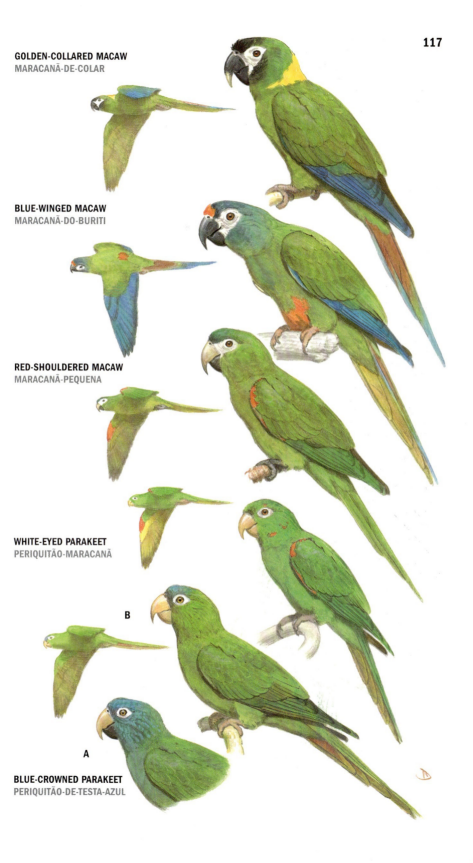

GOLDEN-COLLARED MACAW
MARACANÃ-DE-COLAR

BLUE-WINGED MACAW
MARACANÃ-DO-BURITI

RED-SHOULDERED MACAW
MARACANÃ-PEQUENA

WHITE-EYED PARAKEET
PERIQUITÃO-MARACANÃ

BLUE-CROWNED PARAKEET
PERIQUITÃO-DE-TESTA-AZUL

JANDAYA PARAKEET *Aratinga jandaya* 29 cm | 11.5"

🇧🇷 Uncommon and local in deciduous woodland and clearings in NE of our area (Tocantins). Bare orbital ring gray. *Head and neck bright yellow with orange-red around eyes and over bill*; back and wings contrastingly green, primary coverts blue. *Throat and breast yellow variably suffused with intense orange-red on breast*, belly green. Underwing-coverts red; flight feathers and underside of tail dusky. Nearly unmistakable; does not overlap with the equally beautiful Golden-capped Parakeet. Occurs in flocks, occasionally large; they sometimes fly high, covering large distances, then swooping in to land in a tree or on a snag. Calls are distinctively different from other *Aratinga*, consisting of piercing, *high-pitched* screeches.

GOLDEN-CAPPED PARAKEET *Aratinga auricapillus* 29 cm | 11.5"

🇧🇷 *Rare to uncommon and local in semihumid forest and woodland in SE of our area* (S Goiás, N Minas Gerais). Bare orbital ring gray. Mostly green with *intense red around eye and on forehead, yellow on crown*; primary coverts blue; belly variably washed with red. Underwing-coverts red; flight feathers and underside of tail dusky. Jandaya Parakeet (no overlap) shows much more yellow and red on head, neck, and underparts. Peach-fronted Parakeet ranges in more arid and open habitats; it is much less intensely colored with a brownish foreneck. Behavior much as in the Jandaya, though Golden-capped is more a bird of humid areas. Deforestation has caused a marked reduction in its numbers. Sometimes feeds with White-eyed Parakeets, but seems not to fly with that species. High-pitched vocalizations identical to Jandaya's.

PEACH-FRONTED PARAKEET *Aratinga aurea* 26 cm | 10.5"

Widespread and common in cerrado, deciduous and gallery woodland, gardens, locally even in towns. *Conspicuous yellow-orange feathered eye-ring. Forecrown orange*; otherwise green above, primary-coverts blue. *Foreneck brownish olive*, below yellowish olive. In semiopen areas this is the most frequently encountered parakeet. It tends to occur in pairs or small groups, not the larger flocks of most others; flies in compact groups, swerving between trees and shrubs. Feeds on a variety of fruits and seeds, mainly in trees and shrubs, sometimes on the ground. Usually nests in holes dug into arboreal termitaria (not the tree cavities of most *Aratinga*). Calls are harsh, grating screeches, often given more or less continuously.

🇧🇷 **Caatinga Parakeet** (*A. cactorum*) occurs locally in caatinga woodland and borders in N Minas Gerais. Resembles Peach-fronted and sometimes occurs with it (though favoring more arid areas). Lacks the orange forecrown and has a bare whitish orbital ring. Calls shriller.

MONK PARAKEET *Myiopsitta monachus* 28 cm | 11"

Common to locally abundant in semiopen terrain, woodland borders, and groves (often around buildings) in Pantanal. Long pointed tail. Bill pinkish. *Foreface, throat, and breast pale gray, foreneck scaled darker*. Green above, flight feathers blue; belly paler green. Unmistakable and very conspicuous in Pantanal, omnipresent in some areas. Monk Parakeets are very social and—unique among parrots—they construct bulky, often very large multichambered stick nests in trees (sometimes placed on telephone poles). Sometimes a Jabiru's even more massive nest sits atop a Monk Parakeet colony, the two species coexisting in harmony unless its combined weight causes it all to crash to the ground. Feeds on fruits and seeds, frequently on the ground, and coming to be fed at some ecolodges. Flocks can generate a lot of noise and they call incessantly, even at night when inside their nesting chambers. Calls are shrill and scratchy, e.g., "kreeah-kreeah."

BLACK-HOODED PARAKEET *Nandayus nenday* 31 cm | 12"

Locally fairly common to common (more numerous to S) in seasonally flooded grassland with scattered Copernicia palms and adjacent deciduous and gallery woodland in Pantanal. Long pointed tail. Bare orbital ring gray. Mostly green with *black hood* (some reddish feathers at its margin), *chest powdery blue, thighs red*; flight feathers blue. From below flight feathers and tail blackish, contrasting with green plumage. Not likely confused in the open terrain it favors; *no other parrot here has an obvious black head*. Occurs in small groups that perch in the open (often on fence posts, in which they sometimes nest) and generally fly low. Frequently feeds on the ground, sometimes with Monk Parakeets. Calls are loud and high-pitched screeches, sometimes repeated interminably, given both when perched and in flight.

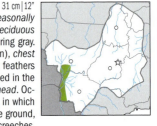

JANDAYA PARAKEET
JANDAIA-VERDADEIRA

GOLDEN-CAPPED PARAKEET
JANDAIA-DE-TESTA-VERMELHA

PEACH-FRONTED PARAKEET
PERIQUITO-REI

MONK PARAKEET
CATURRITA

BLACK-HOODED PARAKEET
PRÍNCIPE-NEGRO

119

PYRRHURA parakeets are *smaller and more slender* than the better known *Aratinga* parakeets. They tend to be more forest-based and less conspicuous, rarely flying much above the canopy, and their plumage patterns are more complex. *Pyrrhura* are almost always in small groups, and apparently are cooperative breeders.

BLAZE-WINGED PARAKEET *Pyrrhura devillei*   25 cm | 10"
Fairly common but local in canopy and borders of deciduous woodland and forest in Mato Grosso do Sul. Regular around Serra da Bodoquena. Bare orbital ring whitish. Mostly green with *crown and auriculars brownish*, throat and breast yellowish green edged brown giving scaly effect, midbelly patch maroon; *tail olive green*; bend of wing bright red, flight feathers edged blue. In flight shows red lesser underwing-coverts, yellow greater underwing-coverts. Green-cheeked Parakeet (no overlap) has a reddish tail (as seen from above) and *lacks* red and yellow on underwing-coverts. Seen in small compact groups that fly swiftly through forest canopy, usually seeming to disappear into the foliage as they alight, only infrequently landing on exposed branches. Perched birds remain inconspicuous, clambering about noiselessly and remaining quiet until they are about to fly, uttering soft alarm notes just before flushing. Most frequent call, given especially in flight, a rather shrill "keey" or "kree," often repeated in series and by flock members more or less at once.

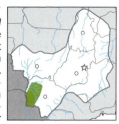

GREEN-CHEEKED PARAKEET *Pyrrhura molinae*   25 cm | 10"
Fairly common but local in canopy and borders of deciduous woodland and forest in extreme W Mato Grosso and Mato Grosso do Sul (especially Corumbá region). Not known to occur with Blaze-winged Parakeet, replacing it just to W. Mostly green with crown brownish and *auriculars a paler ashy brown*, throat and breast faintly scaled, and midbelly patch maroon; *tail maroon-red*; flight feathers blue. Blaze-winged has an olive green tail, as seen from above, and red and yellow on underwing-coverts, visible mainly in flight. Behavior and voice as in Blaze-winged.

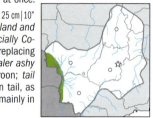

PFRIMER'S PARAKEET *Pyrrhura pfrimeri*   24 cm | 9.5"
VU   A striking, *highly range-restricted* parakeet, *rare* in canopy and borders of forest and woodland in N Goiás and Tocantins (Serra Geral). Bare orbital ring whitish. Mostly green with *pale blue crown and nape* and *maroon-red face and upper throat*; breast scaled grayish, and midbelly patch contrastingly maroon-red; bend of wing red, flight feathers blue; tail reddish. The only *Pyrrhura* parakeet in its (tiny) range, and as such it should be instantly recognizable; cf. various *Aratinga* parakeets, all larger and differing markedly in plumage. Not well known, but behavior and voice presumably much as in Blaze-winged Parakeet.

YELLOW-CHEVRONED PARAKEET *Brotogeris chiriri*   20 cm | 8"
Common and widespread in canopy and borders of deciduous forest and woodland, ranging out into clearings and settled areas, sometimes even towns. A small parrot with a short wedge-shaped tail. Bare orbital ring whitish. Mostly green, darker on back and wing-coverts, with a *conspicuous bright yellow slash on secondary coverts* (visible even on closed wing). Not likely confused; *Pyrrhura* parakeets are larger, more slender, and longer-tailed with more complex plumage patterns. Gregarious, sometimes gathering in large flocks; often flies high and has a distinctive erratic bounding flight, quite unlike other parrots. Tends to remain well above ground, though sometimes coming lower at edge and in trees in clearings. Often perches in the open, but its green plumage can make it amazingly hard to spot when it lands in leafy trees. Feeds in flowering and fruiting trees. Flocks can keep up a shrill chattering, particularly in flight.

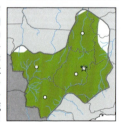

BLUE-WINGED PARROTLET *Forpus xanthopterygius*   12.5 cm | 5"
Uncommon in woodland and forest borders and clearings, mainly in N and E of our area; scarce or lacking in Pantanal. *A very small*, chunky parrot with a *short tail*. Bill whitish. ♂ green, brighter and paler on head and underparts; *greater wing-coverts, secondaries, and rump rich blue*. ♀ lacks the blue. By far the smallest parrot in our area. Usually in small groups that regularly perch in the semiopen, often in *Cecropia* trees (sometimes feeding on their catkins). Often hard to see because of their tiny size and green coloration. Flies in spurts, with fast buzzy wingbeats interspersed with brief moments when the wings are held close to the body. Nests in holes dug into termitaries, also sometimes in hornero nests. Gives a high-pitched chattering, "tzit" or "tzeet," especially in flight, but also when perched.

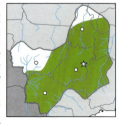

BLAZE-WINGED PARAKEET
TIRIBA-FOGO

GREEN-CHEEKED PARAKEET
TIRIBA-DE-CARA-SUJA

PFRIMER'S PARAKEET
TIRIBA-DE-PFRIMER

YELLOW-CHEVRONED PARAKEET
PERIQUITO-DE-ENCONTRO-AMARELO

BLUE-WINGED PARROTLET
TUIM

PIONUS are stocky parrots with *short, squared tails* and *red undertail-coverts*. They fly with *deep wing-strokes*, the wings sometimes almost touching below the body.

BLUE-HEADED PARROT *Pionus menstruus*   26.5 cm | 10.5"
Fairly common in canopy and borders of forest and woodland in NW of our area. *Unique blue head, neck, and breast*; black patch on ear-coverts, small red patch on foreneck. Otherwise green, with red crissum and base of tail. Scaly-headed Parrot lacks obvious blue on head and neck. Blue-headeds perch on snags in the semiopen and usually move about in groups; feeding birds can be much harder to see. They tolerate disturbed, fragmented forest better than many parrots. Noisy, especially in flight, with most common call a relatively high-pitched "kee-wink, kee-wink...."

SCALY-HEADED PARROT *Pionus maximiliani*   26.5 cm | 10.5"
Fairly common in forest and woodland in most of our area (less numerous in N where partially replaced by Blue-headed Parrot). *Bare ocular ring whitish* in W and S (**A**); or gray and less conspicuous in E (**B**). Mostly green, *head and neck feathers edged darker with a scaly effect* (not always conspicuous); some blue on chest, especially in E birds; *red crissum and base of tail*. Blue-headed Parrot has a blue head and neck (but this can be hard to see in poor light). Amazons are larger with different plumage patterns and flight styles. Behavior and vocalizations much as in Blue-headed, though calls tend to be single-noted: "kreeyk."

YELLOW-FACED PARROT *Alipiopsitta xanthops*   26-28 cm | 10.25-11"
▲ *Rare to uncommon and local in cerrado and gallery woodland*; now mainly in and near protected areas, e.g., Emas and Brasília NPs. Formerly in the genus *Amazona*. Bare orbital ring whitish. *Variable*. Green morph (mostly ♀♀) mostly green, with *bright yellow face, red lores*, yellow scaling below (sometimes extensive on breast and suffused with red and orange). Yellow morph (mostly ♂♂) has *much more yellow on head, neck, and underparts*, variable amount of red on sides. Some (seemingly mostly ♂♂) lose forecrown feathers, revealing "bald" yellow-orange skin. Juvenile all green. The larger Turquoise-fronted Amazon has a red speculum, never as much yellow on face. Occurs in pairs or small groups, usually flying low, sometimes feeding with Turquoise-fronteds. Yellow-faced has a limited vocal repertoire, with a loud "kree-eey" and variants most frequent, given in flight and while perched.

AMAZONA amazons are fine large parrots with *distinctive facial patterns* and *prominent wing speculums* (visible mainly in flight). Gregarious birds, amazons fly with *distinctive stiff, shallow wing-beats*, usually maintaining a steady course; pairs are usually evident within the flock. All too popular as cagebirds, their numbers are depleted in many areas.

ORANGE-WINGED AMAZON *Amazona amazonica*   32 cm | 12.5"
Fairly common in canopy and borders of riparian and deciduous forest and woodland. Bare orbital ring grayish. Mainly green. *Yellow patch on forecrown and cheeks, blue on lores and above bill*. The blue and yellow on the face varies; normally the yellow is extensive, visible at long distances. In flight shows a *conspicuous orange wing speculum* (Turquoise-fronted's is red). Orange-winged favors river-edges and does not seem to associate with Turquoise-fronteds. Engages in regular, conspicuous flights to and from its roosts. Most common flight call a shrill "kee-wik, kee-wik, kee-wik..." or "kwik-kwik, kwik-kwik...," often repeated rapidly. Perched birds give numerous other calls.

TURQUOISE-FRONTED AMAZON *Amazona aestiva*   38 cm | 14.5"
Locally fairly common in gallery forest, woodland, and adjacent open areas (at times out into cerrado); has declined, but remains numerous where protected, e.g., in Pantanal. Mainly green. *Yellow face* (can extend to throat), *blue forehead*; neck, back, and breast feathers edged darker looking scaly. In flight shows *conspicuous red wing speculum* and (**A**; Mato Grosso, S Goiás) yellow or (**B**; elsewhere) *red on shoulders*, sometimes visible on perched birds. Juveniles can be green-headed; older adults can have almost entire head yellow. Orange-winged Amazon shows no color on shoulders, yellow on forecrown (never blue). Mainly occurs in pairs or small groups, but when not breeding can gather in flocks that sometimes raid food crops. Gives a variety of calls, many with a throaty quality (never screechy); most frequent, especially in flight, a repeated "kra-raow."

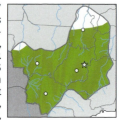

Mealy Amazon (*A. farinosa*) occurs locally in forest in NW of our area. Slightly larger than Turquoise-fronted, it is all green with a prominent pale orbital ring.

BLUE-HEADED PARROT
MAITACA-DE-CABEÇA-AZUL

SCALY-HEADED PARROT
MAITACA-VERDE

YELLOW-FACED PARROT
PAPAGAIO-GALEGO

ORANGE-WINGED AMAZON
CURICA

TURQUOISE-FRONTED AMAZON
PAPAGAIO-VERDADEIRO

123

CUCKOOS & ANIS (Cuculidae) are slender, long-tailed birds found in a variety of habitats, especially forest and woodland. Bills are typically decurved at tip. Most are uncommon and inconspicuous birds, the Guira Cuckoo and both anis being notable exceptions. All are insectivorous, with caterpillars being a favored food of many.

COCCYZUS cuckoos are sleek arboreal birds with *long graduated tails whose feathers are white-tipped*. All are inconspicuous birds, and none is particularly well known in our region.

YELLOW-BILLED CUCKOO *Coccyzus americanus*   28-30.5 cm | 11-12"

Uncommon *boreal winter resident* (Oct-Mar) in deciduous woodland, scrub, and clearings. *Slightly decurved bill black with lower mandible yellow or orange-yellow*; narrow eye-ring yellow. Grayish brown above with darker mask; white below tinged pale gray. *Primaries with rufous edging (prominent as a flash in flight, and usually visible on perched birds); underside of tail black, feathers with large white tips* (visible from below). Juvenile has eye-ring duller (sometimes gray) and often less yellow on bill; underside of tail grayer (consequently the white tail tips contrast less). Pearly-breasted Cuckoo differs in its slightly smaller size and *lack* of rufous in wing. Furtive, found singly as it sneaks about in dense leafy cover at varying heights, occasionally fluttering or running in pursuit of its large-insect prey. Migrants can occur in more open, less wooded terrain where sometimes easier to see though always seems shy. Quiet on its wintering grounds. Breeds in North America.

PEARLY-BREASTED CUCKOO *Coccyzus euleri*   25.5-28 cm | 10-11"

Status uncertain; few records from our region, but can turn up most anywhere. Occurs in forest borders and woodland. Resembles the better known Yellow-billed Cuckoo, with similar *yellow lower mandible*; narrow eye-ring either gray or yellow (as in Yellow-billed, this likely being age-dependent, yellow at least when breeding). Differs in its slightly smaller size and *lack of rufous edging in primaries*; above marginally darker. Despite its English name, *not* appreciably grayer below (though in some lights the foreneck can look slightly more "pearly gray"). Behavior as in Yellow-billed Cuckoo. Pearly-breasted may not breed in our area (occurring just as a transient?) and, as far as known, it is silent here. Song in SE Brazil a simple series of low hollow guttural notes, delivered slowly, "kuow, kuow, kuow...."

DARK-BILLED CUCKOO *Coccyzus melacoryphus*   28 cm | 11"

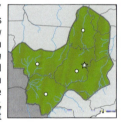

Uncommon summer breeding resident (Oct-Mar/Apr) in shrubby clearings and forest and woodland borders. Slightly decurved *bill all black*; narrow eye-ring usually yellow (sometimes gray). Olive brown above, more grayish on crown with *black mask through eyes and on cheeks* and *gray band from below mask down sides of neck. Below pale buff*, deepest on breast and flanks. Flight feathers show no rufous. Underside of tail black, feathers with large white tips visible from below. The buff on underparts of this handsome cuckoo should be enough to clinch the identification. Usually seen singly, creeping about in dense foliage, generally not too high above ground. Most often silent, its infrequently heard, subdued song is a descending series of guttural notes, "kwo-kwo-kwo-kwo-kolp-kolp," recalling Chotoy Spinetail.

ASH-COLORED CUCKOO *Micrococcyx cinereus*   24 cm | 9.5"

Rare transient in shrubby clearings and borders, occasionally in more open areas; status in our region unclear, with few records, but perhaps breeding in far south. Formerly placed in the genus *Coccyzus*. Slightly decurved bill all black; *iris and narrow eye-ring red*. Grayish brown above, grayer on head. *Throat and chest pale brownish gray*, below white. No rufous in wings. Tail not as graduated as in the *Coccyzus* cuckoos; its underside gray, feathers narrowly white-tipped (visible only from below). The furtive behavior of this little-known species seems to be similar to the *Coccyzus* cuckoos. It is not known to vocalize in our region (but it could). Where it breeds, further south in South America, song a series of 12-16 well enunciated and steadily repeated notes: "kyow-kyow-kyow-kyow-kyow-kyow...."

YELLOW-BILLED CUCKOO
PAPA-LAGARTA-DE-ASA-VERMELHA

PEARLY-BREASTED CUCKOO
PAPA-LAGARTA-DE-EULER

DARK-BILLED CUCKOO
PAPA-LAGARTA-CANELA

ASH-COLORED CUCKOO
PAPA-LAGARTA-CINZENTO

LITTLE CUCKOO *Coccycua minuta*   25.5-28 cm | 10-11"

Widespread but inconspicuous and generally uncommon in shrubby growth, woodland, and forest borders, *mostly near water*. Formerly placed in the genus *Piaya*. Bill greenish yellow; iris reddish with red eye-ring. *Uniform rufous-chestnut above*; throat and breast tawny (paler than upperparts), with belly grayish and crissum blackish. Underside of tail black, feathers tipped white. The longer-tailed Squirrel Cuckoo is so much larger that confusion is unlikely; Little Cuckoo is less arboreal, and is never found inside forest. Little Cuckoos are furtive birds that are usually seen singly as they sneak about in dense vegetation, typically close to the ground. Only occasionally does one rest in the open. Though usually quiet, they have several distinctive vocalizations including a nasal drawn-out "nyaahh, neh-neh-neh-neh-neh" and a much sharper "ek" or "chik! wreeanh."

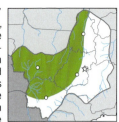

SQUIRREL CUCKOO *Piaya cayana*   40.5-46 cm | 16-18"

Fairly common and widespread in canopy and borders of forest and woodland. *A large*, slender cuckoo with *very long, graduated tail*. Bill greenish yellow; iris red, with *orbital skin red*. Uniform rich rufous-chestnut above, fading to paler vinaceous buff on throat and chest; *lower underparts contrastingly gray*, becoming black on lower belly and crissum. Underside of tail black, feathers with large white tips visible from below. The less arboreal Little Cuckoo is similarly colored but hardly half the size (proportionately shorter-tailed) and lacks the contrast below. Usually in pairs that creep about in foliage and hop along branches, with tail held loosely; sometimes moves with a squirrel-like agility. Generally not hard to see. Rarely flies far, preferring to move about by running and hopping, but sometimes launches across an open area, gliding downward with weak bursts of flapping. Fairly vocal, with a variety of calls. Apparent song a fast series of "kweep" or "kweeyp" notes, often repeated for a protracted period and given from a hidden perch. More often heard are several characteristic calls, notably an abrupt and loud "cheek! kwahh" and a nasal "weeyadidu."

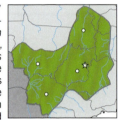

ANIS are *conspicuous long-tailed* birds with *black plumage* and *large, laterally compressed bills*. Their cup-shaped nests are large but well hidden and are sometimes used by more than one pair at once.

SMOOTH-BILLED ANI *Crotophaga ani*   34 cm | 13.5"

Common and widespread in semiopen agricultural and settled areas, lesser numbers in cerrado, etc. *Large black bill laterally compressed and smooth, with arched hump on culmen's basal half.* Iris dark. *All dull black*, sometimes looking quite unkempt. Tail long and rounded, often appearing loosely attached to body. Almost unmistakable, but cf. the much larger Greater Ani. Ranges in small straggling groups of up to 6-10 birds, perching atop bushes and on fences and wires, rarely far above the ground. Often tame; regularly associates with cattle, feeding on insects they flush and sometimes even perching on them; they are thought to pick off ticks but apparently do so only rarely. Flight labored and weak, consisting of a few quick flaps and unsteady glides, landing awkwardly with the long tail often flipping up over back. Frequently heard call a distinctive upslurred "oooo-eeek?" (or "aaaaa-ní") often given in alarm or during flight; also gives a variety of other whining and clucking vocalizations.

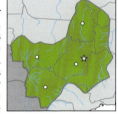

GREATER ANI *Crotophaga major*   46-48 cm | 18-19"

A *very large, long-tailed* ani found in *shrubbery along rivers and lakeshores*; locally numerous in appropriate habitat. *Iris pale straw yellow* (dark brown in juveniles); black bill laterally compressed with arched ridge on base of maxilla *imparting a characteristic "broken-nose" profile*. All glossy blue-black, mantle feathers edged bronzy green and tail glossed purple. Smooth-billed Ani is *much* smaller, has a dark eye, and is nowhere near as glossy. Usually found in groups that forage at varying levels in shrubbery and woodland near water; group size is typically small (5-15 birds) but occasionally a large number gathers together, these sometimes streaming low across water, one by one. Usually occurs independently of other birds, but has been seen accompanying monkeys. Can be quite noisy, with a variety of strange low-pitched growling calls, some abrupt and loud, others given as a long-continued weird bubbling chorus by many birds together. Most characteristic and frequent is a "koro-koro-koro," often given in flight.

STRIPED CUCKOO *Tapera naevia* 28-30 cm | 11-12"

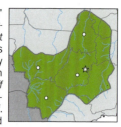

Common but inconspicuous in shrubby clearings and pastures with scattered low trees. Bill yellowish; iris hazel. *Short expressive crest rufescent with blackish streaking;* tail fairly long and graduated, uppertail-coverts elongated. *Pale grayish brown above with blackish streaking;* superciliary buffy whitish, face dusky brown. *Dull whitish below tinged buffy grayish,* with narrow black malar streak. Immature more rufescent above with *large buff spots on crown and mantle;* throat and chest buff with faint scaly markings. Only likely confusion is with much larger Guira Cuckoo, but also cf. much scarcer Pheasant and Pavonine Cuckoos, both forest birds. Striped is a generally secretive cuckoo that creeps unobtrusively in thick vegetation near ground; it seems to occur regularly along dirt roads. A brood parasite, favoring birds that build domed nests. Vocal with a *far-carrying call,* so heard much more often than seen. Most frequent song a pure melancholy 2-noted whistle, "püü-peee;" quite ventriloquial. Also has a 5- or 6-noted song of similar quality, the last several notes short and fast. Both are often given at night. Usually sings from atop a bush or fence post, often ruffling crest and flaring alula at same time.

DROMOCOCCYX are solitary, *secretive,* brownish cuckoos of wooded habitats, both species with *elongated uppertail-coverts* and both *recorded almost entirely by voice.* They are brood parasites, laying their eggs in the cup-shaped nests of other birds.

PAVONINE CUCKOO *Dromococcyx pavoninus* 28-30 cm | 11-12"

Rare and local (overlooked?) in thickets (often bamboo) in forest and woodland. Resembles more numerous Pheasant Cuckoo, with similar strange shape (though its uppertail-coverts are not so elongated), but *considerably smaller* with a proportionately shorter tail. *Narrow postocular stripe buff* and *plain fulvous throat and breast without any spotting or streaking.* Similarly sized Striped Cuckoo has upperparts paler and browner with dusky streaking, more uniform whitish underparts. Exceptionally elusive and skulking, even more so than Pheasant Cuckoo. Almost never noted except when singing. Song resembles Pheasant Cuckoo's; sometimes the two almost seem to switch songs. The typical Pavonine song is a whistled "püü-pee, püü-pi-pi" without Pheasant's trill at end; there is variation, however, and some Pheasant Cuckoos can give *very* close approximations of Pavonine's song.

PHEASANT CUCKOO *Dromococcyx phasianellus* 37-41 cm | 14.5-16"

Uncommon and local (often overlooked) in swampy forest and woodland. *Small-headed* with a short expressive crest; *tail long, wide, and fan-shaped with upper tail-coverts greatly elongated and nearly as long as the tail itself.* Dark brown above, wing-coverts and inner flight feathers with white tips; crown dark with whitish postocular stripe. Whitish below with *dusky spotting and streaking on throat and chest.* The profile of this odd cuckoo is unique, with small head, thin neck, and spread tail feathers. The much more often seen Striped Cuckoo favors more open terrain, is considerably smaller with no streaking on foreneck. Cf. the rarer Pavonine Cuckoo. A secretive bird, Pheasant skulks in dense lower and middle growth, often in vine tangles. Flight labored and rarely long-sustained; even then the tail is usually spread and uppertail-coverts raised resulting in an odd-looking "humped" shape. Song resembles Striped Cuckoo's short song but with a third trilled note added: "püü-peee, pr'r'r'r;" sometimes a rising fourth or fifth note is also added.

GUIRA CUCKOO *Guira guira* 36-40 cm | 14.25-15.75"

Common, widespread, and conspicuous in semiopen and agricultural areas, around towns, and in cerrado. Iris and bill usually yellow, occasionally orange. *Conspicuous shaggy, expressive crest orange-rufous;* head buffy. Dark brown above with whitish streaking, *lower back and rump white;* long tail blackish with white at base and tip. Whitish below, foreneck streaked brown. This unmistakable cuckoo is *one of the more frequently seen birds in semiopen areas.* It occurs in groups of up to 15-20 birds, but usually fewer than ten. In chilly weather they often huddle together as a group; they also often roost communally. Ani-like in overall behavior, Guiras regularly feed on the ground. They sometimes parasitize anis, and are even known to incubate their eggs together with anis! Quite vocal, Guiras give several calls the most distinctive of which is a loud descending series, "kree-yer, kree-yer, kree-yer, kree-yer, kreeyr, kreeyr," and (often in flight) a high-pitched and long-continued rattling.

STRIPED CUCKOO
SACI

PAVONINE CUCKOO
PEIXE-FRITO-PAVÃO

PHEASANT CUCKOO
PEIXE-FRITO

GUIRA CUCKOO
ANU-BRANCO

TYPICAL OWLS (Strigidae) are *nocturnal* predatory birds with large heads, *forward-facing eyes*, strongly hooked bills, and powerful talons; in addition to their superb eyesight, most species also have *exceptionally good hearing*, the better to capture their prey, sometimes in total darkness. They are patterned mainly in shades of brown, and their soft plumage and the serrated leading edge to their outer primaries result in *nearly silent flight*. Most species are infrequently seen, better known from their often *distinctive vocalizations*.

ASIO are large *non-forest* owls with *ear tufts*, brownish with streaking or bold patterning below.

SHORT-EARED OWL *Asio flammeus*   38-40 cm | 15-15.75"
Uncommon and local in campos, cerrado, and adjacent agricultural areas. Iris yellow. Above dusky brown with buff spots and streaks. *Ear-tufts short and inconspicuous* (hence the name); eyebrow whitish, facial area buff becoming blackish around eyes. Below buff *with coarse dusky streaking, especially on chest*. In flight underwing pale with *black carpal patch*, upperwing shows pale patch in primaries. Striped Owl is paler and has much longer ear-tufts; Stygian is much darker generally and also has *long* ear-tufts. Seen most often at dusk, coursing over semiopen grassy terrain with a distinctive loose floppy flight, plunging to ground after prey (usually small rodents). Sometimes perches in the open, e.g., on fence posts, but never much above ground. Roosts by day in tall grass. Usually silent, but gives barks and squeals when breeding.

STRIPED OWL *Asio clamator*   35.5-38 cm | 14-15"
Rare and local in semiopen areas and agricultural terrain. Sometimes placed in the genus *Rhinoptynx*. Iris amber brown. *Very long, buff-edged ear-tufts*. *Cinnamon-buff above* coarsely streaked and vermiculated blackish, *facial area whitish bordered by narrow but conspicuous black rim*; flight feathers and tail grayish buff barred and vermiculated dusky-brown. Throat whitish; *below pale buff with conspicuous but narrow blackish streaking*. Stygian Owl is larger, much darker and sootier, and lacks streaking below. Short-eared Owl is darker overall with short (barely visible) ear-tufts, yellow iris. Roosts by day on the ground, often in tall grass; commences to hunt at or just before dusk, sometimes quartering low over open areas, but more often watching for prey from a fence post or while balancing on a power line. Not very vocal. Gives a low muffled hoot, "whooh," at long intervals, also a much higher-pitched (almost nightjar-like) "keeyr" and other squealing notes.

STYGIAN OWL *Asio stygius*   41-43 cm | 16-17"
Rare and local in semiopen areas with scattered tall trees. A very dark owl with *long blackish ear-tufts*. Iris orange-yellow. *Blackish brown above* with scattered pale spots, facial area blackish outlined with whitish speckling and with *conspicuous whitish area on forehead* (between the "ears"); tail dusky brown barred buff. Below buffy whitish with *blackish blotching across chest*; underparts with a coarse blackish herringbone pattern. Short-eared Owl favors grassy areas (unlike Stygian, it rarely or never roosts in trees); it is almost as dark but has a blurry *streaked* pattern below and its ear-tufts are so short as to be almost invisible. Stygian roosts singly in a leafy tree, sometimes in plantations of exotic trees (even pines), usually close to a trunk where hard to spot. Sometimes active before dusk. Not averse to the vicinity of houses, sometimes feeding around—even perching on—buildings. Most distinctive call a very low-pitched "hooo" repeated slowly at 5- to 10-second intervals; also gives various squeals and screams.

GREAT HORNED OWL *Bubo virginianus*   48-56 cm | 19-22"
Uncommon and local in woodland and groves in semiopen areas; at night often more in the open. *The largest owl in Brazil that shows ear-tufts*, which are *long and well-separated*. Iris yellow. Brown above with buff and whitish mottling; facial area grayish brown bordered with blackish. *Throat white, often puffed and flared out*; below barred dusky brown and whitish. Stygian Owl is darker, smaller, and more slender, with narrower ear-tufts that are set close together, and a coarse herringbone pattern below. Striped Owl is also smaller with narrow ear-tufts and obvious streaking below. Though hunting birds sometimes take prominent perches where they can be seen against the fading light, on the whole Great Horned Owls are inconspicuous. They take a variety of fairly large prey including midsized mammals, e.g., skunks, and even other (small) owls. ♂'s song, given especially at dusk and dawn, a series of deep low hoots with a distinctive cadence, typically "hoo, hoó-hoó, hoo," often followed by the lower-pitched (but otherwise similar) song of ♀.

TROPICAL SCREECH OWL *Megascops choliba* 23.5 cm | 9.25"
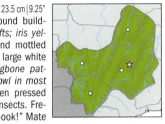
Fairly common and widespread in woodland and groves around buildings. Formerly in genus *Otus*. Short fairly conspicuous ear-tufts; iris yellow. Grayish to cinnamon above, lightly streaked blackish and mottled buff; *brow and facial area whitish*, face *outlined by black rim*; large white spots on scapulars. Below whitish to pale buff with *herringbone pattern of black streaks*. Rufous morph rare. *The only screech owl in most of our area.* Strictly nocturnal; roosts by day in foliage, often pressed against a trunk to escape detection. Feeds mainly on large insects. Frequent song a short trill ending with an abrupt "ook!" or "ook!-ook!" Mate sometimes answers with a soft series of hoots, "tu-tu-tú-tú-tú-tu-tu."

Tawny-bellied Screech Owl (*M. watsonii*), sometimes treated as separate species, Usta Screech Owl (*M. usta*), occurs in forest lower growth in N of our area. Small; *dark-eyed*, browner, and more uniform than Tropical, it lacks the pale facial area; pattern below less evident. Song a quavering trill that swells, then fades; some songs are slower.

FERRUGINOUS PYGMY OWL *Glaucidium brasilianum* 16.5 cm | 6.5"

Widespread and generally common in woodland and forest borders and semiopen areas. *Very small* with a yellow iris, *no ear-tufts*, and a *pair of black "false eyes" on back of head*. Color variable (brown, rufescent, and intermediate morphs). Above grayish to rufous brown, crown with fine pale streaking; scapulars and wing-coverts with white spots. White below with chest band, breast and belly streaking. Tail brown barred whitish to buff. *The only pygmy owl in most of our area.* Often perches in the semiopen, where routinely mobbed by the small birds on which it frequently preys. Flight fast and direct. Commonest song, given day and night, a long series of "pu" notes repeated steadily and rapidly for a minute or more, sometimes starting with a few sharper "wik" notes. Easily whistled; responding birds fly in swiftly, glaring at the "intruder." Also gives various "chirruping" calls.

Amazonian Pygmy Owl (*G. hardyi*) occurs in canopy and borders of forest in N of our area, e.g., Serra das Araras. Resembles Ferruginous, but *head and nape dotted white*. Song a fast, high-pitched series of whistled notes lasting approximately 2 seconds.

BUFF-FRONTED OWL *Aegolius harrisii* 19-20 cm | 7.5-8"

Rare and local in woodland lower growth. Recorded from our region only in Brasília area but likely more widespread. Unmistakable. Iris yellowish. Above dark brown with bold white spots on wings, buff spots on scapulars. *Large area on forecrown, facial area, and underparts orange-buff*, the facial area outlined with black; chin and irregular line across chest also black. Singing birds can be responsive to recordings of their song, a fast, quavering, rather high-pitched trill.

BURROWING OWL *Athene cunicularia* 23 cm | 9"
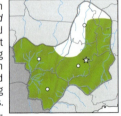
Widespread and locally fairly common in campos and cerrado. Long whitish (feathered) legs. Iris yellow. Brown above, *crown streaked whitish, back and wings spotted white*. Buffy whitish below, with irregular brown spotting and barring. *The only truly terrestrial owl*; nearly unmistakable. Mainly diurnal, but also active in evening. More or less colonial at some sites, with pairs digging burrows into soft, often sandy soil. Frequently perches on the ground (also on fence posts), often adjacent to its burrows, bobbing when approached, and staring intently at the intruder. Not too vocal; occasionally gives a shrieking "kreeey, kik! kik! kik! kik! kik!" with distinctive rhythm; also various cackles.

BARN OWL *Tyto alba* 35.5-40.5 cm | 14-16"

Widespread but only locally common in semiopen and agricultural areas. Classified in separate family, Tytonidae. Iris dark brown. *Distinctive heart-shaped facial disk white outlined by dark rim*. Light morph (illustrated) is grayish and golden buff above, *white below*; usually dotted black and white. Less numerous dark morph grayer above; *buff below*, usually dotted. Underwing whitish, imparting a ghostly appearance to flying birds at night. ♂♂ are paler than ♀♀. Perched birds have a narrow profile, large head, and slender body. In flight cf. Short-eared and Striped Owls. Mostly nocturnal but sometimes active in late afternoon or early morning, Barn Owls hunt while slowly flapping and gliding low over ground, legs often dangled; feeds mainly on rodents, also birds, large insects. Most numerous around houses, sheltering in dark recesses of a barn or tree cavity. Flying birds give a loud raspy shriek or hiss, "sh-h-h-h-h-h!"

TROPICAL SCREECH OWL
CORUJINHA-DO-MATO

FERRUGINOUS PYGMY OWL
CABURÉ

BUFF-FRONTED OWL
CABURÉ-CANELA

BURROWING OWL
CORUJA-BURAQUEIRA

BARN OWL
SUINDARA

SPECTACLED OWL *Pulsatrix perspicillata* 43-48 cm | 17-19"

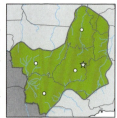

Widespread but uncommon in forest and woodland. *Large; lacks ear-tufts.* Iris yellow. Dark brown above. *White brow extends to lores and around bill* (forming "spectacles"). *Chest band dark brown*, below buff. Juvenile whitish with *large blackish mask*, brownish wings. Nocturnal, but also often encountered by day at roosts, where it seems to flush more readily than other large owls. Hunting birds sometimes come into clearings. Distinctive song a fast series of muffled hoots that start loudly but fade away: "bup-bup-buh-buh-buhbuhbuh." Sometimes duets. Juvenile's call a very different short "woauw."

Crested Owl (*Lophostrix cristata*) occurs in forest and borders at N edge of our area. Large and brown, with long eartufts (flat in repose, flared up when alarmed), rufous facial disks; relatively unpatterned. Distinctive call a low-pitched, far-carrying growl, "groor-r-r-r."

STRIX are fairly large and *round-headed* owls with *dark eyes* (no ear-tufts). Strictly nocturnal, they nest in tree hollows. Formerly placed in genus *Ciccaba*.

MOTTLED OWL *Strix virgata* 30.5-34.5 cm | 12-13.5"

Uncommon and local in forest and woodland; inconspicuous. Iris brown (non-reflective). Light morph *mottled brown above, facial disk outlined buffy whitish, broadest on brow. Belly whitish, streaked dusky.* Dark morph darker above, more rufescent below. Short-eared Owl favors open country, lacks the brow, has short eartufts. By day roosts in dense thickets, often close to ground and not often seen. By night occurs in pairs more than many owls, perching at varying heights. Gives a variety of calls, the most frequent a series of emphatic hoots, "whóh whóh whóh whóh whóh!", slower than Black-banded and nearly evenly pitched; also gives a descending screech, a cat-like "wheeyow?" sometimes followed by a "whowhowho."

BLACK-BANDED OWL *Strix huhula* 38.5 cm | 15"

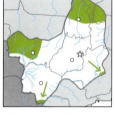

Rare and local, in forest mainly in *N of our area*. Iris brown; bill orange-yellow. *Black narrowly barred white;* facial disks black surrounded by white freckling. Nocturnal; roosts by day in dense cover. Usually perches high, sometimes on open limbs. ♂'s most common call a fast "buh-buh-buh-buh-buh-buh, bwów" with accented last note (sometimes with a final "buh"), often echoed by ♀'s softer, lower-pitched "buh-buh-bó." Also gives a loud resonant "whoóouw" and a screaming, cat-like "keeyow."

POTOOS (Nyctibiidae) are solitary nocturnal birds with *exceptionally large reflective eyes.* They *perch vertically*, motionless by day, relying on their *cryptic coloration* to escape detection.

GREAT POTOO *Nyctibius grandis* 48-53.5 cm | 19-21"

Uncommon in canopy and borders of forest. *Very large; big-headed silhouette* differs from Common Potoo's. Iris brown, *reflecting brilliant orange-red* (visible from great distances). Usually looks very pale above. Buffy whitish below, vermiculated blackish with some blackish mottling on breast. Tail whitish barred dusky. So much larger and paler than Common Potoo that recognition is easy. Common is more streaked, and typically shows a dark gape mark. At night perches high on open branches, sallying after large insects; occasionally lower in clearings. By day roosts in canopy where very hard to spot; can use the same site for long periods. Far-carrying call an explosive, guttural "bwawrrrr" or "bwawrrrru," sometimes just an abrupt "bawr-bü."

COMMON POTOO *Nyctibius griseus* 35.5-40.5 cm | 14-16"

Relatively widespread and numerous in forest and woodland borders and clearings. Iris yellow, reflecting brilliant orange-red. Typical morph grayish brown to brown, mottled and vermiculated blackish. Shoulders usually black, often with small whitish patch on wing-coverts. *Midbreast spotted black*, and usually has a *black malar streak*; tail banded grayish and dark brown. Dark morph dark brown and less patterned, especially below; wings mainly blackish. Upright feathers above eyes impart a slight "horned" effect. Great Potoo is larger and much paler. By day hard to spot, looking like the extension of a broken stub, eyes then held tightly shut. Becomes active at dusk, often flying into clearings and perching on a branch with a commanding view, sallying after large insects. Heard more often than seen. Memorable haunting song, a series of loud wailing notes that descend in pitch, loud at first but then fading, "u-wah, wah, woh, woh, wuh, wüü." What is probably the ♀'s call, shorter and less forceful, often follows ♂.

NIGHTHAWKS & NIGHTJARS (Caprimulgidae) are *cryptically patterned nocturnal* insect-eating birds with *huge gapes*. They are often hard to identify, though calls are usually distinctive.

SHORT-TAILED NIGHTHAWK *Lurocalis semitorquatus* 20 cm | 8"
Uncommon and local in forest canopy and borders. Dark, with notably *short tail* and long, somewhat pointed wings (less so than in *Chordeiles*); *no white in wings or tail*. Mainly dusky speckled rufous; throat white, some whitish marbling on tertials; belly, crissum more rufescent, coarsely barred blackish. Common and Lesser Nighthawks have longer notched tails and narrower, more pointed wings with a band across primaries. *Seen at dawn and dusk, feeding above canopy, clearings, and rivers*; usually found singly or in pairs. *Flight erratic and shifting, somewhat bat-like*, with bursts of shallow wingbeats followed by glides. Roosts by day on branches high in forest canopy. Mostly silent, but gives a sharp stacatto "cu-it."

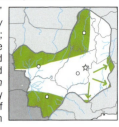

BAND-TAILED NIGHTHAWK *Nyctiprogne leucopyga* 19 cm | 7.5"
Local in woodland near lakes and rivers (*usually blackwater*) in Pantanal and NW of our area. Small and *dark*. Above dusky brown speckled buff, scapulars mottled. Below barred dusky brown and whitish or rufescent, with white slash on sides of throat. In flight *pointed wings dark*; notched tail with *white band near base*. Common Nighthawk is larger with white band across primaries; white on its tail less obvious. Short-tailed has broader wings, much shorter tail; behavior very different. Band-tailed roosts by day in woodland, *emerging at dusk to feed low over water*. Flight fast and erratic with shallow stiff wingstrokes. Breeding ♂♂ give a leisurely "kwoit... ku-woít-kwoit" repeated steadily from low perches near water.

🇧🇷 **Plain-tailed Nighthawk** (*N. vielliardi*) occurs locally along Rio São Francisco (N Minas Gerais). Like Band-tailed Nighthawk but lacks white tail band; behavior similar.

CHORDEILES nighthawks have *long, narrow, pointed wings*. They are more aerial than the more terrestrial nightjars, and become active at dusk, flying with light, easy wingstrokes.

LEAST NIGHTHAWK *Chordeiles pusillus* 16.5 cm | 6.5"
Fairly common but rather local over open grassy areas, usually near water. *The smallest nighthawk*. Above brown mottled and flecked rufous and gray; *white band across primaries, white trailing edge to inner flight feathers*, tail narrowly tipped white. Throat white, chest dark brown spotted white, belly barred gray and brown. Lesser Nighthawk is larger and lacks the trailing edge on wing. Usually seen at dusk, flying with buoyant wingbeats at moderate heights. Roosts under bushes and low trees. Generally silent in flight. Breeding ♂♂ give a fast, repeated "chu-chu-chu-chu-chueé" from a low perch.

Sand-colored Nighthawk (*C. rupestris*) is common along Rio Araguaia, on Ilha do Bananal (elsewhere?). Pale brown above, whitish below, *bold white and black wing pattern obvious in flight*. Roosts on sandbars; at dusk flies low over water, sometimes in large groups.

COMMON NIGHTHAWK *Chordeiles minor* 24 cm | 9.5"
Uncommon boreal migrant (Oct-Mar) to semiopen terrain. Closely resembles Lesser Nighthawk but slightly larger and darker with *longer, more pointed wings* (outermost primary longest). The *white wingband is about midway between tip and bend of wing* (not nearer tip); *underside of inner primaries uniformly blackish*. By day rests lengthwise on tree limbs at varying heights, occasionally on ground. Seen mainly in late afternoon, flying with a languid grace, occasionally in loose flocks, generally higher than Lesser Nighthawk. Usually silent here, but sometimes gives its distinctive nasal "peeeent" call. Breeds in North and Middle America.

LESSER NIGHTHAWK *Chordeiles acutipennis* 21.5 cm | 8.5"
Locally fairly common in semiopen scrub and grassy areas. ♂ blackish and mottled above, below whitish with white throat patch. *At rest wings extend to or slightly beyond tail* (a white primary patch is usually visible); slightly notched tail barred buff, white subterminal band. In flight shows *pointed wings with wide white band across primaries*; underwing-coverts and inner flight feathers (including inner primaries) spotted and barred buff. ♀ has *wingband buff*, buffier throat patch, no white band on tail. Cf. Common and Least Nighthawks. By day rests on low branches; nests are placed on bare ground. Mainly seen coursing about with shallow fluttering wingbeats and brief glides at dusk and dawn, remaining relatively low. Breeding birds give a soft low churring, "ur-r-r-r-r," while on ground.

SHORT-TAILED NIGHTHAWK
TUJU

BAND-TAILED NIGHTHAWK
BACURAU-D'ÁGUA

LEAST NIGHTHAWK
BACURAUZINHO

COMMON NIGHTHAWK
BACURAU-NORTE-AMERICANO

LESSER NIGHTHAWK
BACURAU-DE-ASA-FINA

NACUNDA NIGHTHAWK *Podager nacunda* 28-30.5 cm | 11-12"
Locally and seasonally fairly common in open country, favoring short grass (along roadsides, airstrips). Numbers seem to fluctuate; whether it actually migrates remains unknown. Much *larger than other nighthawks; boldly patterned*, with *comparatively rounded wings*. Above brown with buff and blackish vermiculations. Throat and breast brownish with a white crescent across lower throat; *belly white* (often hidden on resting birds). In flight shows *white underwing-coverts*, blackish primaries *crossed by wide white band*; tail barred dusky buffy brown, ♂ with broad white tail tip. Flight languid and graceful, often well above ground, almost recalling a Short-eared Owl more than a *Chordeiles* nighthawk. By day rests on bare ground (sometimes on stumps), but nonetheless hard to see until it shuffles away or flushes. Most often silent; when breeding, perched birds give a low-pitched guttural "prrrr-pú" (almost owl-like) and clucks.

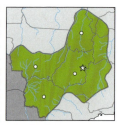

HYDROPSALIS are "fancy" nightjars, ♂♂ with *elongated white outer tail feathers*. They range in semiopen country and, compared to most nightjars, are *notably non-vocal*.

SCISSOR-TAILED NIGHTJAR *Hydropsalis torquata* 25.5-28 cm | 10-11"
Widespread and locally fairly common in cerrado, pastures, shrubby areas, and woodland borders. ♂ *long outer tail feathers* (adding 10-13 cm/4-5" to overall length, though often worn or broken) dusky with *terminal third white*; dusky and brown with buff markings on basal half. Above dusky brown vermiculated and spotted buff and whitish, with *rufous nuchal collar* and whitish malar streak. Below brown, buffier on throat, barred dusky on breast and belly. ♀ has less forked tail showing no white, less rufous on nuchal collar. Shows no white or buff in wing. Spectacular ♂ unmistakable; ♀ less distinctive, but larger than any *Caprimulgus*. Regularly rests on roads, especially soon after dark; ♂'s long tail lies so flat on ground that it often is not noticed until the bird flies. Unimpressive song a steadily repeated, insect-like "tsik, tsik, tsik, tsik...," given while perched.

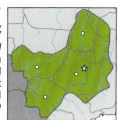

LADDER-TAILED NIGHTJAR *Hydropsalis climacocerca* ♂ 25.5 cm | 10" ♀ 23 cm | 9"
Fairly common in semiopen areas, often islands, along larger rivers and lakes in NW of our area. ♂ *pale sandy grayish buff above* with sparse blackish vermiculation and streaking. Throat whitish, breast buffier barred dusky; belly white. Wings with broad white band across primaries; *strikingly patterned tail long and "double-notched,"* outer feathers longest, mostly white on inner web; inner feathers shorter, also with much white; central feathers grayish. ♀ similar above; below buffyish barred dusky. Wings have *buffyish band across primaries* (narrower than in ♂), tail grayish with dusky barring (no white). Can usually be known by its riverine habitat; ♀ Scissor-tailed Nightjar is larger, lacks pale wing-band. By day roosts on ground near water, occasionally in the open but more often where there is grassy cover or around driftwood piles. Becomes active at dusk, perching near water's edge or on a low branch protruding from water. Sallies after insects, staying low, wings usually held bowed. Infrequently vocal. Flying birds give a squeaky "chweeit;" perched ♂♂ occasionally give a repeated "chip."

PAURAQUE *Nyctidromus albicollis* 27 cm | 10.75"
Widespread and generally common in woodland and borders (not in extensive open terrain or forest interior). *Tail long and rounded, extending well beyond wings at rest*. Gray morph has grayish crown and upper back *contrasting with rufescent cheeks*; grayish brown above with distinctive *"straps" of large black spots on scapulars* and *buff spots on wing-coverts*. Large throat patch white (flaring to sides), chest grayish; below buff narrowly barred dusky. Rufous morph *more cinnamon-rufous* (especially on crown and back, so cheeks contrast less). ♂ has *bold white bar across blackish primaries* and white inner webs of outer tail feathers (both flash in flight); ♀ has narrower wingband buff, less white in tail (only tips of outer feathers). *Longer-tailed than any Caprimulgus*. Rests on roads more often than any *Caprimulgus*, its eyes reflecting fiery red from long distances. Hunts mainly by flying up in pursuit of large insects. Pauraques roost by day on the ground, relying on their cryptic plumage pattern to avoid detection. In breeding season often calls through the night, at other times mainly at dusk and dawn. Most frequent song a hoarse "whe-wheeée-oo," sometimes preceded by "bup" notes (occasionally given alone) or slurred into "por-weeéer."

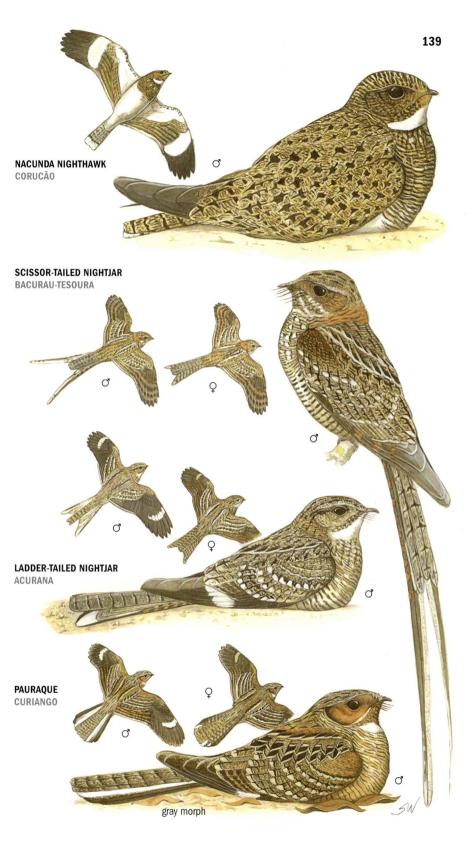

SICKLE-WINGED NIGHTJAR *Eleothreptus anomalus* 19.5 cm | 7.75"
Rare and very local in seasonally flooded campos and adjacent to gallery woodland. *In our area recorded only from Serra da Canastra NP and near Brasília.* ♂ grayish brown; buffyish eyestripe, grayish malar streak; scapulars spotted black, cinnamon patch on wing-coverts; short tail barred buff, outer feathers tipped whitish. *Unique wing shape: primaries strongly curved, blackish with bold whitish tips; secondaries very short.* ♀ browner, with *normal wing shape.* Becomes active at dusk; rests on roads and low perches. Forages by sallying into air and in low fluttery gliding flight with sudden shifts in direction. Often allows a close approach. Not very vocal; gives soft "chip" notes, wing-whirrs during short display flights.

WHITE-WINGED NIGHTJAR *Eleothreptus candicans* 22 cm | 8.5"
EN Rare and very local in cerrado and campo sujo, usually where recently burned. *In our area recently recorded only at Emas NP* where small numbers can be conspicuous, *often resting atop large termite mounds.* Formerly placed in genus *Caprimulgus.* ♂ above pale gray brown, *blackish on crown and face,* whitish eyestripe and malar streak; scapulars with large blackish spots; *tail white from below* (central feathers grayish from above). Throat brown, breast rufous-brown, belly white. In flight shows *large area of white on wings,* black wing tips; primaries curved. ♀ mostly buffyish; *lacks all white,* shows no black wing tips. Occurs with Scissor-tailed Nightjar. Often allows a close approach. Quiet, but gives wing-whirrs during short display flights.

CAPRIMULGUS nightjars have long tails and *rounded wings that do not reach tail tip at rest* (contra *Chordeiles* nighthawks); Pauraque's tail is even longer. They range in semiopen or lightly wooded terrain and can be very *hard to identify,* though *calls are distinctive.*

SPOT-TAILED NIGHTJAR *Caprimulgus maculicaudus* 19.5 cm | 7.75"
Fairly common but local in damp open grassland and pastures, sometimes where seasonally flooded. *Small.* ♂ has *crown and face blackish* with *buffy whitish superciliary and malar stripe* and *rufous nuchal collar;* grayish brown above with *prominent spotting on scapulars and wing-coverts.* Throat buff, breast dusky *spotted buff. No white wingband;* tail grayish barred black, *outer tail feathers tipped white.* White spots on tail underside (hence the name) are almost never visible in the field. ♀ lacks white in tail. *Pale brow and dark face unique among our nightjars.* ♂ Little Nightjar lacks black on crown and the pale brow. Rests by day beneath shrubs and low trees, emerging at dusk to feed. ♂'s far-carrying song a distinctive sharp, high-pitched "pit-sweeeét," given steadily from a low perch.

LITTLE NIGHTJAR *Caprimulgus parvulus* 20 cm | 8"
Locally fairly common on or near ground in woodland and shrubby areas. *Small.* ♂ above grayish brown mottled and streaked gray and buff; vague buff nochal collar. Wings mottled black and buff, coverts with two rows of whitish spots; *rather short tail with white corners.* Throat white, breast brown spotted white. In flight wings show a white bar across blackish primaries. ♀ lacks white in tail; *wing-bar buff.* Spot-tailed Nightjar shows more head pattern; primaries lack bar. Lesser Nighthawk is similar at rest, but its pointed wings, extend past tail tip. Behavior as in Spot-tailed, though Little rarely ventures far from cover; it sometimes sits on roads. ♂'s lovely song a bubbly gurgling, "dree-eé, gur-gur-gur-gur-gur," delivered from a low branch.
Band-winged Nightjar (*C. longirostris*) occurs locally in Minas Gerais, Goiás, and at Chapada dos Guimarães, mainly in *semiopen, rocky areas.* Resembles Little Nightjar though both sexes have a more marbled pattern with rufous nuchal collar, dark throat bordered by a white crescent. Song a far-carrying "pseeu-eét."

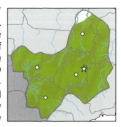

RUFOUS NIGHTJAR *Caprimulgus rufus* 25.5-28 cm | 10-11"
Local in deciduous woodland. *Large.* ♂ dark rufous brown above vermiculated buff and blackish. Throat buff barred blackish, bordered by *buffy whitish crescent of large spots;* breast blackish. Wings profusely barred rufous (*no white wingband*); tail brown barred dusky, *outer third of inner web of outer three feathers whitish.* ♀ lacks whitish in tail. Pauraque has a longer tail showing more white, a prominent white (♂) or buff (♀) wingband, rufescent cheeks, and spots on wing-coverts. Rufous ranges less in the open than many nightjars, not often leaving woodland and rarely resting on roads. At night often perches on branches just above ground. Unmistakable and far-carrying song a fast "chuck, wick-wick-weeeo," given mainly at dawn and dusk but all night long when breeding.

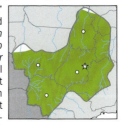

SICKLE-WINGED NIGHTJAR
CURIANGO-DO-BANHADO

WHITE-WINGED NIGHTJAR
BACURAU-DE-ASA-BRANCA

SPOT-TAILED NIGHTJAR
BACURAU-DE-RABO-MANCHADO

LITTLE NIGHTJAR
BACURAU-CHINTÃ

RUFOUS NIGHTJAR
JOÃO-CORTA-PAU

SWIFTS (Apodidae) are *highly aerial, wide-ranging,* insectivorous birds with *long, narrow, pointed stiff wings*; their small bills open wide. They perch on vertical surfaces with short, sharp claws. Compare to swallows.

WHITE-COLLARED SWIFT Streptoprocne zonaris   21 cm | 8.25"

Widespread in *S and E* of our area. Adult blackish with *bold white collar; tail notched*. Juvenile's collar is mostly limited to nape; it can also show faint white scaling on chest. *Much larger than other swifts except Biscutate* (q.v.). Flight fast and powerful on sickle-shaped wings, wingstrokes deep and steady; sometimes soars on stiff outstretched wings. Foraging birds often disperse far from roosting sites and can occur in large groups. Other swifts, especially *Cypseloides*, often accompany them. White-collareds roost and nest in colonies in caves and behind waterfalls. Screeches loudly in flight and while resting on rockfaces; also gives various chippers. At close range the wings can be heard to produce a loud whooshing sound.

BISCUTATE SWIFT Streptoprocne biscutata   20.5 cm | 8"

Uncommon and quite local around its nesting sites, foraging widely. May engage in long-distance movements, but if so this still poorly understood. Regularly seen around Chapada dos Guimarães, sometimes occurring with the similar White-collared Swift. *White collar is reduced to patches on nape and foreneck; tail square or slightly rounded, not* notched (but this hard to discern in the field). Behavior as in White-collared though this species seems mainly to nest in dry caves rather than near waterfalls.

GREAT DUSKY SWIFT Cypseloides senex   18.5 cm | 7.25"

Uncommon and very local around its nesting sites, e.g., Serra da Canastra and Chapada dos Guimarães, but foraging much more widely. *Brown* with a *paler head*, especially on foreface; tail square or slightly rounded; wings long and sweptback. Flight can be very fast, and is interspersed with bouts of gliding. Occurs in large groups where it nests and roosts, there clinging to rockfaces, often near or behind waterfalls. The species is rarely seen elsewhere, perhaps mostly because it flies so high and fast.

CHAETURA are fairly small swifts with *contrasting pale rumps and uppertails*; some have pale throats.

ASHY-TAILED SWIFT Chaetura meridionalis   13.5 cm | 5.25"

Locally fairly common summer resident (Oct-Mar) in semiopen terrain, *often most numerous in settled areas*; possibly resident in N of our area. Formerly named *C. andrei*, and sometimes called Sick's Swift. Rather short-tailed. Blackish brown with *paler brownish gray rump and uppertail-coverts*. Blackish brown below with *paler throat*. Short-tailed Swift is smaller with a shorter tail, differently shaped wings, blackish throat. Usually seen in pairs or small groups, flying in "formations," generally not too high, chippering as they go. Seems not to associate much with other swifts. Nests in palms, sometimes in vents and chimneys.

Gray-rumped Swift (*C. cinereiventris*) occurs around forest at Serra das Araras and Chapada dos Guimarães.

SHORT-TAILED SWIFT Chaetura brachyura   11 cm | 4.25"

Common in clearings and forest borders in *NW* of our area. Sooty black above; *brownish gray rump and uppertail coverts*, the latter covering its *short tail*. Below uniform blackish, crissum and undertail pale brownish. In flight inner primaries longish and secondaries relatively short, creating a "bulge" in rear margin of wing. Easily recognized, its *pale hindquarters* and very short tail obvious. Behavior as in Ashy-tailed; flight loose and floppy, often slow and "rocking." Fast high-pitched, run-together twitters are frequent.

NEOTROPICAL PALM SWIFT Tachornis squamata   13 cm | 5"

Local, *usually near palms*, in semiopen terrain. *Tail long, narrow, and deeply forked, but usually held closed in a point*. Blackish brown above, back feathers edged pale gray. Below whitish, *sides and flanks mottled dusky*. In flight wings narrow, backswept. Usually in small groups, often flying low around clearings; occasionally higher. Generally not with other swifts. Flight fast and direct on stiff, shallow wingbeats. Nests are attached to the underside of palm fronds; many feathers are incorporated, these (at least sometimes) obtained by dive-bombing onto the backs of pigeons, parrots, and other birds. Call a thin, almost buzzy "dzeeee, dzeee-dit" or "dzee-ee-ee-ee-ee-ee."

Lesser Swallow-tailed Swift (*Panyptila cayennensis*) occurs along N edge of our area. *Deeply forked tail* usually held closed in a point; black with *white bib, narrow collar, and patch on lower flanks*.

WHITE-COLLARED SWIFT
ANDORINHÃO-DE-COLEIRA

adult   juv.

BISCUTATE SWIFT
ANDORINHÃO-DE-COLEIRA-FALHA

GREAT DUSKY SWIFT
TAPERUÇU-VELHO

ASHY-TAILED SWIFT
ANDORINHÃO-DO-TEMPORAL

SHORT-TAILED SWIFT
ANDORINHÃO-DE-RABO-CURTO

NEOTROPICAL PALM SWIFT
ANDORINHÃO-DO-BURITI

ACV

HUMMINGBIRDS (Trochilidae) are enchanting *tiny* birds whose wings beat so rapidly as to appear blurred; flight is exceptionally maneuverable. Many boast *glittering iridescent colors* (produced by microscopic feather structure, not pigmentation) that depend on light angle to be seen; at the "wrong" angle they look black. They feed mainly on nectar with their slender pointed bills and are important pollinators for many plants; they also consume many insects. Some, especially ♀♀, *can be hard to identify*; others, especially ♂♂, are *gorgeous*. Elsewhere hummingbirds come regularly to specialized feeders that have been filled with sugar water, providing a wonderful spectacle. One hopes that this will become more frequent in our area.

RUFOUS-BREASTED HERMIT Glaucis hirsuta   10-11 cm | 4-4.25"

Locally fairly common in undergrowth of forest and woodland in *NW of our area*. Bill long and decurved, lower mandible mostly yellow. Metallic green above, cheeks dusky outlined buffy whitish. *Dull cinnamon-rufous below*, throat duskier. *Tail rounded*, central two feathers bronzy green *tipped white*, others *rufous* with black subterminal band and white tip. ♀ slightly paler and brighter below. Other hermits have *elongated* white-tipped central tail feathers. Rufous-breasted favors vicinity of water and *Heliconia* thickets where it is solitary, usually remaining low in vegetation, not often seen. Sometimes one is spotted while feeding at *Heliconia* flowers, or gleaning for insects from underside of leaves. Flight call a sharp, thin, inflected "swyeep!" Fast, high-pitched territorial song a "tzee-tzee-tzi-tzi-tzit."

PHAETHORNIS hermits comprise a distinctive group of dull-plumaged, undergrowth-inhabiting hummingbirds with *long decurved bills* and (most species) *elongated central tail feathers*. All have a dusky to dark mask bordered by pale lines. Breeding ♂♂ gather in leks where they emit squeaky "songs" to attract ♀♀ which, as in all hummingbirds, raise young on their own. Otherwise hermits tend to be inconspicuous, regularly noted in fast direct flight but not all that often seen perched.

PLANALTO HERMIT Phaethornis pretrei   14-14.5 cm | 5.5-5.75"

Fairly common and widespread in undergrowth of deciduous and gallery woodland, adjacent gardens and semiopen areas. Largely absent from Pantanal. Lower mandible basally reddish. Metallic green above, *uppertail-coverts bright rufous*; central tail feathers tipped white, much elongated. Below uniform rich cinnamon-buff. Cf. much smaller Buff-bellied and Cinnamon-throated Hermits. Planalto Hermit is the only *Phaethornis* on the planalto, where it is regular in semiopen areas and occurs locally even in towns. It feeds at a variety of flowers, and also comes to hummingbird feeders. ♂'s song at leks a repeated "tsu tsi-tsi, tsu tsi-tsi...."

BUFF-BELLIED HERMIT Phaethornis subochraceus   11-12 cm | 4.25-4.75"

Uncommon in undergrowth of deciduous and gallery woodland in N Pantanal, e.g., along the Transpantaneira. *Small and dull-plumaged*. Metallic green above, feathers of rump and uppertail-coverts edged rufous; central tail feathers elongated, *all tail feathers conspicuously tipped white*. Below dull grayish buff, duskier on throat, buffier on belly. Planalto Hermit is larger, brighter on rump and below. Cinnamon-throated has lateral tail feathers *tipped rufous*. Buff-bellied is an inconspicuous hermit, rarely emerging from its woodland haunts. ♂'s song at leks a repeated "tseu-tsi, tseu-tsi...."

CINNAMON-THROATED HERMIT Phaethornis nattereri   10-11 cm | 4-4.25"

Uncommon and local in undergrowth of forest and woodland in N and NW of our area, e.g., at Chapada dos Guimarães. Above bronzy green, rump and uppertail-coverts rufescent; central tail feathers elongated and white-tipped, *lateral ones tipped cinnamon-rufous*. Buff below. ♀ longer tailed. Planalto Hermit is considerably larger and has all tail feathers tipped white. Cinnamon-throated does not occur with Buff-bellied Hermit, also with *all* tail feathers tipped white. Behavior as in Buff-bellied; likewise not well known. ♂'s song at leks a fast series of high-pitched notes: "tsi-tsi-tsi-tsi-tsiu-tsiu," ending with several jumbled notes.

REDDISH HERMIT Phaethornis ruber   7.5-8 cm | 3-3.25"

Uncommon in undergrowth of forest and woodland in N of our area. Lower mandible mostly orange-yellow. ♂ bronzy green above with *rufous rump*. Uniform cinnamon-rufous below with *narrow black band across breast*. *Wedge-shaped tail* bronzy green, feathers narrowly tipped rufous-buff, central feathers (*not especially elongated*) tipped white. ♀ paler below (black band reduced or absent). Usually found alone as it darts and weaves in undergrowth, pausing to feed at flowers (often *Heliconia*). Displaying ♂♂ seem to sing solitarily, and their song sometimes descends into a jumble: "tsii-tsii-tsiitsiiteeteeteeu."

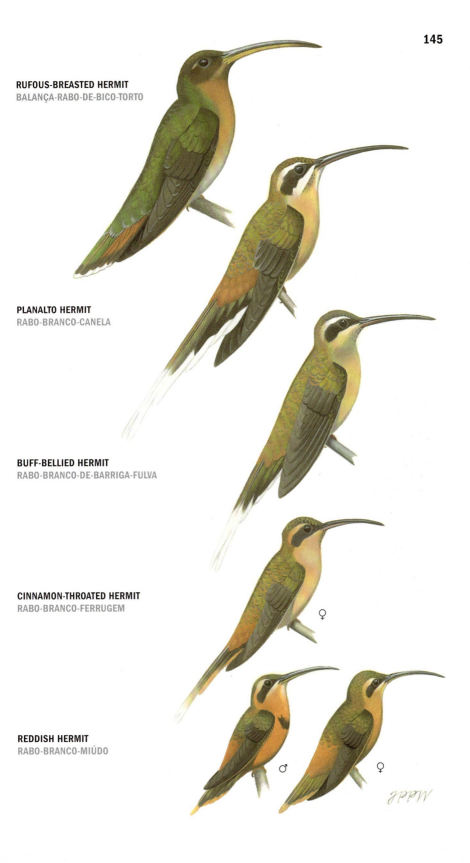

RUFOUS-BREASTED HERMIT
BALANÇA-RABO-DE-BICO-TORTO

PLANALTO HERMIT
RABO-BRANCO-CANELA

BUFF-BELLIED HERMIT
RABO-BRANCO-DE-BARRIGA-FULVA

CINNAMON-THROATED HERMIT
RABO-BRANCO-FERRUGEM

REDDISH HERMIT
RABO-BRANCO-MIÚDO

SWALLOW-TAILED HUMMINGBIRD *Eupetomena macroura* 15-18 cm | 6-7"
Common, conspicuous, and widespread in semiopen areas (including cerrado), woodland, forest borders, gardens. Bill slightly decurved (2.1 cm/1"). *Head, neck, breast, and deeply forked tail violet-blue* (tail 7-9 cm/2.75-3.5"); otherwise shining green; crissum blue. Outer primaries thickened and flat, as in the sabrewings. ♀ slightly smaller and duller. Nearly unmistakable, though conceivably molting birds with short tails might be confused with ♂ Blue-tufted Starthroat (also large and equally dark, but with a much less deeply forked tail, etc.). Searches for insects and flowers in many different situations though hardly ever in the interior of forest and woodland. Regular in gardens, and pugnacious, dominating other hummingbirds at reliable food sources such as hummingbird feeders. Common call a husky "tchu," often rapidly repeated in succession.

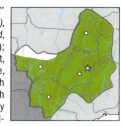

GRAY-BREASTED SABREWING *Campylopterus largipennis* 13-14 cm | 5-5.5"
Uncommon and local in humid and deciduous forest and woodland. Bill slightly decurved. Shining metallic green above with *prominent white postocular spot. Uniform gray below*. Central tail feathers bluish green above, others blackish, *outer feathers with broad white tips*. Outer primaries thickened and flat (hence the term "sabrewing"). A *large, relatively dull* hummingbird. Sombre Hummingbird is smaller and lacks sabrewing's obvious white tail corners. ♀ Fork-tailed Woodnymph is considerably smaller and has much smaller white tail corners; it *lacks* a postocular spot. The sabrewing forages at flowers at varying levels, most often in lower growth and at edge. Not very vocal; ♂♂ give a repeated, sharp tanager-like "tchip" note every 1-2 seconds, sometimes in small groups.

SOMBRE HUMMINGBIRD *Aphantochroa cirrhochloris* 12 cm | 4.75"
Local and uncommon in lower growth of humid and gallery forest and woodland in S Goiás and Minas Gerais. Bill slightly decurved. Above bronzy green with *a small white postocular spot* and slight coppery sheen on uppertail-coverts; *tail all blackish*. Below dull gray, throat and chest with some inconspicuous green spotting. Well-named, this *drab* hummingbird can often be known from its *lack* of good field characters. ♀ Fork-tailed Woodnymph is much smaller, lacks the white postocular and green on foreneck (but shows some on flanks), and has narrow white tail tipping. Gray-breasted Sabrewing is larger with conspicuous white tail corners. Found singly and, at least in the heart of its range (SE Brazil), regularly in clearings and gardens. ♂'s rather sweet song a repeated "chee-wee," often doubled.

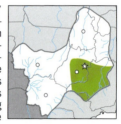

WHITE-VENTED VIOLETEAR *Colibri serrirostris* 12-13 cm | 4.75-5"
Fairly common and widespread in cerrado, woodland borders, and gardens (but largely absent from Pantanal). Bill slightly decurved. Mostly shining green with *conspicuous violet-purple ear tufts*; tail with wide subterminal band steel blue; *lower belly and crissum contrastingly white*. Glittering-bellied Emerald, often with it, is much smaller. Forages mostly in the open, often quite close to the ground; also hawks insects regularly. Aggressive toward other hummingbirds, especially smaller species (but in turn the violetear is dominated by Swallow-taileds). Song a series of "tsit" notes, often paired, interspersed occasionally by a more chortled note, delivered (often interminably) from a prominent perch.

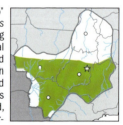

BLACK-THROATED MANGO *Anthracothorax nigricollis* 11 cm | 4.5"
Fairly common and widespread in semiopen areas, gardens, woodland and forest borders. Bill slightly decurved. ♂ shining metallic green above. *Throat, breast, and midbelly black*, throat and chest bordered by glittering blue and green, flanks green. *Tail purplish maroon*, feathers tipped black, central feathers blue-green. Nearly unmistakable ♀ like ♂ above but *white below with conspicuous black stripe down median underparts*; tail feathers tipped whitish. ♂, which looks dark and unpatterned in poor light, recognized by combination of its fairly large size, decurved bill, purplish tail, and black below. ♀ not likely confused. Mangos forage mainly at flowering trees in the semiopen, less often closer to the ground in plantations. They are frequently seen capturing insects in the air, sometimes well above the ground. Not especially vocal, though foraging birds give a "tsik" note.

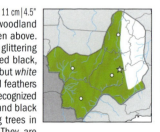

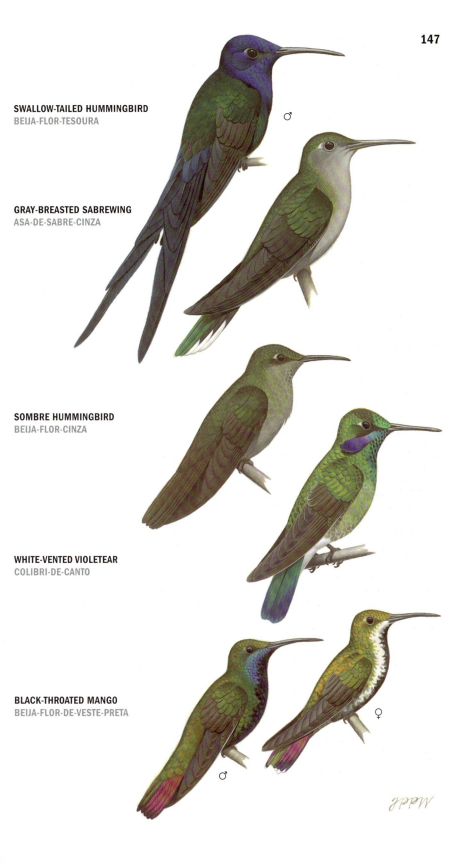

SWALLOW-TAILED HUMMINGBIRD
BEIJA-FLOR-TESOURA

GRAY-BREASTED SABREWING
ASA-DE-SABRE-CINZA

SOMBRE HUMMINGBIRD
BEIJA-FLOR-CINZA

WHITE-VENTED VIOLETEAR
COLIBRI-DE-CANTO

BLACK-THROATED MANGO
BEIJA-FLOR-DE-VESTE-PRETA

HORNED SUNGEM *Heliactin bilopha* 9.5-11 cm | 3.75"-4.25"

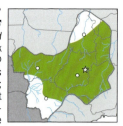

Uncommon in cerrado and campo rupestre. Bill short, straight. Beautiful ♂ has *dark shining blue crown with short lateral tufts glittering red, blue, and gold*; above bronzy green. *Lower face, throat, and chest black*; sides of neck and underparts white. *Tail long and narrow, all but central feathers white*. ♀ lacks tufts and blue on head. Throat dusky, *sides of neck and underparts white*, flanks greenish. Tail as in ♂ but slightly shorter. *Tiny but long-tailed*; not likely confused. Even ♀ should be easily known in its restricted habitat where there are few other hummingbirds (the most common being the Glittering-bellied Emerald). Usually found singly, foraging at flowers close to the ground; infrequent at flowering shrubs or trees. Occasionally hawks insects.

AMETHYST WOODSTAR *Calliphlox amethystina* 6.5 cm | ♀ 2.5"- ♂ 3"

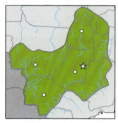

Uncommon and local (or seasonal?) in canopy and borders of forest and woodland, sometimes adjacent clearings. Bill straight. ♂ bronzy green above with *small white patch on sides of back extending behind wings to lower flanks*. Gorget glittering amethyst, bordered by *white pectoral collar extending up sides of neck* almost to white postocular spot; belly dusky green with some rufous on lower flanks and crissum. *Tail long and deeply forked*, dark bronzy purple. ♀ like ♂ above. Throat whitish spangled violet and green, bordered by *fairly distinct whitish pectoral collar; lower underparts rufous* with some intermixed white on midbelly; *flank patch as in ♂*; tail shorter, less forked, dull green with black subterminal band. Most likely confused with ♀ coquettes (q.v.). Usually found singly, most often while feeding at flowering trees; regularly perches on twigs high above ground. Flight slow and weaving, almost bee-like, with tail usually held partially cocked. Displaying ♂ arcs, pendulum-like, over a perched ♀.

COQUETTES are *tiny short-tailed hummingbirds with pale rump bands, ♂♂ ornately plumed*. Their flight is slow, often seeming to "float" in the air, tails often partially cocked.

FRILLED COQUETTE *Lophornis magnificus* 7 cm | 2.75"

Uncommon and apparently seasonal and local in canopy and borders of forest and woodland, sometimes adjacent clearings and gardens. At most rare in Pantanal. ♂ has *mainly red bill. Crown and bushy crest rufous* with glittering green forehead and throat; above metallic green with *conspicuous white rump band; fan-like elongated feathers on sides of neck white tipped dark green*; middle tail feathers bronzy green, others rufous edged green. Below grayish green. ♀ lacks crest and fan on neck; *forecrown pale buff dotted green*; pale rump band and tail as in ♂. Throat whitish dotted green; below whitish washed green on sides. ♂ of even fancier Dot-eared Coquette has longer crest feathers and neck plumes tipped with round green spots; ♀ Dot-eared has rufous forecrown and throat (*no dots*). Forages at all heights, often in flowering trees, e.g., *Inga* spp.

DOT-EARED COQUETTE *Lophornis gouldii* 7 cm | 2.75"

Rare and seemingly seasonal and local in canopy and borders of forest and woodland in NW of our area. Seems less numerous than the previous species, but the two have been found together at a few sites, e.g., around Chapada dos Guimarães. Resembles Frilled Coquette. ♂ has *rufous crest feathers longer and more pointed; white plumes on sides of neck even longer and tipped with round dark green spots*. ♀ has *throat rufous (no dots)* and more rufescent forecrown (*no dots*). Behavior as in Frilled Coquette.

RUBY TOPAZ *Chrysolampis mosquitus* 8-9 cm | 3.25-3.5"

Uncommon in deciduous and gallery woodland, scrubby areas, borders, gardens. Seems local and/or seasonal. ♂ has *crown and nape shining ruby red, throat and chest glittering gold*, these separated by a black band; back and underparts sooty blackish; *tail bright rufous tipped black*. ♀ above shining green, often with small white postocular spot; *tail mostly rufous with black subterminal band and white tip*. Below uniform pale grayish. Immature ♂ like ♀ but with some glittering orange on throat. Gorgeous in good light, but often ♂ looks just blackish; its distinctive silhouette, with *flat forecrown ruffled to the rear*, may then be more helpful. The drab ♀ can be confusing; watch for flashing rufous in its tail (best seen in flight); head shape can echo ♂. Usually seen feeding in flowering trees (often high), sometimes with other hummingbirds.

HORNED SUNGEM
CHIFRE-DE-OURO

AMETHYST WOODSTAR
BEIJA-FLOR-ESTRELINHA

FRILLED COQUETTE
TOPETINHO-MAGNÍFICO

DOT-EARED COQUETTE
TOPETINHO-PONTILHADO

RUBY TOPAZ
MOSQUITINHO

WOODNYMPHS are *forest* hummingbirds with slightly decurved bills. Colorful ♂♂ have *long, deeply forked tails*.

VIOLET-CAPPED WOODNYMPH *Thalurania glaucopis* ♂ 10-11 cm | 4-4.25" ♀ 9 cm | 3.5"
Uncommon and local in forest and woodland in S of our area. ♂ has *crown glittering violet-blue*; above dark shining green; *blue-black tail long and deeply forked*. Below shimmering golden green. ♀ shining green above; tail less forked. *Below whitish*. ♀ whiter below than ♀ Fork-tailed Woodnymph, tail tipping less obvious. Behavior similar, though Violet-capped more apt to forage in clearings and gardens.

FORK-TAILED WOODNYMPH *Thalurania furcata* ♂ 9.5 cm | 3.75" ♀ 8 cm | 3.25"
Fairly common and widespread in lower growth and borders of forest and woodland. ♂ shining green above. Glittering green forehead, blackish crown; *violet-blue on shoulders and across back*; *blue-black tail long and deeply forked*. Throat and chest glittering green; glittering violet-blue lower underparts. ♀ shining green above; tail less forked than ♂'s, outer feathers tipped whitish. *Below uniform pale grayish*. ♀ Glittering-bellied Emerald is smaller with whitish postocular, blackish mask. Cf. Violet-capped Woodnymph. Seen singly in forest or at edge, regularly foraging at *Heliconia* flowers, also shrubs. Tends to remain in lower growth, but sometimes ventures into subcanopy (usually ♂). Infrequent song a series of weak "tsip" notes.

AMAZILIA emeralds are "typical" hummingbirds with rather plain plumage, virtually straight bills, and little or no sexual dimorphism. They range in *semiopen terrain*.

GLITTERING-THROATED EMERALD *Amazilia fimbriata* 9-10 cm | 3.5-4"
Common and widespread in forest and woodland borders, clearings, and gardens. Lower mandible pinkish. ♂ shining green above. *Foreneck glittering green* (can look bluish in some lights), flanks shining green, with *obvious white stripe from lower belly on underparts to midbreast*. Tail blue-black, central feathers greener. ♀ with *white median stripe extending up to throat*. Versicolored Emerald *lacks* the white stripe on underparts. A rather conspicuous hummingbird that usually forages fairly close to the ground, but also ascends into the canopy of flowering trees; regularly hawks and gleans for insects.

Sapphire-spangled Emerald (*A. lactea*) occurs in SE of our area, e.g., Serra da Canastra area. Resembles Glittering-throated, including white median stripe, but foreneck glittering blue.

VERSICOLORED EMERALD *Amazilia versicolor* 8-9 cm | 3.25-3.5"
Uncommon at edge of forest and woodland, and in clearings and gardens (apparently not in Pantanal; but status unclear). Lower mandible reddish. Shining green above; tail bronzy green. Foreneck glittering bluish green, belly whitish. ♀ similar, but outer tail feathers narrowly tipped white. Similar to Glittering-throated Emerald but *lacks its white midbelly stripe*. Behavior also similar, but in our area Versicolored never seems to be as numerous. Call a fast, repeated "tsit-tsit-tsit...."

GLITTERING-BELLIED EMERALD *Chlorostilbon aureoventris* 8-9.5 cm | 3.25-3.75"
Common in semiopen (cerrado, borders, clearings, gardens); our most numerous and widespread hummingbird. ♂'s bill *mostly red*. Glittering golden green above, forked tail steel blue. Throat and chest glittering bluish green, belly glittering bronzy green. ♀ has less red on bill, *whitish postocular line, and blackish mask*; tail as in ♂ but outer feathers tipped grayish. Throat white, grayish below. The gorgeous ♂ has an overall shimmering effect, ♀ a unique (in our area) facial pattern. Found singly, usually feeding at flowers close to the ground. Song a rapidly repeated "tzz, tzz, tzz...."

WHITE-CHINNED SAPPHIRE *Hylocharis cyanus* 9 cm | 3.5"
Uncommon in canopy and borders of forest and adjacent clearings *in NW of our area*. ♂'s bill *coral red tipped black*. Head, throat, and chest glittering violet-blue; white chin (hard to see). Shining green above, uppertail-coverts more coppery. Below shining blue-green. *Tail steel blue*. ♀'s mandible pinkish tipped dusky. Shining green above, uppertail-coverts more coppery. Below grayish white, sides of throat spotted, chest glittering blue-green. Tail as in ♂, outer feathers tipped whitish. ♀ Glittering-bellied Emerald shows more facial pattern; ♀ Fork-tailed Woodnymph has all-black bill, no spotting on its darker and grayer underparts. Feeds at varying heights but inside forest remains high. ♂♂ sing in loose assemblies of 2-3 birds. Song a jumbled series of Bananaquit-like notes, "sweesee-see-see-seé."

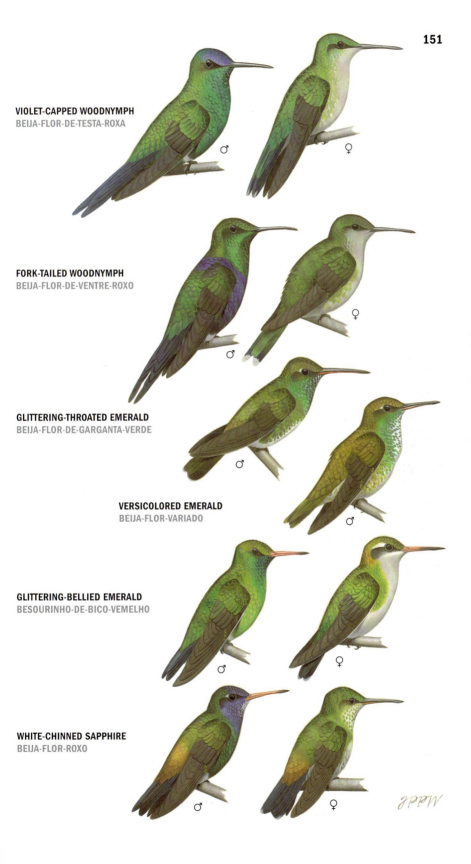

GILDED SAPPHIRE Hylocharis chrysura   9-10 cm | 3.5-4"
Common and widespread in forest and woodland borders, shrubby areas, and gardens in *S of our area*; particularly numerous in the Pantanal. Bill coral red tipped black. *Uniform shining golden green* with very small white postocular spot; *tail glittering coppery gold*. ♀ duller with less golden sheen, bill pinkish with dusky tip. Generally the most frequently seen hummingbird in the Pantanal, and should not be confused with any other species there. Feeds at various flowering plants at varying heights, but typically not too high above the ground. Song a series of high-pitched jumbled notes.

Rufous-throated Sapphire (*H. sapphirina*) occurs in forest borders in NW of our area, e.g., Serra das Araras, occasionally further S (Poconé area). ♂ has coral red bill, glittering violet-blue foreneck, rufous on chin, and *mostly chestnut tail*. ♀ has whitish underparts with green spangling, rufous on chin; tail as in ♂. White-chinned Sapphire has blue-black tail.

WHITE-TAILED GOLDENTHROAT Polytmus guainumbi   10 cm | 4"
Uncommon and somewhat local in shrubby areas with tall grass, most often near marshy areas or open water. Bill rather long and *decurved*, mostly reddish pink. ♂ above bronzy golden green with *white postocular and malar lines*; rather long, rounded tail mostly green, *outer feathers increasingly white toward base* (conspicuous in flight). Below glittering golden green (brighter than upperparts). ♀ buffier below with some green spangling. Tends to remain close to the ground, the white in its tail flashing prominently in flight, especially as it hovers. Feeds at various flowers (generally not in flowering trees), and also gleans tiny insects from foliage and twigs.

`STARTHROATS` are *large* hummingbirds with *long, nearly straight bills*. They range in semiopen country and none is particularly numerous.

LONG-BILLED STARTHROAT Heliomaster longirostris   11.5 cm | 4.5"
Uncommon at forest and woodland borders and in clearings in *NW of our area*. Bill very long and straight. ♂ shining green above with *glittering pale blue crown*, white postocular spot, and partially concealed *white stripe on lower back and rump*. Gorget *glittering reddish purple* bordered by white malar streak; below pale grayish with some green on sides, sometimes showing a white flank patch. Tail bronzy green, outer feathers tipped white. ♀ has less blue on crown and its gorget is smaller and duller, often scaled grayish. Both sexes of Stripe-breasted Starthroat differ in having a striking white midventral line; ♂'s gorget flares more to the sides. Cf. Blue-tufted Starthroat. Usually found in the semiopen, often perching high on dead branches. Feeds at flowering trees, often congregating with other hummers. Sometimes captures tiny insects, most often while hovering in the open.

STRIPE-BREASTED STARTHROAT Heliomaster squamosus 11-11.5 cm | 4.25-4.5"
Rare to uncommon in gallery woodland, caatinga woodland and scrub, and gardens in *SE of our area*. Bill long and straight. ♂ *rather dark bronzy green above*, crown more glittering, with a white postocular spot and moustachial streak; tail forked, mostly dull bluish green. Gorget and tuft at sides of neck glittering violet; below dark green with *narrow but conspicuous white midventral* stripe. ♀ duller, lacking the gorget and tuft; throat dark green scaled white and outlined by white moustachial streak; below grayish with *white midventral stripe* (less evident than in ♂); tail less forked than in ♂, outer feathers tipped white. Dark-looking ♂ should be easy to recognize, while the drabber ♀ can be known by its midventral stripe (*lacking* in the other starthroats). Behavior much as in Long-billed Starthroat.

BLUE-TUFTED STARTHROAT Heliomaster furcifer   12-13 cm | 4.75"-5"
Uncommon in scrub, woodland borders, and gardens in *S of our area*. Bill very long and straight. Beautiful ♂ shining bronzy green above with glittering bluish green crown. Gorget *glittering violet-magenta; underparts and long tuft on sides of neck glittering blue*. Tail deeply forked, bronzy green. ♀ bronzy green above with small white postocular spot. *Dingy pale grayish below*, whitest on narrow malar streak and midbreast and midbelly. Tail forked, bronzy green, outer feathers tipped white. The stunning full-plumaged ♂ resembles no other hummingbird. Much drabber ♀ can be confused with both the other starthroats, though her relatively large size, long bill, and essentially uniform underparts should identify her. Behavior similar to Long-billed Starthroat, though Blue-tufteds can be commoner at least seasonally.

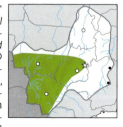

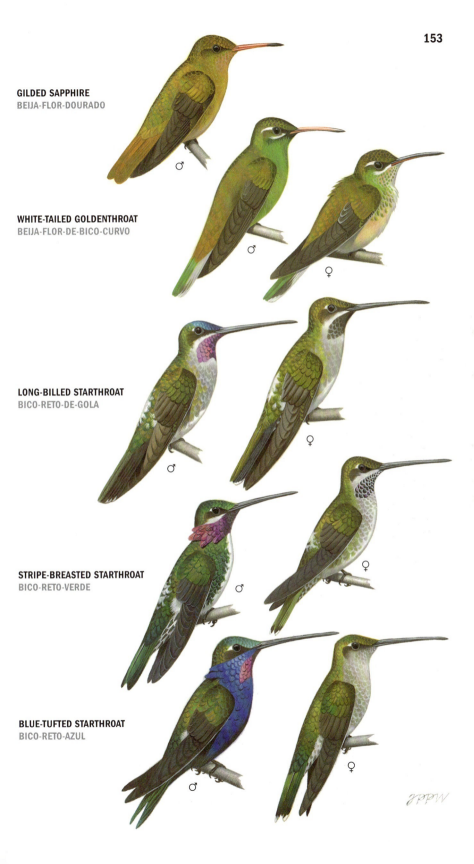

GILDED SAPPHIRE
BEIJA-FLOR-DOURADO

WHITE-TAILED GOLDENTHROAT
BEIJA-FLOR-DE-BICO-CURVO

LONG-BILLED STARTHROAT
BICO-RETO-DE-GOLA

STRIPE-BREASTED STARTHROAT
BICO-RETO-VERDE

BLUE-TUFTED STARTHROAT
BICO-RETO-AZUL

TROGONS (Trogonidae) are *colorful* birds with soft lax plumage, short legs, heavy bills that are slightly hooked at tip, and *long, graduated, square-tipped tails*. They range in a variety of forest and woodland habitats, where they perch upright and (in spite of their often gorgeous coloration) are rather *inconspicuous aside from their frequently given vocalizations*. Many species gather in singing assemblies, both sexes participating. Their diet consists of a mixture of fruit and larger insects, both usually captured on the wing. They nest in cavities, often dug into arboreal termitaries.

BLACK-TAILED TROGON Trogon melanurus   30.5 cm | 12"

Uncommon in forest and borders in *NW of our area*, e.g., Serra das Araras. A large, red-bellied trogon. ♂ has *bright yellow bill* and red eye-ring; ♀'s bill yellow below. ♂ shining bluish green above, bluest on rump and upperside of tail, with black-and-white vermiculated wing-coverts. *Underside of tail black*. Face and upper throat black, lower throat and chest shining green with narrow white chest band; *breast and belly bright red*. ♀ gray above and on throat and chest, *breast and belly red*, sometimes with a little white separating the gray and red. Underside of tail blackish, outer feathers often with white tips and some barring. Blue-crowned Trogon is smaller, has obvious black-and-white barring on tail. Surucua Trogon occurs only in the SE. A phlegmatic bird, tending to remain well above the ground; usually inconspicuous, heard more often than seen. Mostly ranges in pairs, perching on larger limbs, often motionless for long periods but periodically hurtling out to pluck a fruit or pick off a large insect, continuing on to another perch. Sometimes gathers in small groups that displace and chase each other, the ♂♂ singing and the ♀♀ calling softly. Song a long-continued series of up to 20-30 resonant, quite low-pitched "cow" notes that starts softly: "cuh-cuh-cuh-cuh-cow-cow-ców-ców-ców...." Call a soft clucking, often given as the bird raises and lowers its tail.

AMAZONIAN WHITE-TAILED TROGON Trogon viridis   28 cm | 11"

Common in midlevels and subcanopy of forest and borders in NW of our area. *Pale blue eye-ring in both sexes*, unique among our trogons. ♂ mostly shining bluish green above, hindcrown and nape more violet-blue; forehead, face, and throat black. Underside of tail black, but *appearing mostly white because outer feathers are so broadly white-tipped*. Chest violet-blue; breast and belly rich yellow. ♀ gray above and on throat and chest, with some faint white vermiculations on wing-coverts and barring on inner flight feathers. Underside of tail blackish, *outer feathers barred and broadly tipped white*. Breast and belly yellow. ♀'s undertail like that of smaller Violaceous, but latter differs in its whitish eye crescents and whitish band separating gray and yellow on underparts. Often in pairs, usually staying well above the ground but coming lower at edge; there they can be tame, peering around sluggishly. Small groups gather during courtship. Song a fast, fairly even series of 15-20 "cow" or "cowp" notes, higher-pitched than in Black-tailed Trogon. Both sexes also give soft "chuk" notes and a nasal scolding: "kwa kwa kwo-kwo-kwo-kwo-kwo," often accompanied by raising and lowering tail.

SURUCUA TROGON Trogon surrucura   27 cm | 10.5"

Locally fairly common in deciduous and gallery forest and woodland in SE of our area. ♂ has *red eye-ring*; ♀ has *feathered white crescents in front and behind eye*. ♂ has *shining violet-blue head, neck, and chest* with black forehead, face, and throat; shining green above with black-and-white vermiculations on wing-coverts and inner flight feathers. *Underside of tail looks mostly white. Breast and belly bright red*. ♀ gray above with narrow white barring on wing-coverts and inner flight feathers. Underside of tail blackish with *outer feathers broadly tipped and edged white*. Throat dark gray, becoming paler on breast; belly red. Blue-crowned Trogon is smaller; ♂ has orange eye-ring, underside of tail boldly barred, and a white chest band; ♀ also has white chest band and shows some barring on outer tail feathers. Black-tailed Trogon occurs only in far northwest. Behavior similar to Amazonian White-tailed Trogon. Song a measured series of up to 20-30 "cow" notes, in the second half their pitch sometimes dropping and the pace slightly speeding up. Also gives various "cluk's" and "chirr's."

BLACK-TAILED TROGON
SURUCUÁ-DE-RABO-PRETO

AMAZONIAN WHITE-TAILED TROGON
SURUCUÁ-GRANDE-DE-BARRIGA-AMARELA

SURUCUA TROGON
SURUCUÁ-VARIADO

BLUE-CROWNED TROGON *Trogon curucui* 24 cm | 10"
Fairly common in midlevels and borders of deciduous and gallery forest and woodland. The only trogon in the Pantanal. ♂ has yellow-orange eye-ring, ♀ feathered white crescents in front of and behind eye. ♂ shining green above, with *crown and nape shining dark blue*; wing-coverts vermiculated black and white. *Underside of tail black, feathers barred and tipped white*. Forehead, face, and throat black. ♂'s Chest shining blue, *bordered below by white band*; breast and belly bright red. ♀ gray above and on throat and breast, with white barring on wing-coverts and inner flight feathers. Underside of tail black, *outer webs of outer feathers barred and tipped white*. Breast and belly red, separated from chest by a white band. ♂ Collared Trogon (only far NW) has green crown and chest; ♀ Collared is predominantly brown. ♀ Black-tailed Trogon, also gray with red belly, has little or no white on its tail underside. Cf. Surucua Trogon. Usually in pairs and phlegmatic even for a trogon; more often in the semiopen than the others. Song a fast, even repetition of a relatively high-pitched "cow" note, ending abruptly; higher-pitched than Amazonian White-tailed's. Both sexes give a "churrr" call.

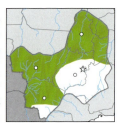

COLLARED TROGON *Trogon collaris* 24 cm | 10"
Uncommon in midlevels of forest and woodland in NW of our area. Bill yellow (with black culmen in ♀); ♂ has red eye-ring, ♀ feathered white crescents in front of and behind eye. ♂ *shining green above*, wing-coverts and inner flight feathers vermiculated black and white; forehead, face, and throat black. Underside of tail black, feathers barred and broadly tipped white. *Chest shining green*; a *prominent white band* separates it from bright red breast and belly. ♀ brown above and on throat and chest, more rufous-chestnut on upperside of black-tipped tail. *Underside of tail whitish freckled and mottled dusky*. ♂ Blue-crowned Trogon has blue crown and chest (not green); ♀ Blue-crowned is gray where Collared is brown. Collared usually ranges in pairs and is easily overlooked, as it tends to remain *within* forest. Song a short fast "karow-ków-ków-ków," sometimes up to 5 "ców's." Both sexes give a descending "churrr," often while raising and lowering tail.

VIOLACEOUS TROGON *Trogon violaceus* 22.5 cm | 8.75"
Uncommon in midlevels and subcanopy of forest and woodland *in NW of our area*. ♂ has *orange-yellow eye-ring*, ♀ feathered white crescents in front of and behind eye. ♂ shining bluish green above, *crown and nape shining violet-blue*; forehead, face, and throat black. Upperside of tail bluish; underside barred black and white, outer feathers tipped white. An indistinct white band separates *violet-blue chest* from rich yellow breast and belly. ♀ gray above and on throat and chest, with faint white barring on wing-coverts and inner flight feathers. *Underside of tail blackish, outer webs of outer feathers barred and tipped white*. Breast and belly yellow. Most likely confused with Amazonian White-tailed Trogon, both sexes of which have a blue eye-ring and show more white on their undertails (especially in ♂). ♂ Black-throated has green chest; ♀ Black-throated is predominantly brown (not gray). Behavior similar to Amazonian White-tailed. Song a fast but rather short series of clipped "cow" notes, the notes often becoming doubled: "cadow-cadow-cadow...."

BLACK-THROATED TROGON *Trogon rufus* 25 cm | 10.25"
Rare and local in lower growth of forest in NW and far S of our area. In NW (illustrated) ♂ has yellow eye-ring, ♀ white crescents in front of and behind eye. ♂ shining green above, wing-coverts and inner flight feathers vermiculated black and white; forehead, face, and throat black. Upperside of tail shining coppery bronze; underside barred black and white, outer feathers tipped white. *Chest shining green*, bordered below by diffused white band; *breast and belly rich yellow*. In SW, eye-ring pale blue and tail upperside bronzy green. ♀ brown above and on throat and chest, wing-coverts vermiculated buff and blackish. Upperside of tail rufous-chestnut; underside as in ♂. Breast and belly yellow with diffused white chest band. ♂ *is only Brazilian trogon combining green chest and yellow underparts*. ♀ *is the only "brown" trogon with yellow underparts*. Behavior similar to Collared Trogon, but is even more likely to remain *within* forest and favors the vicinity of streams. Song in NW a slow "cuh, cwuh-cwuh," sometimes with an extra note or two added; in S a much faster series of 6-7 evenly paced "coh" notes. Both sexes give a "churrr" call.

BLUE-CROWNED TROGON
SURUCUÁ-DE-BARRIGA-VERMELHA

COLLARED TROGON
SURUCUÁ-DE-COLEIRA

VIOLACEOUS TROGON
SURUCUÁ-PEQUENO

BLACK-THROATED TROGON
SURUCUÁ-DE-BARRIGA-AMARELA

KINGFISHERS

KINGFISHERS (Alcedinidae) are *fish-eating* birds with *powerful long and pointed bills*. They nest in burrows dug into banks, almost always near water.

RINGED KINGFISHER *Megaceryle torquata*   39.5 cm | 15.5"
Widespread and generally common along rivers, larger streams, and lakes; most numerous in Pantanal. Our largest kingfisher, with distinctive bushy crest. Massive bill. ♂ *blue-gray above* with nuchal collar white; tail feathers blackish sparsely barred white. *Below chestnut-rufous.* In flight shows white patch at base of primaries; underwing-coverts white. ♀ has *chest band blue gray*, underwing-coverts rufous. A conspicuous bird that perches at varying levels, often higher than other kingfishers, sometimes raising and lowering its tail. Plunges for fish from a perch, often from a considerable height. The only kingfisher that regularly flies high above ground. Noisy, with a variety of loud calls including a harsh "krek!" (normally given when flying high), a loud, fast rattle (especially when flushed), and sharp chatters.

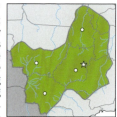

AMAZON KINGFISHER *Chloroceryle amazona*   28 cm | 11"
Fairly common along rivers, larger streams, lakes, and ponds; most numerous in Pantanal. *The largest "green" kingfisher, with heavy blackish bill.* ♂ *dark, shiny, oily green above* with narrow nuchal collar white; outer tail feathers have some white banding. *White below with broad rufous chest band.* ♀ has *chest band broken dark green.* Green Kingfisher is much smaller with a slighter bill and white wing-spotting; on flying birds the white in its tail is more obvious. Amazons are seen singly or in pairs, often perching in the open on low branches. Their flight is fast and direct, usually low over the water. Plunges for fish from perches, less often after hovering briefly. Noisy, giving a loud "cak!" or "chat!" call, sometimes doubled, and a descending series of sputtering or squeaky notes that can end in a rattle.

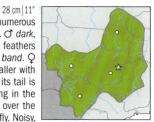

GREEN KINGFISHER *Chloroceryle americana*   19 cm | 7.5"
Fairly common and widespread along rivers, streams, margins of lakes and ponds. Bill black. ♂ *dark, shiny, oily green above* with narrow nuchal collar white; *two rows of white spots on flight feathers; outer tail feathers mainly white* (flashing in flight). White below with a broad rufous chest band. ♀ with *two green-spotted chest bands.* Smaller than similarly plumaged Amazon Kingfisher. Green has more white spotting on wings and in flight it shows more white in tail. Cf. American Pygmy Kingfisher. Green occurs more often than Amazon *along small flowing streams.* Rests more often on rocks than other kingfishers; always perches low. Flight is fast and direct, usually low over water. Dives for fish from perches, sometimes first hovering. Flight call a raspy "dzeet" or "treet," often doubled or trebled, like two pebbles being struck together. Also gives a descending series of "tsu" notes.

GREEN-AND-RUFOUS KINGFISHER *Chloroceryle inda*   21.5 cm | 8.5"
Uncommon and inconspicuous along sluggish wooded streams and shady lake margins in N of our area and Pantanal. Bill proportionately heavy, black. ♂ dark oily green above with buff streak above lores, wings and tail lightly speckled white or pale buff. *Collar on sides of neck and entire underparts rich orange-rufous;* paler on throat. ♀ has green-and-white chest band. American Pygmy Kingfisher, though similarly colored, is much smaller with a slighter bill; its midbelly is white. Green-and-rufous is unobtrusive and usually solitary, perching on twigs and branches low over water, generally remaining within cover where hard to see. Often seems wary. Dives directly into the water without hovering. Flight swift and direct, low over water; flushed birds usually zoom off and disappear. Flight call a sharp, grating "dreet" or "dzreet," often trebled.

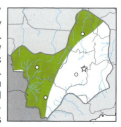

AMERICAN PYGMY KINGFISHER *Chloroceryle aenea*   13 cm | 5"
Uncommon and inconspicuous along small, usually sluggish streams and swampy backwaters, most numerous in N of our area and Pantanal. *Our smallest kingfisher.* Bill blackish. ♂ dark oily green above with rufous streak above lores. *Collar on sides of neck and underparts rich orange-rufous,* with *white midbelly and crissum.* ♀ has narrow green-and-white chest band. Green-and-rufous and Green Kingfishers are much larger (but confusion is possible in fast flight). American Pygmy is found singly and generally perches close to the surface of shallow water, and sometimes even feeds at stagnant pools; it can be confiding. Flight low and darting, often so fast that it can be hard to follow to its next perch. Flight call a sharp but rather weak "tzit" or "tyeet."

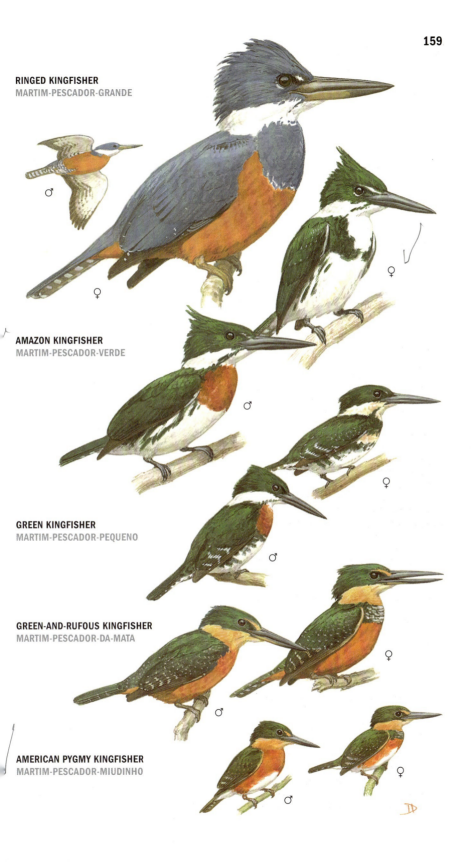

RINGED KINGFISHER
MARTIM-PESCADOR-GRANDE

AMAZON KINGFISHER
MARTIM-PESCADOR-VERDE

GREEN KINGFISHER
MARTIM-PESCADOR-PEQUENO

GREEN-AND-RUFOUS KINGFISHER
MARTIM-PESCADOR-DA-MATA

AMERICAN PYGMY KINGFISHER
MARTIM-PESCADOR-MIUDINHO

MOTMOTS (Momotidae) are attractive but *inconspicuous* birds that favor the forest interior and nest in burrows dug into banks. They *often switch their racket-tipped tails back and forth*.

RUFOUS-CAPPED MOTMOT *Baryphthengus ruficapillus*   42 cm | 16.5"

Uncommon and somewhat local in lower growth of forest and woodland in SE of our area. Tail long and slender, terminal rackets usually present but sometimes not well developed. Iris reddish brown. *Crown and nape rufous*; green above with broad black mask through eyes; primaries edged blue-green. Below bright olive green with black chest spot, *cinnamon-rufous band across upper belly*, bluish green lower belly. Nearly unmistakable; lacks Blue-crowned Motmot's obvious blue in crown, etc. Usually in pairs that seem more active at dawn and dusk. Call a brief fast hooting, "hoo-oo-oo-oo-oo," guttural and vaguely owl-like; vocalizes primarily in predawn darkness and at dusk.

BLUE-CROWNED MOTMOT *Momotus momota*   40 cm | 15.75"

Fairly common and widespread but inconspicuous in lower growth of forest, woodland, and borders. Iris red. *Tail long and slender, terminal rackets usually obvious* (except when molting). *Blue crown encircling black midcrown patch turquoise in front, more violet to rear*; broad black mask through eyes. Above green. Below bronzy olive, belly more tawny-rufous. Rufous-capped Motmot has rufous in crown and nape, *no blue*. Found singly or in pairs, perching quietly in shady lower growth; in early morning sometimes more in the open. Feeds on larger insects, lizards, snakes, and fruit, often dropping to the ground. Most frequent call, given especially at dawn and dusk but also during the day, a fast, hollow "hooo-do." Also has a rolling series of more tremulous hoots.

JACAMARS (Galbulidae) are distinctive slender birds with *long straight bills* and graduated tails. They nest in burrows dug into banks or arboreal termitaries.

BROWN JACAMAR *Brachygalba lugubris*   16.5 cm | 6.5"

Uncommon and local in borders and canopy of deciduous and gallery woodland; absent from Pantanal but present in Chapada dos Guimarães area. *Long slender bill*, blackish with *lower mandible at least basally yellowish*; iris pale bluish. *Mostly sooty brown*, face and throat freckled and streaked white; wings and *rather short tail* glossy bluish black. *Belly contrastingly paler, buff to buffy whitish*. ♀ richer brown, less mottled. Smaller, shorter-tailed, and duller-plumaged than other jacamars, with an almost hummingbird-like silhouette. Occurs in pairs or loosely associated groups that perch in the open on twigs and small branches, watching alertly with rapid head movements for passing insects, making long sallies to pursue them. Calls shrill and piercing and include a single sharp "piii" or "peeey" and a thin, somewhat descending and slowing "pididideedeedee-dee-dee-dew."

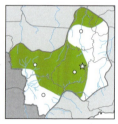

RUFOUS-TAILED JACAMAR *Galbula ruficauda*   24 cm | 9.5"

Fairly common and widespread in lower growth of forest and woodland borders, often (but not always) near water. *Bill very long and slender*. ♂ shining metallic golden green above; upperside of tail green, *underside mostly rufous*. Throat white, chest shining golden green; *lower underparts rich rufous*. ♀ has throat buff, lower underparts paler. *The only jacamar in the Pantanal*. Unmistakable; Paradise and Brown Jacamars (q.v.) are very different. Found in pairs that perch, often close together, on open branches, looking around alertly. Sallies into air after insects, e.g., wasps, butterflies (even *Morphos*!), and dragonflies, usually removing stingers and wings before swallowing. Notably sedentary. Most frequent call, given by both sexes, a sharp "peeyeek!" sometimes rapidly repeated. ♂'s song a long, high-pitched, ascending series of "pee" notes ending in a trill.

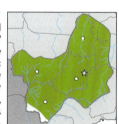

PARADISE JACAMAR *Galbula dea*   29.5 cm | 11.5"

Uncommon in *canopy and borders of forest in NW of our area*. Unmistakable; *large* with a *long slim tail*. Bill very long and slender. Mostly dark bluish black, wings and tail more bronzy green; tail underside blackish, feathers tipped whitish. *Large throat patch white*. ♂ has crown mottled brownish, darker in ♀. No other jacamar is as long-tailed or dark overall. Occurs in pairs that perch on exposed snags from which they make long looping sallies into the air in pursuit of insects, usually large ones. Easiest to see at forest borders. Call a sharp "pik!" or "peeyk!" sometimes given in a short series. ♂'s song a rather slow series of 10–12 well-enunciated notes that gradually descend in pitch: "peeyr-peeyr-peeyr-peer-peer-peer-pur-pur-pur-pr."

PUFFBIRDS (Bucconidae) are chunky birds with large heads, short necks, and stout hooked bills. *Nystalus* are midsized puffbirds with *nuchal collars*; most have *brightly colored bills*. Our species range primarily in semiopen and wooded areas, and all have *characteristic loud vocalizations*.

WHITE-EARED PUFFBIRD *Nystalus chacuru* 20 cm | 8"

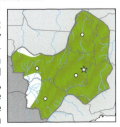

Fairly common in cerrado and at edge of *deciduous and gallery woodland*; absent from Pantanal. *Bill mostly bright orange*. Iris brownish orange. *Boldly patterned; nearly unmistakable*. Brownish rufous above with *complex black-and-white facial pattern* and whitish nuchal collar. Buffy whitish below with black scaling on sides. Caatinga and Chaco Puffbirds lack the bold facial pattern, have streaking below. Found singly or in loosely associated pairs, usually perching low and in the open, sometimes on fences or even phone wires. Feeds on larger insects and small lizards, often captured on the ground. Song a loud, far-carrying "truu, truturay, truturay, truturay," often given as a duet and with a distinctive rollicking cadence that almost has a laughing quality. It is often one of the first bird sounds heard at dawn.

CAATINGA PUFFBIRD *Nystalus maculatus* 19 cm | 7.5"

Fairly common in *deciduous and gallery woodland and scrub*; occurs S to Poconé area, but *not in actual Pantanal*. *Bill orange-red*; iris yellow. Brown above with *crown and face streaked dusky*, obscure nuchal collar buff, mantle and tail barred blackish. Below whitish with *orange-rufous band across lower throat, coarse black spotting on breast and down flanks*. Chaco Puffbird (occurring to S) is streaked on its lower underparts. Cf. also White-eared Puffbird (with much bolder face pattern, *no* streaking, essentially plain whitish underparts). Occurs singly and in pairs, perching in the open, and often tame to the point of being stolid; a little less conspicuous than the White-eared Puffbird (with which Caatinga can occur, though latter is more of a woodland bird). Mainly consumes large insects, captured both on the ground and from foliage. Song, as loud and far-carrying as the White-eared's, a ringing "cherowee, cherowee, cho-ró," sometimes given as a duet.

CHACO PUFFBIRD *Nystalus striatipectus* 19 cm | 7.5"

Fairly common in deciduous and gallery woodland and scrub in *Mato Grosso do Sul*. Resembles Caatinga Puffbird, differing most notably in being *streaked* (not spotted) with *black on breast and sides*; paler nuchal collar and throat band, more streaked face. As far as known, Chaco does not occur with the Caatinga; until recently the two were considered conspecific. Behavior and voice similar to Caatinga's, but at least typically its song is a little shorter: "toh-wr-ee, toh-wr-ee, tw-oh;" it likewise is sometimes given as a duet.

STRIOLATED PUFFBIRD *Nystalus striolatus* 20.5 cm | 8"

Uncommon and local in subcanopy and borders of forest in NW of our area, e.g., Serra das Araras. Bill olive green with dark tip; iris buff. Above dark brown spotted and edged rufous buff; lores whitish, *nuchal collar cinnamon-buff*; tail dusky barred cinnamon-buff. Mostly whitish below *narrowly streaked black*, breast tinged buff. Though rather drab, this arboreal puffbird is unlikely to be confused in its range and habitat. Caatinga Puffbird is not a forest bird; it has an orange-rufous chest band and spotting below. Striolated is typically found alone or in loosely associated pairs, perching stolidly on open forest branches quietly surveying its surroundings; it sallies abruptly to leaves and branches for prey, mostly large insects. Heard much more often than seen, its soft but far-carrying call a melancholy whistled "whip, whi-wheeu, wheeeeeuu" with a characteristic cadence.

RUSTY-BREASTED NUNLET *Nonnula rubecula* 14.5 cm | 5.75"

Uncommon and very local in lower and middle growth of forest and woodland in SE of our area; often in viny growth. *Long, slightly decurved bill* bluish gray with blackish culmen and tip. Rather short tail. Crown grayish with *buffy whitish loral area* and *prominent eye-ring white or bluish white*; face and upperparts dull brown. *Below mostly orange-buff*, midbelly contrastingly white. Obscure in plumage and retiring by nature, this nunlet occurs singly or in pairs, perching quietly and nearly motionless on branches where it usually is overlooked except when singing. Whistled song a simple series of 10–20 steadily repeated "wheer" notes.

Rufous-capped Nunlet (*N. ruficapilla*) occurs very locally in lower growth of forest in NW of our area, e.g., Serra das Araras. Overall much as in Rusty-breasted Nunlet, but has rufous crown contrasting with gray face, neck, and sides.

WHITE-EARED PUFFBIRD
JOÃO-BOBO

CAATINGA PUFFBIRD
RAPAZINHO-DOS-VELHOS

CHACO PUFFBIRD
RAPAZINHO-DO-CHACO

STRIOLATED PUFFBIRD
RAPAZINHO-ESTRIADO

RUSTY-BREASTED NUNLET
MACURU

SWALLOW-WINGED PUFFBIRD *Chelidoptera tenebrosa* 16.5 cm | 6"
Common and conspicuous in forest and woodland borders in N of our area; not in Pantanal. *Chunky; short-tailed.* Short black bill. Blue-black above with *white patch on lower back and rump* (obvious in flight, though usually hidden at rest); *underwing-coverts white.* Throat and breast dull black, *belly cinnamon-rufous,* crissum gray. In flight broad wings and short tail impart an unmistakable silhouette. Perched birds can look vaguely like Gray-breasted Martins, though the latter are not so short-tailed; the "square-headed" shape can even recall a parrot. Pairs or small groups perch on high open branches, even in mid-day heat. They make long sallies after insect prey, flying out rapidly, then swooping back with a slower long glide. Nest in holes dug into sandy banks, sometimes in level ground. Generally silent, occasionally twittering weakly.

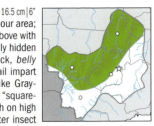

WHITE-NECKED PUFFBIRD *Notharchus hyperrhynchus* 25 cm | 10"
Uncommon in canopy and borders of forest and woodland in NW of our area. *Broad heavy bill.* Black above with *prominent white forecrown and nuchal collar.* White below with *wide black breast band,* flanks barred black. Not likely confused. Generally solitary, less often in pairs, especially in the early morning perching for long periods on high exposed branches. Sallies abruptly to capture large insects from leaves and branches, often grabbing prey with an audible snap, beating it against a branch before swallowing it. Infrequently heard song an evenly pitched trill that lasts 3-5 seconds, sometimes given sequentially by members of a pair.

Pied Puffbird (*N. tectus*) occurs in small numbers in NW of our area. Considerably smaller than White-necked, with crown black dotted white and a white scapular patch.

SPOTTED PUFFBIRD *Bucco tamatia* 18.5 cm | 7.25"
Rare and local in lower and middle growth of forest and woodland in NW of our area. Iris red. Brown above with *orange-rufous forehead and short superciliary,* and white malar streak; *large black patch on sides of neck.* Throat orange-rufous; below white *with irregular coarse black spotting.* Boldly patterned, not likely confused. Stolid in demeanor, perching for long periods without moving, sometimes fully in the open on an unobstructed branch but usually hard to spot. Mainly solitary, less often in pairs. Though usually quiet, its distinctive song (usually given in the early morning) is loud and far-carrying, a long series of monotonous and inflected "kueeép" notes that starts softly and gradually becomes louder, continuing for 10-15 seconds.

NUNBIRDS are *mainly gray* puffbirds with *distinctive coral red bills.* They occur in forest and woodland, and nest in burrows dug into sloping ground, with "helpers" (presumably relatives) often being present.

BLACK-FRONTED NUNBIRD *Monasa nigrifrons* 26.5 cm | 10.5"
Common and widespread in lower and midlevels of riparian, gallery, and deciduous forest and woodland. *Bill coral red.* Slaty gray, darker on foreparts; black on forehead and upper throat. White-fronted Nunbird occurs only in NW of our area, is strictly confined to terra firme forest; Black-fronted is darker and lacks white on foreface. Occurs in groups of up to 4-6 individuals that perch upright, often in the open on horizontal branches or lianas. They rest quietly, then sally to foliage or branches (occasionally to the ground) after insect prey. Flight is slow, a few shallow flaps and a glide. Other species are regularly with them. Foraging Black-fronted Nunbirds are usually quiet, giving only an occasional soft or querulous note, but then they may abruptly break into a rollicking, gabbling chorus that can continue for several minutes, most often given by two birds (a pair?) that perch side by side.

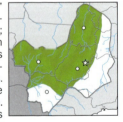

WHITE-FRONTED NUNBIRD *Monasa morphoeus* 25.5 cm | 10"
Uncommon in midlevels and subcanopy of forest and borders in *NW of our area.* Bill coral red. Mostly gray; paler on belly, blacker on face and throat; *forecrown and chin white.* Only likely confusion is with Black-fronted Nunbird, which lacks white on face and is restricted to areas near water. Cf. also the all-gray Screaming Piha. Behavior as in Black-fronted, though White-fronted is harder to see, descending close to the ground much less often. Group size is usually smaller; mixed flocks often coalesce around them. Voice similar to Black-fronted's but less jumbled, a loud gabbling chorus given together by a pair, less often several birds at once.

SWALLOW-WINGED PUFFBIRD
URUBUZINHO

WHITE-NECKED PUFFBIRD
MACURU-DE-TESTA-BRANCA

SPOTTED PUFFBIRD
RAPAZINHO-CARIJÓ

BLACK-FRONTED NUNBIRD
BICO-DE-BRASA

WHITE-FRONTED NUNBIRD
TANGURUPARÁ-DE-CARA-BRANCA

TOUCANS (Ramphastidae) are spectacular large birds that use their *long, laterally compressed, usually colorful bills* to stretch for fruit, their primary diet. The bills are hollow and light. They mainly range in the canopy of forest and woodland, the Toco in much more open country. Most toucans are *quite vocal*, and give a variety of croaking, grunting, or even squeaky sounds.

GOULD'S TOUCANET *Selenidera gouldii* 33 cm | 13"

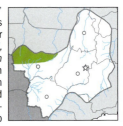

Rare and local in canopy and midlevels of forest in *NW of our area*. Iris yellow with a "horizontal" pupil; large bare ocular area bluish green. Outer bill orange-yellow; base of maxilla black, of mandible whitish. ♂ has *head, neck, and underparts black* with *conspicuous tuft of golden feathers on ear-coverts* and a yellow nuchal collar. Olive green above, tail duskier with central feathers tipped rufous. Flank patch mixed yellow and rufous, crissum red. ♀ has *rufous replacing ♂'s black*, and ear-tufts duller. In its limited range nothing really resembles this *ornately plumaged* small toucan. Unlike most toucans, this toucanet is furtive and inconspicuous, tending to remain hidden in foliage and rarely perching in the open. It usually ranges in pairs, less often small (family?) groups. Distinctive call a low-pitched guttural growl, almost frog-like, "groww-groww-groww-groww…," repeated at a rate of about one note per second for 8-10 seconds; calling birds often seesaw up and down, bowing head and raising tail up over back.

ARAÇARIS are slender, boldly patterned, *long-billed* toucans with *long, graduated tails*. They range widely in forest and woodland, usually in small groups, and are noisy and conspicuous. Nests are placed in tree cavities (sometimes in holes made by woodpeckers). They roost communally, sometimes in the same holes where nesting occurs.

LETTERED ARAÇARI *Pteroglossus inscriptus* 34 cm | 13.5"

Fairly common in canopy and borders of forest and woodland in *N of our area* (including Chapada dos Guimarães). Iris dark red; bare ocular area mostly turquoise, darker below eye. Bill brownish yellow with dark culmen, tip, and *vertical "tooth" marks on the maxilla (the "letters")*. ♂ dark green above with black head, neck, and throat; rump red (hidden at rest). *Below pale yellow (no bands)*; thighs mainly brown. ♀ differs in its chestnut brown face and throat. Lettered is our only araçari *without a red band on underparts*. Usually in groups, sometimes up to 6-8 birds, that forage in canopy, subcanopy, and borders where they hop on larger branches. They often perch in the open early in the morning, less so at other times, and feed mainly on fruit (also some insects, small lizards, eggs, and nestlings). Flight is level on rapidly beating wings, ending with a short glide just before landing; often a group straggles along single file, following much the same flight path. Quiet for an araçari, with call consisting of an often long-continued series of guttural "cha" notes.

CHESTNUT-EARED ARAÇARI *Pteroglossus castanotis* 44 cm | 17.5"

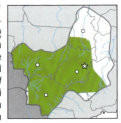

Fairly common and widespread in canopy and borders of forest and woodland; the only araçari occurring in the majority of our region, including Pantanal. Maxilla orange-yellow deepening to rich brown basally, culmen black; mandible black. Iris white, bare ocular area blue. Dark olive above with red rump (hidden at rest) and black crown; sides of head, throat, and nuchal collar maroon-chestnut. Chest patch black; below yellow *crossed by a broad red band*; thighs brown. Behavior similar to Lettered Araçari but as Chestnut-eared occurs in more open habitats, it tends to be a more conspicuous bird. Call a harsh, querulous squeal: "skreeeyip!" often repeated many times.

RED-NECKED ARAÇARI *Pteroglossus bitorquatus* 37 cm | 14.5"

Rare in canopy and borders of forest in *NW of our area* (Serra das Araras). A *colorful* araçari. Maxilla greenish yellow, mandible mostly black. Bare ocular area red with a little blue in front of eye. Head and throat black, *nape and upper back red*; otherwise olive above with red rump (hidden at rest). *Broad breast band bright red*; lower underparts yellow. No other araçari shows anywhere near as much red as this rather small and beautiful species. Its behavior is similar to the others, though Red-necked is more a forest-based bird and hence can be inconspicuous. Most frequent call a loud, repeated "kree-yeéik, kree-yeéik, kree-yeéik…;" also various chatters and bill clappings.

GOULD'S TOUCANET
ARAÇARI-DE-GOULD

LETTERED ARAÇARI
ARAÇARI-DE-BICO-RISCADO

CHESTNUT-EARED ARAÇARI
ARAÇARI-CASTANHO

RED-NECKED ARAÇARI
ARAÇARI-DE-PESCOÇO-VERMELHO

RAMPHASTOS toucans are well known from their *extraordinary long bills* (longer in ♂♂ and reaching an extreme in the spectacular Toco) and *bold color patterns*, with black predominating. Primarily frugivorous, they also consume some insects, vertebrates, and bird eggs. Nests are placed in tree cavities.

CHANNEL-BILLED TOUCAN *Ramphastos vitellinus* 44 cm | 17.5"

Fairly common in canopy and borders of forest in *N of our area* (including Chapada dos Guiamarães). *Bill "keeled"* (with a slightly protruding ridge along culmen), *black with yellow tip, culmen, and base of maxilla*; base of mandible blue. Bare ocular area blue. Mostly black with yellow uppertail-coverts and red crissum. *Throat and chest white*, bordered below by a narrow red band (usually inconspicuous), sometimes washed with yellow on chest. Identical in plumage and bill coloration to the larger White-throated Toucan (see below). Usually in pairs, less often small groups that forage at all levels in forest, most often high. Hops in canopy and subcanopy with springy bounds and surprisingly agile while feeding, using its long bill to reach out for fruit. Sometimes accompanies groups of other larger birds. Typical undulating flight seems weak, but these toucans are capable of traversing fairly long distances, crossing rivers and openings with relative ease. Flight consists of several quick flaps and a glide; often the bird seems gradually to be losing altitude. Oft-heard call a rapidly repeated frog-like croaking, "kreeek, kreeek, kreeek...," with each note often being accompanied by an upward toss of head. Calling birds sometimes take to high exposed perches.

White-throated Toucan (*R. tucanus*) also occurs in NW of our area. Likewise a forest bird, it regularly occurs with Channel-billed Toucan and here generally is outnumbered by it. In plumage identical to Channel-billed, and only to be distinguished by its slightly larger size and longer, "smoother" bill (with no keel). The two are best told apart by voice, White-throated's call being a loud yelping phrase: "kreeo-keh-keh" (very different from Channel-billed's croak).

GREEN-BILLED TOUCAN *Ramphastos dicolorus* 45 cm | 18"

Uncommon and local in canopy and borders of forest and adjacent clearings in *SE of our area*, e.g., Serra da Canastra NP. Sometimes called Red-breasted Toucan. *Bill greenish* with black base. Bare ocular area orange-red; iris pale blue. Black above with red uppertail-coverts. *Throat and breast yellow*, stained orange on chest; *lower underparts red*, with some black on lower belly. Overlaps little or not at all with similarly sized Channel-billed Toucan, which in any case differs in its mainly black bill and white bib. Otherwise unmistakable. Behavior much as in Channel-billed. Call a slow guttural croaking, "reenh... reenh... reenh...."

TOCO TOUCAN *Ramphastos toco* 59 cm | 23"

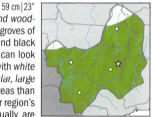

Fairly common and widespread in deciduous and gallery forest and woodland; foraging birds often range far out into cerrado and to isolated groves of trees (including palms). *Very long bill flame orange* with a large round black area at tip of maxilla; in some particularly long-billed ♂♂ the bill can look as long as the body! Bare ocular area orange, eye-ring blue. *Black with white throat and chest* and uppertail-coverts, red crissum. This spectacular, large toucan is essentially unmistakable; it occurs in much more open areas than any other. A conspicuous bird, the splendid Toco Toucan is one of our region's emblematic species. It occurs in pairs and small groups that usually are easy to observe as they fly from tree to tree or hop among larger branches. They often perch on high snags, almost as if to draw attention to themselves and their unique, flamboyant bill. Tocos sometimes fly surprisingly high, with bouts of short flaps interspersed by long glides, and they are capable of crossing extensive open areas. In some areas Toco Toucans come to feeders around ecolodges, there becoming quite tame. Call a short series of notably deep, low-pitched croaks, but in general less vocal then its congeners.

CHANNEL-BILLED TOUCAN
TUCANO-DE-BICO-PRETO

GREEN-BILLED TOUCAN
TUCANO-DE-BICO-VERDE

TOCO TOUCAN
TUCANUÇU

WOODPECKERS (Picidae) are well known for their clinging to trunks and branches, using stiffened tail feathers for bracing. Many "drum," hammering their chisel-like bills rapidly, to advertise territory.

BAR-BREASTED PICULET *Picumnus aurifrons*   9 cm | 3.5"

Uncommon in lower and middle growth of forest and borders in *NW of our area*. ♂'s *crown black with forecrown dotted red (can look solid), rearcrown dotted white*. Mostly olive above. Below pale yellowish, breast barred and belly streaked brownish. ♀'s crown dotted white. White-wedged Piculet has an obvious chevron pattern below. Occurs singly or in pairs, working up and down slender stems and in vines; often hangs sideways, pausing to peck or hammer into dead wood. Regularly with mixed flocks. High-pitched call a well-enunciated series of 3 descending notes: "tsi-tse-tsu," hummingbird-like.

Rusty-necked Piculet (*P. fuscus*) occurs in small numbers in várzea forest around Guaporé. Not well known, but distinctive: brown above, *tawny-buff below*; black crown dotted red in ♂.

🇧🇷 **Spotted Piculet** (*P. pygmaeus*) is fairly common in deciduous forest in N Minas Gerais (Peruaçú/Januária area). Brown; *bold white spotting below*, fainter spots above.

WHITE-WEDGED PICULET *Picumnus albosquamatus*   9.5 cm | 3.25"

Fairly common in deciduous and gallery forest and woodland; sometimes in more open areas. In Pantanal (**A**) ♂'s crown black, forecrown spotted red, rearcrown dotted white; above grayish brown. Below whitish, belly buffier, *throat and breast scalloped blackish*; tail black, outer feathers edged white. ♀'s entire crown dotted white. Elsewhere (**B**) slightly larger and darker with *heavy black chevrons below* (may look solid black), wing-coverts tipped white. *The only piculet in most of our area*; not likely confused. Behavior as in Bar-breasted Piculet, but White-wedged is more apt to forage in the open. High-pitched call a simple descending trill, "tsi-si-si-si-si."

White-barred Piculet (*P. cirratus*) occurs in deciduous woodland and scrub in S Pantanal. It resembles White-wedged but its underparts are *uniformly barred blackish*. The two species can occur together.

VENILIORNIS woodpeckers are *small* and *relatively dull*-plumaged woodpeckers that are olive above and barred or banded below. See p. 173 for Checkered Woodpecker.

LITTLE WOODPECKER *Veniliornis passerinus*   15.5 cm | 6"

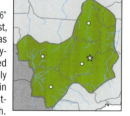

Fairly common and widespread in deciduous, riparian, and gallery forest, woodland, and adjacent semiopen areas. *Small* and *dull-plumaged*. ♂ has head grayish, *hindcrown red*. Yellowish olive above, tail dusky. Below grayish olive with *obscure pale barring*. ♀ lacks red on hindcrown. Red-stained Woodpecker has a golden-yellow nape, bolder barring below; it occurs only in NW, in forest. Found singly or in pairs, foraging at varying levels, usually in the open and easy to observe. Generally not with flocks. Call a series of rattled "wik" or "kik" notes, starting slowly, often given from a prominent perch.

White-spotted Woodpecker (*V. spilogaster*) occurs in small numbers in gallery forest of the Brasília area. Larger than Little Woodpecker with whitish postocular and malar streaks, bold yellowish white spotting and banding on wings.

RED-STAINED WOODPECKER *Veniliornis affinis*   17 cm | 6.75"

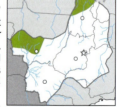

Uncommon in canopy and borders of forest in *NW of our area*. ♂ bright yellowish olive above with crown red (brown intermixed) and *nuchal collar golden yellow*. Below barred olive and buffy whitish. ♀ has crown brown. Little Woodpecker is smaller, less barred below; lacks yellow collar. Found singly or in pairs; generally remains well above ground, so less conspicuous than Little. Often accompanies mixed flocks. Call a fast series of up to 12-14 nasal, high-pitched "kee" notes that can sound like a Bat Falcon.

GOLDEN-GREEN WOODPECKER *Piculus chrysochloros*   21 cm | 8.25"

Uncommon in deciduous and gallery woodland in Pantanal, and also locally in N of our area. ♂ has crown and malar streak red; long *yellow streak across lower cheeks*; otherwise olive above. Below *boldly banded yellowish and dark olive*. ♀ differs in its olive crown, blackish malar. No other woodpecker in our area is so conspicuously banded below; *much larger* than any *Veniliornis*. Found singly or in pairs, foraging at all levels, in open areas regularly quite low. Relishes ants and termites, sometimes feeding on them for long periods. Infrequent call a shrill "shreeyr," doubled or tripled in quick succession.

Yellow-throated Woodpecker (*P. flavigula*) occurs locally in deciduous and riparian forest and woodland in N of our area, e.g., Ilha do Bananal. Superficially like Golden-green Woodpecker, but with yellow face, coarsely scalloped underparts.

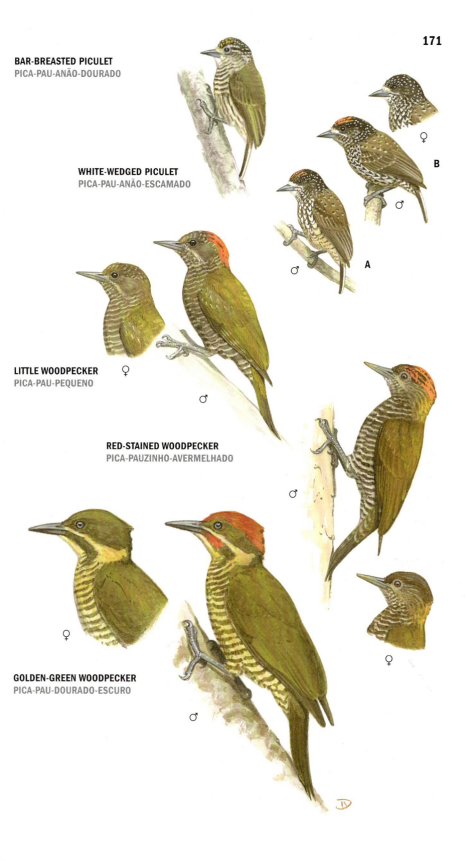

BAR-BREASTED PICULET
PICA-PAU-ANÃO-DOURADO

WHITE-WEDGED PICULET
PICA-PAU-ANÃO-ESCAMADO

LITTLE WOODPECKER
PICA-PAU-PEQUENO

RED-STAINED WOODPECKER
PICA-PAUZINHO-AVERMELHADO

GOLDEN-GREEN WOODPECKER
PICA-PAU-DOURADO-ESCURO

171

YELLOW-TUFTED WOODPECKER *Melanerpes cruentatus* 19 cm | 7.5"
Common in canopy and borders of forest and adjacent clearings *in NW of our area*. Mainly black; *pale yellow eye-ring extends back as a white postocular stripe reaching a golden yellow nuchal band*; midcrown red (lacking in ♀). Rump white, midbreast and belly red, flanks and crissum barred black and whitish. Colorful and not likely confused. *No other woodpecker is so fond of dead trees and snags*. Conspicuous and social, usually in groups of 3-5 birds (sometimes 12 or more); nesting is communal. Generally perches high, often in the open, even at mid-day. Feeds mainly by probing wood like other woodpeckers, but also sallies for insects and eats fruit. Upon alighting, usually holds wings outstretched for a moment. Noisy. Most frequent call a distinctive "krr-rr-rr-rr-krénh-krénh" with variants, but always with the rolled beginning. Sometimes several birds call together, their wings flaring and heads bowing.
Yellow-fronted Woodpecker (*M. flavifrons*) occurs locally in forest and borders in S Goiás and W Minas Gerais. Even gaudier than the Yellow-tufted Woodpecker, with yellow forecrown and throat, bright red crown and nape (lacking in ♀).

WHITE-FRONTED WOODPECKER *Melanerpes cactorum* 18 cm | 6.75"
Uncommon and local in deciduous woodland and scrub in Pantanal, often (but not always) in association with cactus. Previously thought an austral migrant here, but recent evidence confirms resident status. Mostly black above; *prominent white forecrown and nape patch*, red spot on forecrown (lacking in ♀); midback striped whitish, wings boldly barred white. *Throat pale yellow, white malar stripe*; below grayish, flanks and crissum barred dusky. Not likely confused, especially not in the semiopen scrub this species favors, where there are few woodpeckers. Conspicuous, often perching in the open, even on cactus. Quite social, most often in groups of 3-5 (presumably related?) birds. Calls are less forceful than in many woodpeckers and have an odd, almost squeaky quality: "skwee-kyup." As with the Yellow-tufted, calling birds often raise their wings and bow their heads.

CHECKERED WOODPECKER *Veniliornis mixtus* 15 cm | 6"
Uncommon and local in cerrado and light woodland, but only very locally in Pantanal. *Small*. Above *blackish spotted and barred white*, with a *long white superciliary and malar stripe* and small red area on hindcrown (lacking in ♀). Below whitish, breast with sparse and narrow blackish streaking. Only possible confusion is with Little Woodpecker (plain above, etc.). Occurs at inexplicably low densities and not often encountered even in seemingly "perfect" habitat. Forages by gleaning and tapping on smaller branches and twigs in low trees. Infrequently heard call a fairly loud, descending series of sharp notes: "kweh-weh-weh-weh-weh-weh."

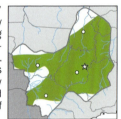

GREEN-BARRED WOODPECKER *Chrysoptilus melanochloros* 28 cm | 11"
Fairly common and widespread in deciduous and gallery woodland, cerrado, and clearings with scattered trees. Sometimes placed in the genus *Colaptes*. *Above golden olive boldly barred black, forecrown black, hindcrown and malar stripe red (the latter black in ♀), face creamy whitish*. Throat whitish streaked black; *below yellowish with prominent black spotting*. Campo Flicker is larger, shows no red on crown, has bright yellow face and chest. This fancy woodpecker is found singly or in pairs, foraging in trees and also on the ground (but less often on ground than Campo Flicker); apparently feeds primarily on ants. Most frequent call a loud ringing "kip!" often repeated several times; song, often given from a commanding perch, a faster series of "keeu" notes.

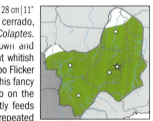

CAMPO FLICKER *Colaptes campestris* 31 cm | 12"
Common and widespread in open terrain. More than most birds found originally in campos and cerrado, this species has adapted well to agricultural terrain. *Crown and throat black; face, sides of neck, and chest bright yellow*; inconspicuous malar streak flecked reddish (♂) or whitish (♀). *Above boldly barred black and whitish*; below whitish with black scaling. Not likely confused, but cf. Green-barred Woodpecker (with spotted breast, whitish face, etc.). A very conspicuous bird, foraging primarily on the ground for ants and termites; also perches freely in low trees and often rests for long periods on fence posts or termite mounds. Nests in holes dug into banks, trees, or sometimes termite mounds. Gives various loud ringing calls, often in repetition: "kyu! kyu! kyu-kyu," frequently accompanied by wing flaring.

WHITE WOODPECKER *Leuconerpes candidus* 25.5 cm | 10"
Fairly common, widespread, and conspicuous in cerrado, campos, and open agricultural terrain (especially where there are groves of trees). Sometimes placed in the genus *Melanerpes*. Iris whitish; bare orbital area yellow. *Mainly white with black mantle* connecting to streak behind eye; tail black; yellow nape patch and midbelly. Unmistakable; *no other woodpecker shows as much white*. A conspicuous woodpecker that almost always is in the open; often perches on snags. Rather social, occurring in small groups of up to 4-5 individuals and seeming to occupy large home ranges. Regularly seen in long flights high above the ground and, notably for a woodpecker, its flight paths are relatively level (not bounding). Eats mainly fruit, also some insects. Call ultra-distinctive and far-carrying, typically a sharp nasal "kweeyr," often given in flight, heralding the approach of a group of these spectacular, unique woodpeckers.

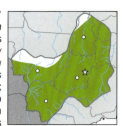

CELEUS woodpeckers have *bushy crests* and *greenish-yellow bills*. They range in forests of various types and tend to occur at low densities. Usually inconspicuous aside from their loud, often ringing vocalizations.

CREAM-COLORED WOODPECKER *Celeus flavus* 26 cm | 10.25"
Uncommon in canopy and borders of forest and woodland in Pantanal and NW of our area. Iris dark red, bill *yellow*. Obvious bushy pointed crest. *Yellow to creamy yellow* with red malar streak (lacking in ♀); *wings dusky brown*, coverts with yellowish edging (some individuals have cinnamon-rufous in flight feathers); tail dusky. Forages at all levels, hitching along trunks and major branches. Feeds especially by breaking open arboreal ant nests or termitaries, often then lingering at them for protracted periods. Most frequent call a loud, ringing, descending "peeyr, peeyr, peeyr, puh."

KAEMPFER'S WOODPECKER *Celeus obrieni* 27 cm | 10.5"
Rare and very local in bamboo patches within gallery forest in N of our area; rediscovered in 2006 at Goiatins in N Tocantins, after having long been known only from a single specimen from Piauí. Recently recognized as a species separate from Rufous-headed Woodpecker (*C. spectabilis*) of W Amazonia. *Head uniform rufous* (with red malar stripe in ♂); *above creamy buff* with scattered black chevrons and spots, flight feathers edged rufous; tail black. *Chest black*; creamy buff below. Should be easily recognized in its very limited range. Behavior presumably similar to most other *Celeus*, but hardly known in life. Distinctive call a loud squeal followed by several muffled notes: "squeeah! kluh-kluh-kluh-kluh-kluh." Apparently gravely threatened by highway construction and agricultural development.

PALE-CRESTED WOODPECKER *Celeus lugubris* 26 cm | 10.25"
Uncommon in deciduous and riparian forest and woodland in Pantanal and around Guaporé. *Mostly rufescent brown, head yellowish buff* except for brown around eye and red malar stripe (latter brown in ♀), rump also yellowish buff; feathers on back and wing-coverts faintly scaled buff. Blond-crested Woodpecker (occurring to E and N) has body plumage much more blackish, and a *dark rump*. Behavior much as in Cream-colored Woodpecker. Call a loud ringing "kree-kree-kree."

Chestnut Woodpecker (*C. elegans*) occurs in forest in NW of our area, e.g., Serra das Araras. Mostly rich rufous, with rump paler and yellower, black tail, and (in ♂) a red malar stripe.
Ringed Woodpecker (*C. torquatus*) also occurs at Serra das Araras. Strikingly patterned, with buff head and (in ♂) red malar, black chest and upper back; rufous above with black scaling, whitish below with black chevrons.

BLOND-CRESTED WOODPECKER *Celeus flavescens* 28-29 cm | 11.25"
Locally fairly common in gallery, deciduous forest, and woodland (does not occur in Pantanal, where Pale-crested Woodpecker takes over). *Very long, expressive bushy crest*. In N of our area (illustrated) yellowish buff of head extends down over upperparts (which are extensively marked with blackish chevrons), and more sooty below. In most of our area mainly *blackish except for contrasting yellowish buff head* (with red malar stripe in ♂); feathers on back and wing-coverts scaled yellowish buff, rump also yellowish buff. A spectacular bird, not likely confused, but cf. Pale-crested (which occurs to W of this species' range) and the rare Kaempfer's Woodpeckers. Blond-crested's behavior is similar to that of other *Celeus* but it seems a more numerous bird. Call a loud and fast repetition of 3-5 ringing notes: "kree-kree-kree-kree."

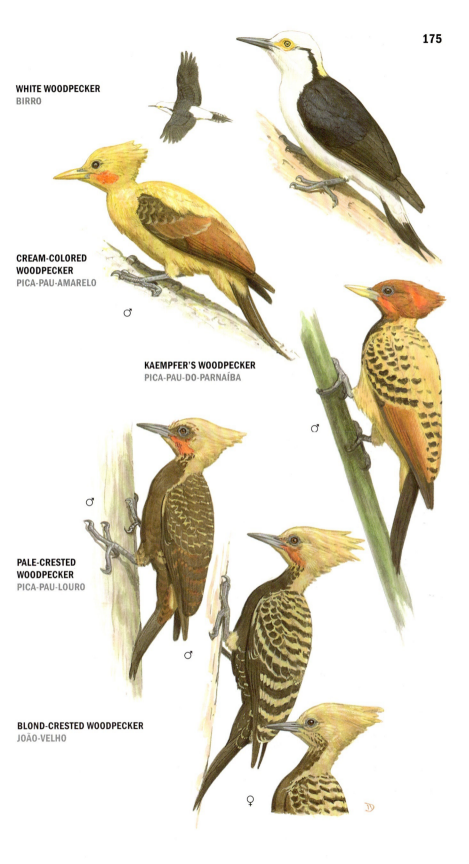

WHITE WOODPECKER
BIRRO

CREAM-COLORED WOODPECKER
PICA-PAU-AMARELO

KAEMPFER'S WOODPECKER
PICA-PAU-DO-PARNAÍBA

PALE-CRESTED WOODPECKER
PICA-PAU-LOURO

BLOND-CRESTED WOODPECKER
JOÃO-VELHO

LINEATED WOODPECKER *Dryocopus lineatus* 34.5 cm | 13.5"
Fairly common and widespread *in forest, woodland, and adjacent clearings. The most frequently seen large woodpecker.* Iris pale yellow; bill blackish. Conspicuous pointed crest. ♂ black above with crown, crest, and malar streak bright red, gray face, and a *distinctive narrow white stripe from base of bill across lower face and down sides of neck; two parallel white stripes down sides of back also distinctive.* Throat and chest blackish, below buffy whitish with blackish barring. ♀ has a black forehead and malar streak. ♂ Crimson-crested Woodpecker differs in its all-red head, while ♀ has black along entire front of crest; in both, the *white back stripes converge on lower back.* Found singly or in pairs, foraging at all levels, most often on trunks and larger limbs. Though forest-based, Lineated regularly flies out to trees in clearings. Flight deeply undulating. Pries off pieces of bark and probes deeply into rotting wood, searching for beetles and larvae; also takes many ants and consumes some fruit. Common calls include a "wik-wik-wik-wik-wik" (often up to 15-20 "wik's") that starts softly but quickly becomes loud and far-carrying, and a "keép-grrrr." Also gives a loud drum of 6-8 notes.

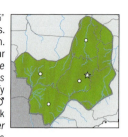

CAMPEPHILUS are *spectacular* and *large* woodpeckers with long bills and *long pointed crests.*

CRIMSON-CRESTED WOODPECKER *Campephilus melanoleucus* 36.5 cm | 14.5"
Uncommon but widespread *in forest, woodland, and borders.* Iris yellow; bill mainly dark horn. Conspicuous pointed crest. ♂ mainly black above with *head bright red; whitish patch at base of bill,* black-and-white spot on ear-coverts; a *white stripe* extends down sides of neck onto sides of back, *almost converging into a "V" on lower back.* Throat and chest black; breast and belly barred blackish and buff. ♀ has forehead and front of crest black extending to around eye, and lower neck stripe extending forward as a *broad stripe across lower face.* Both sexes of Lineated Woodpecker differ in their head patterns, neither showing as much red as ♂ Crimson-crested, while ♀ Lineated's white facial stripe is much narrower. Its back stripes do *not* converge. Behavior as in Lineated, but Crimson-crested is more of a forest-based bird, moving less often out into clearings. It often forages in pairs, occasionally in small (family?) groups, hitching up trunks and larger branches at all levels. Sometimes remains for protracted periods in the same tree or snag, whacking off pieces of bark and chiseling deeply into wood. Often heard is a loud, far-carrying drumming that consists of a very loud rap followed by several weaker strokes (at a distance often just the first two can be heard). Also gives several calls, a slightly nasal and fast "ski-zi-zik" or "skik-skik-ski-zi-zik," and a down-slurred "kiarrh."

RED-NECKED WOODPECKER *Campephilus rubricollis* 34.5 cm | 13.5"
Uncommon *in forest in NW of our area* (including Chapada dos Guimarães). Bill greenish horn; iris yellow. Conspicuous pointed crest. ♂ has *head, neck, and chest bright crimson,* becoming *rufous on remaining underparts;* a small black-and-white spot on lower ear-coverts. Above black, with *rufous patch in primaries* (conspicuous in flight, and sometimes even visible on perched birds). ♀ similar but lacks the ear-spot, this replaced by a *broad white malar stripe* narrowly bordered black. Not likely confused, this splendid woodpecker *lacks* the prominent white neck and back stripes shown by Crimson-crested and Lineated Woodpeckers. Usually found in wide-ranging pairs, tending to feed on trunks and larger branches of bigger trees, most often rather low; infrequent at edge and hardly ever out into clearings. Gives a far-carrying, two-noted rap, the first stroke much louder than the second. Call a loud, nasal "kiahh."

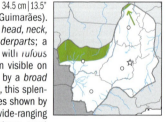

Cream-backed Woodpecker (*C. leucopogon*) occurs very locally in woodland in SW Mato Grosso do Sul. ♂ has mainly red head and neck, and black body with *large creamy buff patch on back and rump.* ♀ has black forehead and front of crest (as in ♀ Crimson-crested) and white stripe back from bill.

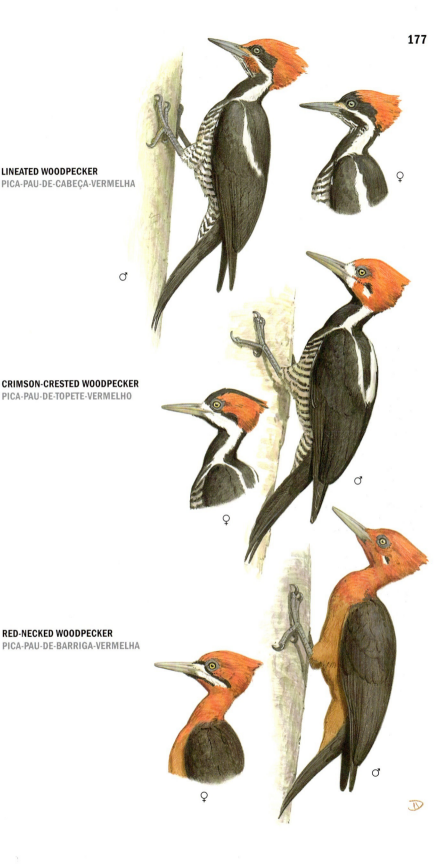

LINEATED WOODPECKER
PICA-PAU-DE-CABEÇA-VERMELHA

CRIMSON-CRESTED WOODPECKER
PICA-PAU-DE-TOPETE-VERMELHO

RED-NECKED WOODPECKER
PICA-PAU-DE-BARRIGA-VERMELHA

WOODCREEPERS (Dendrocolaptidae) are inconspicuous insectivorous birds that *hitch* up tree trunks and branches using their *stiff tails for support*, probing bark and epiphytes with bills. *Their loud calls* are often distinctive. *All species have rufous wings and tail*. They nest in holes in trees, natural cavities being preferred.

OLIVACEOUS WOODCREEPER Sittasomus griseicapillus 15 cm | 6"
Common and widespread in humid, deciduous, and gallery forest and woodland. Bill short, straight. In most of our area (**A**) *head, neck, and underparts grayish to olive grayish*; back more olive; flight feathers, tail bright rufous. In SE (**B**) *bright ochraceous olive above* (lacking gray). Combined small size and *lack of streaking* distinctive. Climbs along open trunks and larger branches at varying levels, most often fairly high. Regularly accompanies mixed flocks. Song, usually given at long intervals, a series of 6-14 successively higher-pitched, gradually louder notes: "pu-pu-pew-pew-peh-peh-peé-peh" with a slight drop at end. In SE song differs: a series of 8-10 enunciated "weep" notes dropping in pitch, ending in a stutter.

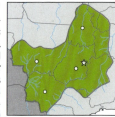

PLAIN-BROWN WOODCREEPER Dendrocincla fuliginosa 20.5 cm | 8"
Fairly common in lower and middle growth of forest and woodland in NW of our area. The most *uniformly brown* woodcreeper, with *paler grayish lores and ear-coverts*, vague buff postocular, *dusky malar stripe*, and whitish throat. Sometimes with mixed flocks, but mostly seen attending army ant swarms. At large ones six or more may gather, sallying after prey. Sometimes perches "normally" across branches. Song a descending series of rattled notes, slowing toward end, lasting 2-3.5 seconds. Less often gives a protracted series of "keé" or "keh" notes can continue for 30 seconds or more.

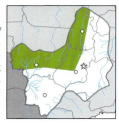

XIPHORHYNCHUS woodcreepers constitute a confusing group elsewhere in the Neotropics, but in our region only four species occur, each of them quite different. All are *streaky* with *relatively heavy bills*. See p. 181 for Buff-throated Woodcreeper.

STRAIGHT-BILLED WOODCREEPER Xiphorhynchus picus 20.5 cm | 8"
Generally common and widespread in várzea, riparian woodland, and borders in N of our area, including N Pantanal. Pale ivory to dull pinkish bill, *straight and dagger-shaped*. Head and nape dusky, narrowly buff-streaked, with indistinct whitish superciliary; above otherwise rufous. Throat whitish, *chest with large black-edged whitish squamate spots*; below brown. Frequently with mixed flocks, sometimes in the semiopen, most often near ground or water. Song a *descending* trill, usually with an upturn at end; sings more at mid-day than most other woodcreepers.
Striped Woodcreeper (*X. obsoletus*) is also found in várzea and riparian forest in N of our area, but is less numerous than Straight-billed. Striped has a more decurved and darker bill, and more streaking on its back and underparts. Its similar song *ascends*.

LESSER WOODCREEPER Xiphorhynchus fuscus 17.5 cm | 7"
Uncommon in forest, woodland, and borders in SE of our area. Formerly in genus Lepidocolaptes. Slender, somewhat decurved bill. Brown above, crown and nape duskier with buffyish spotting; vague pale eye-ring and postocular; back with faint buff streaking. Throat buffy whitish; below dusky-olive with blurry buffy streaking. Narrow-billed Woodcreeper is larger and longer-billed, with a much more prominent superciliary; it favors semiopen terrain. Lesser's behavior is typical for a woodcreeper. Song commences with a series of stuttered notes, then moves to a faster trill before ending with more stuttered notes: "chit, chit, chit, chee-ee-ee-ee-ee-ee-ee, chit, chit-chit."
Scaled Woodcreeper (*Lepidocolaptes squamatus*) occurs locally in forest patches in NE of our area (NW Minas Gerais and adjacent Tocantins). Resembles Lesser Woodcreeper but has an even more slender pinkish bill, is brighter rufous above, has bolder streaking below.

NARROW-BILLED WOODCREEPER Lepidocolaptes angustirostris 20.5 cm | 8"
Fairly common and widespread in lighter woodland, scrub, cerrado, and trees in agricultural areas. *Long, narrow, decurved pale bill*. Crown blackish with narrow pale streaking, *bold whitish superciliary*; above bright rufous brown. *Below buffy whitish*. Long bill, bold head pattern, and pale underparts should preclude confusion. This handsome woodcreeper favors more open situations than most others. Cf. Lesser Woodcreeper, much more a forest bird. Song a loud series of sharp, well-enunciated notes that either accelerate and fade away, "peeé, pee-pee-pee-pee-peepeepeepeepupupu," or simply descend, "peer, peer, peer, peeer, peeeer, pweeeer."

OLIVACEOUS WOODCREEPER
ARAPAÇU-VERDE

PLAIN-BROWN WOODCREEPER
ARAPAÇU-PARDO

STRAIGHT-BILLED WOODCREEPER
ARAPAÇU-DE-BICO-BRANCO

LESSER WOODCREEPER
ARAPAÇU-RAJADO

NARROW-BILLED WOODCREEPER
ARAPAÇU-DO-CERRADO

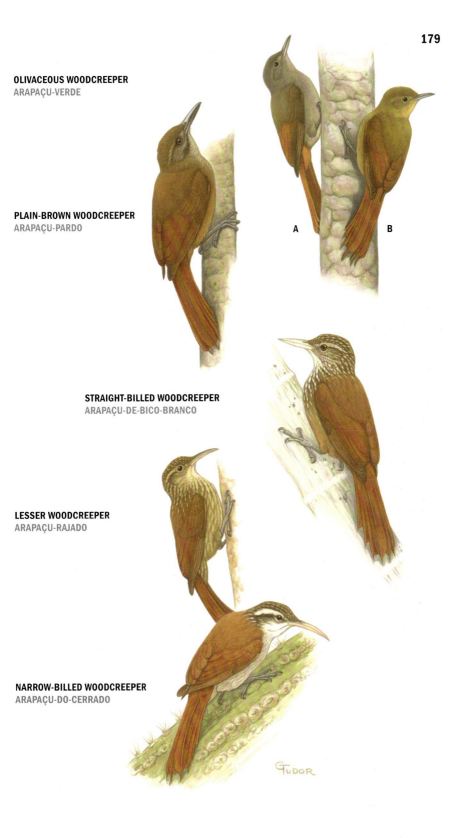

BUFF-THROATED WOODCREEPER *Xiphorhynchus guttatus* 26.5 cm | 10.5"
Common and widespread in forest borders, gallery woodland, and semi-deciduous woodland. Bill long, pale dusky to grayish horn. Above brown, duskier on crown and nape, with buff spotting on crown and back. *Throat pale buff*; below brown with buff streaking, belly plain. Due to its large size can be confused with similarly sized *Dendrocolaptes*. Black-banded Woodcreeper has a streaked throat and barring on belly; cf. also Planalto Woodcreeper. Conspicuous for a woodcreeper, foraging with mixed flocks at varying levels, often high, frequently rummaging in dead leaves. *Very vocal*, more so than most woodcreepers. Most frequent song, given mainly at dawn and dusk but also throughout the day, a series of evenly paced, loud ringing whistled notes starting slowly, becoming louder. Also gives a shorter fast series of descending "laughing" notes, and a doubled "wheeyer, wheeyer." Oft-heard call a loud "kyoow."

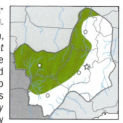

DENDROCOLAPTES are large, *heavy-billed* woodcreepers found in forest and woodland where they are relatively inconspicuous.

BLACK-BANDED WOODCREEPER *Dendrocolaptes picmnus* 27 cm | 10.5"
Rare and local in lower growth of semihumid forest and woodland in Pantanal. Bill dusky. Above rufous brown, streaked buff on head, more sparsely on back. *Throat and breast buffy brown streaked whitish, belly buff weakly barred blackish*. A large woodcreeper with distinctive combination of forepart streaking and belly barring (though the bars can be hard to see in field). Buff-throated has a *plain* buff throat; it *lacks any* barring below. Rather stolid, often perching low on trunks. Song a fast, slightly descending series of liquid-sounding "winh" notes with a "laughing" quality. When agitated also gives various whines and chatters.

PLANALTO WOODCREEPER *Dendrocolaptes platyrostris* 25.5 cm | 10.25"
Uncommon in lower growth of forest and woodland, but largely absent from Pantanal. *Bill black*. Brown above, *head with buff streaking. Throat whitish with dusky streaking*; below brown, breast with buffy whitish streaking, belly vaguely barred blackish. White-throated Woodcreeper is larger with a notably longer, heavier, arched bill and a pure white throat. Black-banded Woodcreeper (no known overlap, but comes close) is paler and more rufous, with a dusky (not black) bill. Song a fast series of sharp "whik" notes, sometimes fading toward the end (Black-banded's song is shorter and less shrill).

XIPHOCOLAPTES are *heavy-bodied, massive-billed* woodcreepers found at low densities in forest and woodland. Their *loud vocalizations* draw attention to them, but they are not all that often seen.

WHITE-THROATED WOODCREEPER *Xiphocolaptes albicollis* 28 cm | 11"
Uncommon in forest and gallery woodland in SE of our area. *Long, stout, somewhat decurved black bill*. Above olivaceous brown, head blacker with whitish superciliary and supramalar. *Throat pure white bordered by blackish malar stripe*; below pale brownish, breast streaked whitish, belly barred blackish. Planalto Woodcreeper has a shorter, straighter bill, shows streaking on throat, no malar pattern. Found singly and in pairs, most often not with mixed flocks. Regularly forages close to ground, mainly on trunks but also sometimes on fallen logs. Often stolid and tame. Far-carrying song, given mainly at dawn and dusk (sometimes in near-darkness), a leisurely series of piercing notes, each slightly lower-pitched than its antecedent and preceded by soft hiccupping: "mc-wheer, mc-wheer, mc-wheer, mc-wheer, mc-wheer." It often starts with a snarl. Also gives whining or snarling calls independently.

MOUSTACHED WOODCREEPER *Xiphocolaptes falcirostris* 28 cm | 11"
🇧🇷
VU *Rare and local in deciduous and gallery forest in N Minas Gerais and Tocantins. Bill pale dusky to horn*. Mainly olivaceous brown with *pale buff superciliary, supramalar, and throat*; pale streaking fine or even absent. No similar woodcreeper occurs in this species' restricted range. White-throated Woodcreeper (no known overlap, occurring to south) differs in its obviously black bill, blacker head with more pale streaking. Moustached's behavior and voice are similar to White-throated Woodcreeper; individual notes of song are less clearly bisyllabic.

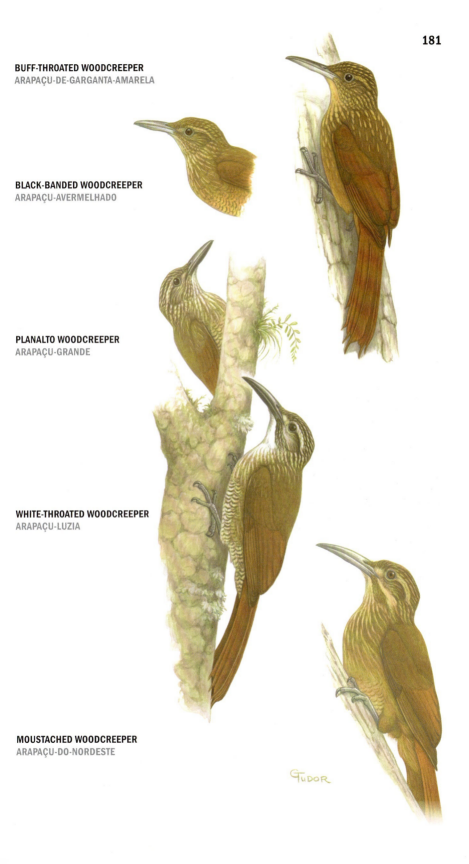

BUFF-THROATED WOODCREEPER
ARAPAÇU-DE-GARGANTA-AMARELA

BLACK-BANDED WOODCREEPER
ARAPAÇU-AVERMELHADO

PLANALTO WOODCREEPER
ARAPAÇU-GRANDE

WHITE-THROATED WOODCREEPER
ARAPAÇU-LUZIA

MOUSTACHED WOODCREEPER
ARAPAÇU-DO-NORDESTE

GREAT RUFOUS WOODCREEPER *Xiphocolaptes major* 28-30.5 cm | 11-12"
Uncommon in deciduous and gallery woodland in Pantanal, sometimes venturing out to trees in adjacent cleared areas. Very large and *mainly rufous*. Long, heavy, somewhat decurved bill grayish to horn-colored. Above mostly bright rufous; paler and browner on head with dusky lores; *paler rufous below*. A few individuals show faint shaft streaking on breast and dusky barring on belly. Because of its massive size (for a woodcreeper) not likely confused. Black-banded Woodcreeper is markedly smaller and has streaked foreparts. Though conspicuous, the Great Rufous Woodcreeper never seems to be very numerous and its territories are so large that it's not encountered all that often. Forages by hitching up trunks and along larger limbs, but also drops to the ground where it hops about rather clumsily, sometimes even rummaging in leaf litter. Usually found singly or in pairs. At times seems quite fearless. Far-carrying and unmistakable song a series of double-stopped whistled notes, each pair dropping a little in pitch and interspersed with soft whining or chattered notes.

LONG-BILLED WOODCREEPER *Nasica longirostris* 35.5 cm | 14"
Uncommon in midlevels, subcanopy, and borders of várzea and riparian forest along *NW edge of our area*, e.g., Ilha do Bananal; *rarely far from water*. Spectacular and unmistakable, with *very long, nearly straight ivory white bill and long thin neck*. Crown blackish narrowly streaked buff with narrow white postocular stripe; *above bright rufous-chestnut. Throat snowy white*; below brown with black-edged white lanceolate streaks on sides of neck and breast. Found singly or in pairs, often foraging in the semiopen along edges of lakes, streams, and rivers; inconspicuous except by voice. Generally not with flocks. Frequently probes into bromeliads and other epiphytic plants, also into tree cavities. Loud and far-carrying song an easily recognized series of 3-4 eerie plaintive whistled notes: "twooooóoo... twooooóoo... twoooooóoo." Birds responding to tape playback or a whistled imitation also often utter various chuckled calls.

RED-BILLED SCYTHEBILL *Campylorhamphus trochilirostris* 24-26.5 cm | 9.5-10.5"
Uncommon in lower growth of deciduous, gallery, and riparian forest and woodland; seemingly most numerous in Pantanal. *Unmistakable*, with *spectacular long (7 cm/3") reddish or reddish brown decurved bill* (in some birds quite bright). In most of range (**A**) rufescent brown above, somewhat duskier on crown; head and back streaked buff to whitish; wings and tail rufous-chestnut. Throat whitish; below brown with dark-edged buff streaking. Birds in Pantanal and nearby (**B**) are larger and *even longer-billed*, and their plumage is paler and more rufescent. Despite their spectacular bills, scythebills forage much like other woodcreepers, hitching up trunks and along larger lateral branches at various levels but most often quite low, probing into bark crevices and epiphytic plants. They often accompany mixed flocks. Song a series of musical notes with an antbird-like quality, variable in phraseology, a descending and gradually slowing series of upslurred notes: "tuwee-tuwee-toowa-tew-tew." Also gives an ascending "twee-twee-twee-twi?-twi?" sometimes adding a trill at the start. Agitated birds can give a loud semimusical chipper.

GREAT RUFOUS WOODCREEPER
ARAPAÇU-GIGANTE

LONG-BILLED WOODCREEPER
ARAPAÇU-BICUDO

RED-BILLED SCYTHEBILL
ARAPAÇU-BEIJA-FLOR

MINERS & LEAFTOSSERS (Scleruridae) These two genera have recently been recognized as a separate family from Furnariidae. They place their nests at the end of long burrows dug into the ground. See p. 195 for our one leaftosser.

CAMPO MINER *Geositta poeciloptera*   12.5 cm | 5"

VU   Uncommon, erratic, and local in campos and open cerrado. Favors recently burned areas, and though it can be numerous when conditions are favorable seems intolerant of agricultural land. *Short-tailed*. Brown above with narrow buff superciliary; wings dusky with *rufous in flight feathers* (obvious in flight), *tail rufous with broad black subterminal band*. Throat white; below dull buff, breast flammulated brown. Confusion is possible with the more uniform *and much more numerous* Rufous Hornero, but the latter's habitat and habits differ markedly. Miners occur in pairs and are terrestrial and inconspicuous; they hug the ground, crouching when approached, then flush abruptly, usually flying some distance away. Song, most often given during a hovering display flight, a simple repeated "zhliip" or "zh-zh-zh-leép."

OVENBIRDS (Furnariidae) A large, diverse group of insectivorous birds found in both forested and open environments. Ovenbirds are subtly colored in browns, grays, and rufous. Many are skulking birds, though vocalizations can be loud and distinctive. Nest structure and placement vary, some being constructed of sticks while a few are made of mud and shaped like a Dutch oven with a side entrance; others place theirs in holes.

BAND-TAILED HORNERO *Furnarius figulus*   16.5 cm | 6.5"

Locally fairly common in semiopen areas and woodland borders, *usually near water*, in NE of our area. Legs grayish. Cinnamon-rufous above with *contrasting dark brown crown and cheeks*, bold white superciliary; primaries blackish, *tail feathers variably tipped black*. Throat whitish; drab buffyish below. Pale-legged Hornero is brighter-plumaged, shows a more contrasting white throat, paler legs; it has *no* black on tail, but *more* black along wing's leading edge. This hornero walks on the ground, often in pairs, head nodding. It feeds by picking at the ground, most often where moist. Song an explosive series of notes that gradually slows and descends in pitch.

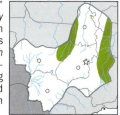

PALE-LEGGED HORNERO *Furnarius leucopus*   17 cm | 6.75"

Fairly common in gallery woodland and forest borders in Pantanal and N of our area. Favors areas near water. Legs pale, dull fleshy gray; bill pale. *Bright orange-rufous above* with contrasting dusky crown and bold white superciliary. Throat white; rich cinnamon-buff below. The similar Band-tailed Hornero is duller and less cinnamon below, and shows blackish on tail-tip. Behavior as in Band-tailed; much less conspicuous than Rufous Hornero. Song a fast descending series of loud notes that slows toward end.

RUFOUS HORNERO *Furnarius rufus*   18 cm | 7"

Common and widespread in open and semiopen areas, often around houses and in agricultural lands. Very plain and uniform. Above rufous brown, crown slightly grayer. Below paler buffy brown; throat and midbelly whiter. Vaguely thrush-like, much more conspicuous than our other two horneros, both with white superciliary. Struts about boldly on ground and roads, perching freely on fences and in trees. Uses its conspicuous mud nests only once; afterward Saffron Finches, Brown-chested Martins, and other birds take them over before they gradually disintegrate. Song a loud, abrupt series of harsh notes that rise, then subside. Pairs often sing more or less in unison.

GRAY-CRESTED CACHOLOTE *Pseudoseisura unirufa*   23 cm | 9"

Fairly common in borders of gallery and deciduous woodland and near buildings in Pantanal. Iris yellow. *Large*. Uniform cinnamon-rufous with expressive bushy crest usually showing gray. No other rich rufous bird has such an obvious crest. Though mainly arboreal, cacholotes sometimes drop to the ground to feed, there hopping or walking with an unsteady, almost lurching gait. Pairs often linger around their large, conspicuous stick nests. Their vocalizations are loud and unmistakable, and are usually given in a duet. One gives a descending series of "chup" notes that accelerates into a long-continued churring; its mate gives a fast series of "che" notes. The overall effect is complex and sounds jumbled, before eventually winding down to a stuttered stop.

Caatinga Cacholote (*P. cristata*) occurs in woodland borders and semiopen areas in N Minas Gerais and Tocantins. It resembles Gray-crested Cacholote (recently split), but Caatinga shows no gray in crest. Its loud raucous vocalizations are similar though given at a slightly slower pace.

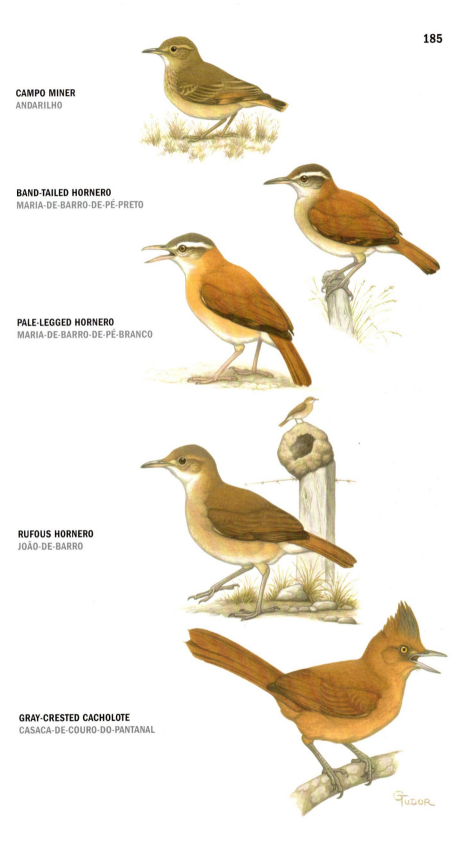

CAMPO MINER
ANDARILHO

BAND-TAILED HORNERO
MARIA-DE-BARRO-DE-PÉ-PRETO

PALE-LEGGED HORNERO
MARIA-DE-BARRO-DE-PÉ-BRANCO

RUFOUS HORNERO
JOÃO-DE-BARRO

GRAY-CRESTED CACHOLOTE
CASACA-DE-COURO-DO-PANTANAL

SYNALLAXIS spinetails are slender *skulking* furnariids found in low dense growth. Most have a rufous crown and wings, and *long "double-pointed" tails*. They are hard to identify and see, *but have generally distinctive calls*. Nests are globular structures with a side entrance.

SPIX'S SPINETAIL *Synallaxis spixi* 16 cm | 6.25"
Uncommon in woodland borders and shrubby areas in far SE. Olive brown above with an *all-rufous crown (no dark frontlet)*, rufous wing-coverts, and grayish tail. *Mostly gray below*, throat flecked black. Cinereous-breasted Spinetail has a shorter, less pointed tail and dusky frontlet; voice very different. Pale-breasted Spinetail favors more open habitats; it too has a frontlet, is paler below. Rufous-capped Spinetail has a yellowish streak below its rufous crown; shorter rufous tail. Spix's creeps about within dense cover, usually in pairs, emerging mostly when it is singing. Song a constantly reiterated "whít, di-di-dit."

PALE-BREASTED SPINETAIL *Synallaxis albescens* 16.5 cm | 6.5"
Fairly common in cerrado and shrubby areas, also locally in marshes. Olive grayish brown above with dusky frontlet and rufous crown; *wing-coverts rufous*, remainder of wing and *rather long tail brownish*. Below whitish with dusky flecking on throat. Cinereous-breasted Spinetail is darker and grayer below; it has a very different voice. Usually skulking, remaining in bushes and tall grass. Characteristic, often incessantly repeated song a nasal, husky "kweé-beyrr," generally given from concealment.

SOOTY-FRONTED SPINETAIL *Synallaxis frontalis* 16.5 cm | 6.5"
Fairly common in undergrowth, borders of low woodland, and scrub. Olive brown above with sooty frontlet and *rufous crown*; wings and rather *long tail also rufous*. Throat whitish flecked black, breast pale grayish, midbelly whitish. Pale-breasted Spinetail (often sympatric) and the more local Cinereous-breasted have brownish tails. Sooty-fronted favors more wooded environments than Pale-breasted. Forages in pairs, with typically furtive *Synallaxis* behavior. Song a sharp "ka-kweé," very different from our other spinetails.

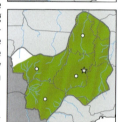

CINEREOUS-BREASTED SPINETAIL *Synallaxis hypospodia* 15.5 cm | 6"
Uncommon and local in low shrubby and grassy areas and borders, usually near water. Can be locally numerous in N Pantanal. Olive brown above with sooty frontlet and *rufous crown*; wings mainly rufous, *tail dull dark brown*. Throat whitish flecked black, breast pale gray, midbelly whitish. Sooty-fronted Spinetail has a longer, obviously rufous tail. Pale-breasted Spinetail, also brownish-tailed, is whiter below. *All of these species' voices differ dramatically.* Behavior much as in Sooty-fronted. Song a fast series of chippered notes that coalesce into trill: "chew, chew-chee-chee-chee-ee-ee-ee-ee-ee-ee-ee-ee-ee-eu."

RUFOUS-CAPPED SPINETAIL *Synallaxis ruficapilla* 16 cm | 6.25"
Uncommon in undergrowth and borders of forest in SE of our area. Crown bright rufous, bordered below by a buffy yellowish streak and then dusky cheeks; above brown. Most of wings and *rather short tail rufous-chestnut*. Throat silvery grayish, breast pale grayish, belly pale ochraceous. A relatively colorful spinetail, not likely confused; cf. Spix's Spinetail, sometimes nearby. Usually in pairs, remaining near the ground in dense undergrowth and *favoring bamboo*. Occasionally accompanies mixed flocks. Oft-heard song a somewhat nasal, fast "di-di-di-reét" phrase, sometimes repeated interminably.

WHITE-LORED SPINETAIL *Synallaxis albilora* 16 cm | 6.25"
Locally common in undergrowth of deciduous and gallery woodland and riparian shrubbery in Pantanal. Very plain. Head grayish with *prominent white lores*; back brown, wings and tail contrastingly rufous. *Below mostly bright ochraceous.* Ochre-cheeked Spinetail has a more rufous back, black throat patch, much more prominent superciliary. Forages in pairs, on or near the ground, keeping in contact through their frequent calling. Associates with flocks only rarely. Song a sharp, piercing "keeeu, kit-kweeit," given at a leisurely pace.

Araguaia Spinetail (*S. simoni*) occurs locally along Rio Araguaia, favoring **VU** riparian thickets adjacent to oxbow lakes and rivers. Resembles White-lored Spinetail but more uniform rufous on back and whiter below (showing less ochre). Behavior similar. Song has a more nasal, whining tone; often simply a repeated "kweeu" (not sliding up as in White-lored).

OCHRE-CHEEKED SPINETAIL *Synallaxis scutata* 15.5 cm | 6"
Uncommon and local in undergrowth of deciduous forest and woodland. From Goiás E and N (illustrated), crown and upper back grayish brown with *white superciliary; otherwise rufous above*. Chin white, *prominent black patch on lower throat*; mostly buff below. In Mato Grosso, back and rump olive brown; only the wings and tail are rufous. White-lored Spinetail *lacks* obvious superciliary, has more ochraceous underparts, *no* black on throat. Forages in pairs, hopping in thickets on or near ground, not accompanying mixed flocks. Usually detected by its frequent song, a bisyllabic, shrill "tweeeyt, weét?" with a leisurely cadence; often repeated for protracted periods.

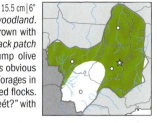

⬤ **Red-shouldered Spinetail** (*Gyalophylax hellmayri*) is a large, dark spinetail of caatinga scrub in extreme NE. Iris yellow-orange. Brownish gray with black throat patch; wing-coverts rufous-chestnut; long tail blackish.

CHOTOY SPINETAIL *Schoeniophylax phryganophilus* 19-22 cm | 7.5-8.75"
Fairly common in semiopen areas with scattered trees and bushes, gallery woodland, and gardens in SW of our area; also in far NE. *Tail extremely long, "spiky," and graduated*. Crown and lesser wing-coverts chestnut; superciliary white; *above sandy brown, boldly streaked blackish*. Chin yellow; patch on lower throat black; malar area white; *chest band cinnamon*; below mainly whitish. NE birds smaller. Much longer-tailed and more arboreal than any *Synallaxis* (none of which is streaked above). Pairs and small groups usually clamber inside leafy cover; they are often around houses, and construct conspicuous large stick nests. Named for its most frequent call, a distinctive, low-pitched gurgling or chortling "cho-cho-cho-cho-chchchchchchchch."

CRANIOLEUCA are small spinetails of wooded areas, *more arboreal* than most *Synallaxis* (Rusty-backed Spinetail less so). They forage by inspecting limbs and epiphytes. Nests are ball-shaped and often large, with entrance at bottom.

GRAY-HEADED SPINETAIL *Cranioleuca semicinerea* 14 cm | 5.5"
⬤ Uncommon and local in subcanopy and borders of deciduous forest and woodland in S Goiás and N Minas Gerais. Bill pinkish. *Head and neck pale ashy gray*, forehead and superciliary whiter, becoming *dingy pale grayish below*; above otherwise rufous. Goiás birds are darker on crown and ear-coverts and have a more prominent superciliary. Should be distinctive, but cf. Pallid Spinetail (with obvious rufous crown, etc.). Usually in pairs, regularly foraging with loose flocks. Its frequently given shrill and high-pitched song accelerates into a trill.

PALLID SPINETAIL *Cranioleuca pallida* 14 cm | 5.5"
⬤ Uncommon in canopy and borders of forest and woodland in SE part of our area, (e.g., around Brasília, Serra da Canastra). *Crown rufous with bold whitish superciliary*; olive brown above, wings and tail rufous. Throat whitish, below pale buffy brownish. May occur very locally with Gray-headed Spinetail though tending to occur in more humid regions. Rusty-backed Spinetail is much more tied to water and is decidedly less arboreal; it is entirely rufous above. Behavior and high-pitched shrill song resemble Gray-headed's.

RUSTY-BACKED SPINETAIL *Cranioleuca vulpina* 14.5 cm | 5.75"
Fairly common in shrubby thickets and lower growth of gallery forest and woodland; often in seasonally flooded areas. *Uniform rufous above*; cheeks grayish. *Below drab buffy grayish*. Yellow-chinned Spinetail is much whiter below. Pairs of Rusty-backeds forage in dense tangled growth where they usually remain hard to see. Song a fairly long, fast series of descending nasal chortling notes, initially enunciated "kwee-kwee-kwee-kweh-kweh-kweh-kwakwakwakwakwakwa," often delivered together by a pair.

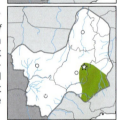
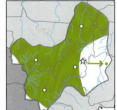

YELLOW-CHINNED SPINETAIL *Certhiaxis cinnamomeus* 14.5 cm | 5.75"
Common and widespread in marshes, adjacent grassy, shrubby areas, and around pond and lake margins. Above bright reddish brown, wings and tail rufous, with indistinct pale grayish superciliary, dusky lores and postocular line. *Below white with inconspicuous pale yellow chin spot*. This clean-cut, attractive spinetail looks bicolored, rufous and white. Sometimes with it, the Rusty-backed Spinetail is notably dingier below. Conspicuous for a marsh bird, often feeding in the open on floating vegetation or in low bushes near water's edge. Notably vocal, calling at intervals through the day, with most frequent song a harsh, loud churring rattle that recalls certain *Laterallus* crakes.

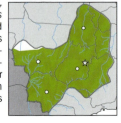

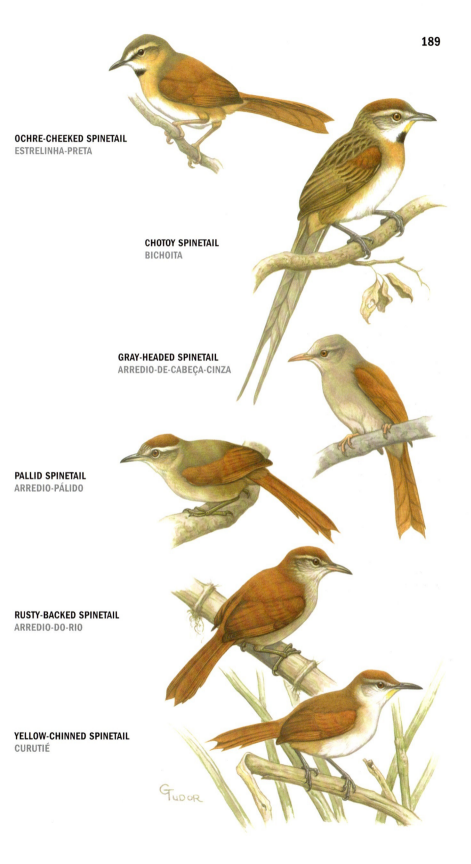

OCHRE-CHEEKED SPINETAIL
ESTRELINHA-PRETA

CHOTOY SPINETAIL
BICHOITA

GRAY-HEADED SPINETAIL
ARREDIO-DE-CABEÇA-CINZA

PALLID SPINETAIL
ARREDIO-PÁLIDO

RUSTY-BACKED SPINETAIL
ARREDIO-DO-RIO

YELLOW-CHINNED SPINETAIL
CURUTIÉ

THORNBIRDS are plain unstreaked furnariids that range in semiopen areas and are *best known from their large stick nests*. Both of our species have slightly decurved bills and obvious *rufous on forecrown*. They are inconspicuous aside from their *loud vocalizations*.

RUFOUS-FRONTED THORNBIRD Phacellodomus rufifrons 16.5 cm | 6.5"

Locally common in a variety of semiopen and lightly wooded habitats, often in agricultural areas and ranch country, mainly in Pantanal and E of our area. Uniform drab brown above with *rufous forecrown* and a pale buffyish superciliary. Whitish below, buff on flanks. In Minas Gerais has some rufous on shoulders, flight feathers, and basally on tail. Greater Thornbird is larger and is notably yellow-eyed. Ranges in pairs or small groups; usually arboreal, but sometimes drops to the ground. Seen particularly often near their large stick nests, which are usually placed at the tips of large branches. Very vocal, with typical song a loud, abrupt series of forceful "cheh" or "chit" notes that start slowly, then accelerate and drop off. Often both members of a pair call in sequence, or the ♀ contributes a chatter.

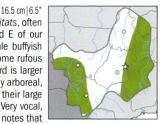

GREATER THORNBIRD Phacellodomus ruber 20.5 cm | 8"

Fairly common at edge of gallery woodland and shrubby areas, *usually near water*. *Iris bright yellow*. *Large*. Brown above, *more rufous on crown, wings, and tail*; lores pale and *showing little or no superciliary*. Whitish below, foreneck sometimes scaled dusky. Rufous-fronted Thornbird is smaller and dark-eyed, and it shows less rufous on crown and essentially none on wings and tail. Usually in pairs, and quite skulking, especially when not singing; vocalizes frequently, then often more in the open. Song a series of loud, arresting notes that start explosively and accelerate, but then gradually become less forceful: "kur-cheé-chee-chee-chee-che-che-che-chew-chew-chu-chu-chu-chuchuchu." Some songs are shorter, and pairs regularly duet.

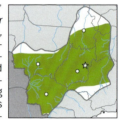

CIPÓ CANASTERO Asthenes luizae 17 cm | 6.75"

🇧🇷 *Uncommon and local on rocky slopes with scattered bushes and ground bromeliads in central Minas Gerais on Serra do Cipó and Campina do Bananal*. An isolated species, related to a widespread group in the Andes and Patagonia. Above grayish brown with narrow whitish superciliary; long tail dusky brown, outer feathers rufous-chestnut. *Chin patch white with fine black streaks*; below grayish. A *drab and obscure* bird, likely to be missed unless specifically searched out. Few if any birds are likely to be confused with it. Inconspicuous and mainly terrestrial, it hops and runs with partially cocked tail on rocky slopes, often disappearing among boulders. Song, often delivered from atop a rock, a rather musical series of 10-15 descending notes.

FIREWOOD-GATHERER Anumbius annumbi 19.5 cm | 7.75"

Locally fairly common in grassy cerrado and pastures with scattered trees, at edge of light woodland, and around ranch and farm buildings, *mainly in SE of our area*. Above sandy brown, crown and back streaked blackish with rufous forehead and whitish superciliary; flight feathers edged rufescent; *tail long and graduated, feathers pointed and broadly tipped white*. Throat white, often outlined with black spots; below pale buffyish. Vaguely recalls a thornbird or Chotoy Spinetail, *though neither of these shows the obvious white in tail*. Long-tailed silhouette also recalls Wedge-tailed Grassfinch. Feeds mainly on ground or within cover so inconspicuous except when near their nests, around which they can linger for protracted periods. These enormous nests are constructed from thorny, often surprisingly long twigs; each year a new nesting chamber is added. Most frequent song, often given while perched on or near nest, a fast gravelly "chit, chit, chit, chr-chr-r-r-r-r-r."

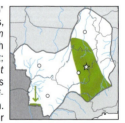

POINT-TAILED PALMCREEPER Berlepschia rikeri 21.5 cm | 8.5"

Local *in buriti palm stands (almost entirely Mauritia flexuosa) in N of our area*. A *boldly patterned*, unmistakable *furnariid with long straight bill*. Head, neck, and entire underparts black boldly streaked white; otherwise bright rufous-chestnut. Strictly arboreal and inconspicuous unless vocalizing, palmcreepers rummage in the bases of large, fan-shaped palm fronds, usually remaining well hidden. They occur in widely dispersed pairs and rarely associate with other birds. Infrequent song an unmistakable, far-carrying series of fast, ringing notes, "dedede-kee!-kee!-kee!-kee!-kee!-kee!-kee!" One sometimes will vocalize while perched in the open on a palm's spiky growing stalk.

RUFOUS-FRONTED THORNBIRD
JOÃO-DE-PAU

GREATER THORNBIRD
GRAVETEIRO

CIPÓ CANASTERO
LENHEIRO-DO-CIPÓ

FIREWOOD-GATHERER
COCHICHO

POINT-TAILED PALMCREEPER
LIMPA-FOLHA-DO-BURITI

XENOPS are small arboreal furnariids with *laterally compressed bills* and a *silvery malar streak*. Widespread in wooded areas, they nest in small holes excavated in soft wood.

STREAKED XENOPS *Xenops rutilans*   12 cm | 5"

Fairly common and widespread in canopy and borders of deciduous and gallery forest and woodland. *Lower mandible strongly upturned*. Rufous brown above *lightly streaked buff on crown and upper back*; superciliary whitish, *prominent silvery white malar streak*. Wings mostly rufous (a wing-band shows in flight); tail rufous, inner webs of inner tail feathers with black. Throat whitish; *below olive brown streaked whitish*. Plain Xenops is much less streaked, *showing none at all above* and only a little on chest. Found singly or in pairs, often with mixed flocks, foraging at various levels though usually not in lower growth. Works along or under twigs and in vine tangles, swiveling from side to side, tapping at dead wood or flaking off bark, not using its tail for support. Song a fast series that first rises, then descends, consisting of 4-7 shrill notes: "swee-swee-swee-swee-swee."

PLAIN *XENOPS Xenops minutus*   12-12.5 cm | 4.75-5"

Fairly common in lower, middle growth, and borders of forest and woodland in *N and far SE of our area*. *Lower mandible strongly upturned*. Above olive brown with pale buff superciliary and *prominent silvery white malar streak*; wings mainly rufous (a wing-band shows in flight); rump and tail rufous, black on lateral tail feathers. Throat whitish; pale brown below with whitish chest streaking. SE birds slightly smaller. Streaked Xenops is much more streaked, especially below. Plain's habits are similar to Streaked, though it forages much more at lower levels. Song much as in Streaked, a series of shrill notes that first rises and then drops: "swee-swee-swee-swee-swee."

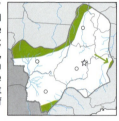

PHILYDOR foliage-gleaners are *relatively slender and long-tailed* furnariids that range in forested and wooded habitats and often accompany mixed flocks. They show little streaking, but do have a bold superciliary. Nests are usually placed in holes in snags.

BUFF-FRONTED FOLIAGE-GLEANER *Philydor rufum*   19 cm | 7.5"

Uncommon and somewhat local in subcanopy and borders of forest and woodland in *S of our region*. *Forehead and broad superciliary buff, gray crown and postocular stripe*; olive brown above with rufous wings and tail. Below uniform buff. Confusion is most likely with Ochre-breasted Foliage-gleaner, but note the latter's *all-gray* crown (no buff). Forages singly or in pairs, regularly with mixed flocks; frequently lingers in the semiopen, hence easy to see, hopping and twisting along horizontal limbs, sometimes in terminal foliage. Song a fast descending series of sharp, metallic, *Veniliornis* woodpecker-like notes, "whi-ki-ki-ki-ke-ke-ke-kuh-kuh."

OCHRE-BREASTED FOLIAGE-GLEANER *Philydor lichtensteini*   18 cm | 7"

Uncommon and local in midlevels and subcanopy of forest and woodland in *SE of our area*. *Crown and nape grayish* with slight scaly effect, *broad buff superciliary* and gray postocular stripe. Above brown, wings and tail rufous. Below ochraceous. Buff-fronted Foliage-gleaner (sometimes with it) is larger and should show a prominent buff forehead. Cf. Planalto Foliage-gleaner. Behavior much as in Buff-fronted, also mainly foraging with mixed flocks but tending more to inspect vine tangles and dead leaves (less often on terminal branches). Song similar.

PLANALTO FOLIAGE-GLEANER *Syndactyla dimidiata*   17.5 cm | 6.75"

Uncommon and local in lower and middle levels of gallery forest and woodland; seems to be absent from actual Pantanal. *Above uniform rufous brown* with long superciliary and sides of neck rich ochraceous; tail brighter rufous. *Below uniform rich ochraceous*. Buff-fronted and Ochre-breasted Foliage-gleaners both have gray on crown, are paler and buffier below. Both of those species are more arboreal and are more restricted to actual forest. Cf. also Henna-capped Foliage-gleaner. Found singly or in pairs, most often foraging independently of mixed flocks, inspecting epiphytic plants on branches. Song a series of strongly emphasized harsh metallic notes usually preceded by a softer accelerating chatter.

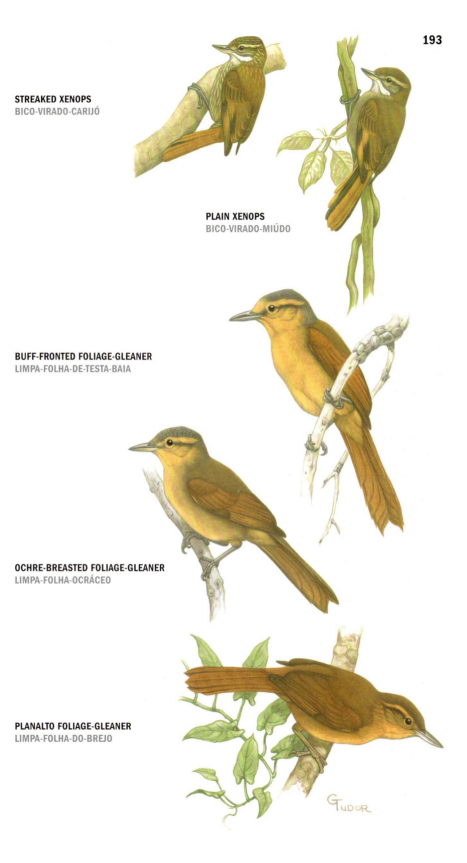

STREAKED XENOPS
BICO-VIRADO-CARIJÓ

PLAIN XENOPS
BICO-VIRADO-MIÚDO

BUFF-FRONTED FOLIAGE-GLEANER
LIMPA-FOLHA-DE-TESTA-BAIA

OCHRE-BREASTED FOLIAGE-GLEANER
LIMPA-FOLHA-OCRÁCEO

PLANALTO FOLIAGE-GLEANER
LIMPA-FOLHA-DO-BREJO

GREAT XENOPS Megaxenops parnaguae 16 cm | 6.25"
🇧🇷 Uncommon and local in caatinga woodland and scrub in N Minas Gerais and Tocantins. Unmistakable. *Heavy bill with lower mandible sharply upturned, pinkish at base. Bright cinnamon-rufous above*, somewhat dusky around eye; below somewhat paler cinnamon-rufous, *throat contrastingly white.* Found singly or in pairs, usually independent of the small mixed flocks found in its range. Generally forages like a xenops, hitching along branches, pecking or prying off small bark pieces, but also sometimes gleans more like a foliage-gleaner. Distinctive song a bubbly series of closely spaced notes that become louder and higher-pitched before trailing off at end.

WHITE-EYED FOLIAGE-GLEANER Automolus leucophthalmus 19 cm | 7.5"
Uncommon and local in undergrowth of forest and woodland in S of our area. *Iris white.* Above rufescent brown; rump and tail bright cinnamon-rufous. *Throat white* (often puffed out); below ochraceous, midbreast whiter. The only *Automolus* foliage-gleaner in our area and as such should be easily recognized; the white eye and white throat are both conspicuous. Superficially similar Greater Thornbird is yellow-eyed, and is found in a totally different, semiopen habitat. Seen singly or in pairs, often accompanying mixed flocks of understory birds. Not particularly shy. Clambers in dense tangled habitat, often hanging upside-down while rummaging in epiphytes and suspended dead leaves. Nests in holes dug into banks. Distinctive song a loud and fast "ki-deee, ki-dee, ki-dee, ki-dee, ki-dee, ki-dee, ki-dee" (sometimes varied to "ki-trrr, ki-trrr…").

HENNA-CAPPED FOLIAGE-GLEANER Hylocryptus rectirostris 20.5 cm | 8"
▲ *Rare and local on or near ground in gallery and deciduous woodland*; seems absent from actual Pantanal. Iris brownish yellow. *Crown and nape orange-rufous* (henna), *back golden brown*; wings and tail also orange-rufous. *Below mostly bright pale ochraceous.* Planalto Foliage-gleaner is smaller, shorter-billed, and more uniformly rufescent above. Inconspicuous, foraging mainly in leaf litter on the ground; usually in pairs, not often joining flocks. Distinctive song a loud staccato chattering: "cuk-cuk-cuk-cuk, cuh-cuk, cuk-cuk," at times sounding uncannily like a clucking domestic chicken.

RUFOUS-BREASTED LEAFTOSSER Sclerurus scansor 18 cm | 7"
Rare and local on or near ground in forest and woodland, *mainly in E of our area*. Bill slender. Above dark brown with rufous-chestnut rump and black tail. *Throat whitish scaled dusky, contrasting with rufous chest*; below grayish brown. Sharp-tailed Streamcreeper is obviously spotted below. A furtive bird, the leaftosser is *mainly terrestrial* and solitary. It favors damp places, there hopping and shuffling on the ground on its short legs, flicking aside leaves and other debris with its bill and often lingering in the same spot. Not particularly wary, it may suddenly flush from close by, giving a sharp "tseeeét." Song a sharp descending chippered trill that sometimes ends in a chatter: "chee-ee-ee-ee-ee-ee-ee-ee-eu, cht, cht."

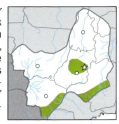

SHARP-TAILED STREAMCREEPER Lochmias nematura 15 cm | 6"
Uncommon and local *along or near streams* in more humid forest and woodland, *mainly in hilly terrain* but also locally in gallery woodland along streams in cerrado. Inconspicuous and apt to be overlooked; present at Brasília NP. Bill long, slender, and slightly decurved. Legs dull grayish pink. Dark brown above with *white-spotted superciliary*, mantle more chestnut; tail black. *Below dark brown profusely spotted white.* Though leaftosser-like in shape, the streamcreeper is smaller and has unique bold white spotting below. Hops on or near the ground and usually hard to see, though sometimes one will feed more in the open, even on rocks out in the middle of a stream. It probes damp soil and also tosses leaves aside with its bill. Song a series of dry unmusical notes that starts slowly, often preceded by a single sharp note, and accelerates into a chipper.

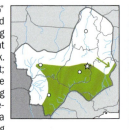

GREAT XENOPS
BICO-VIRADO-DA-CAATINGA

WHITE-EYED FOLIAGE-GLEANER
BARRANQUEIRO-DE-OLHO-BRANCO

HENNA-CAPPED FOLIAGE-GLEANER
FURA-BARREIRA

RUFOUS-BREASTED LEAFTOSSER
VIRA-FOLHA

SHARP-TAILED STREAMCREEPER
JOÃO-PORCA

ANTBIRDS (Thamnophilidae) comprise a large, varied group of small to midsized insectivorous forest birds species. None actually eat ants, but some follow army ant swarms to catch fleeing insects. Favoring dense undergrowth, antbirds tend to be restless, fast-moving, and hard to see well. Most are sexually dimorphic and have distinctive plumage patterns. Their *loud and frequent vocalizations* attract attention. Nests are cup-shaped.

GLOSSY ANTSHRIKE *Sakesphorus luctuosus* 17.5 cm | 7"

Fairly common in riparian thickets, lower growth of várzea woodland, and river islands in N of our area, e.g., Ilha do Bananal. *Long shaggy crest is often fully raised*, especially when singing. ♂ uniform deep black, scapulars edged white (looks like a bar), outer tail feathers narrowly tipped white. ♀ similar but *sootier black with contrasting chestnut crown*. This antshrike should be distinctive in its limited range. Pairs hop about methodically in vine tangles and bushes, sometimes mounting higher into trees; the tail is often slowly wagged, invariably when singing. Song a series of slow, measured, nasal notes that speeds up toward end: "caw, caw, caw, caw-caw-caw-ca-ca-ca-cacaca."

Silvery-cheeked Antshrike (*S. cristatus*) ranges in lower growth of arid scrub and woodland in N Minas Gerais and Tocantins. Handsome, crested ♂ has black foreparts and chest contrasting with whitish cheeks and sides. ♀ rufous above, wings dusky with two buffy whitish bars.

GREAT ANTSHRIKE *Taraba major* 20 cm | 8"

Common and widespread in undergrowth of shrubby clearings, woodland, and thickets in forest borders. Heavy hooked bill; *iris bright red*. Slight shaggy crest. ♂ *black above* with white edging on wing-coverts; tail with white banding. *White below*. ♀ has *rufous* replacing ♂'s *black*; wings and tail *lack* white. No other antshrike is so boldly black, or rufous, and white. Pairs frequent thick undergrowth, where they sometimes accompany loose flocks. Though not shy, heard more often than seen. Song an accelerating series of hoots with a bouncing-ball effect, trogon-like aside from its snarled "nyaah" ending (which occasionally is left off). Also gives various gravelly or rattled calls.

THAMNOPHILUS antshrikes are engaging, midsized antbirds with *strongly hooked bills* that are found in scrubby habitats and the lower growth of forest and woodland. Usually in pairs, some species regularly ranging with flocks. All species are notable for their *oft-given vocalizations*.

BARRED ANTSHRIKE *Thamnophilus doliatus* 16-16.5 cm | 6.25-6.5"

Common and widespread in thickets, dense lower growth of borders, clearings, scrub, and gardens. Expressive loose crest. Iris yellow. ♂ *above black barred white*; crown black, often showing its semiconcealed white; sides of head and hindneck more streaked. *Below white barred black*. ♀ bright *cinnamon-rufous above*; sides of head and hindneck streaked buffy whitish and black. Below ochraceous buff. Though often numerous, this striking bird usually remains hidden and hard to see in the dense vegetation it so loves. Forages in pairs, hopping and peering about in search of insects, generally not with mixed flocks. Heard much more often than seen, with song a fast, accelerating series of nasal notes, "wah-wah-wa-wa-wawawawawawawawa-wánh," with distinctive emphasized final note. The singing ♂ usually raises its crest and spreads its tail, and sometimes bobs stiffly; the ♀ may echo with her own song, higher-pitched and usually shorter.

RUFOUS-WINGED ANTSHRIKE *Thamnophilus torquatus* 14 cm | 5.5"

Uncommon in scrub, lighter woodland, and cerrado, but absent or very local in Pantanal. Iris orangey. ♂ has *black crown*, grayish face and neck, brownish back, and *contrasting rufous-chestnut wings*; tail black with white barring. Below whitish with *black barring across breast*. ♀ olivaceous brown above with *rufous crown, wings, and tail*; below dingy buffy whitish. Confusion is possible only with the much commoner Barred Antshrike; ♀ lacks ♀ Barred's streaking on face. An unobtrusive bird, thus rarely seen except when vocalizing; pairs seems usually to forage independently of mixed flocks. Song a rather high-pitched and nasal "renh, renh, renh, reh-reh-reh-reh-rénh."

GLOSSY ANTSHRIKE
CHOCA-D'ÁGUA

GREAT ANTSHRIKE
CHORÓ-BOI

BARRED ANTSHRIKE
CHOCA-BARRADA

RUFOUS-WINGED ANTSHRIKE
CHOCA-DE-ASA-AVERMELHADA

"GRAY" *THAMNOPHILUS* antshrikes are confusingly similar, sexually dimorphic antbirds with white or buff markings on their wings. They are typical members of mixed-species flocks in undergrowth.

AMAZONIAN ANTSHRIKE *Thamnophilus amazonicus* 15 cm | 6"

Uncommon and local in lower and middle growth of várzea woodland in NW of our area. ♂ gray above with black crown and considerable black mottling on back; *wings black with bold white markings*; tail and upper tail-coverts black, *feathers tipped white*. More colorful ♀ has *orange-rufous head, neck, and underparts*; back and rump olive brown, wings and tail as in ♂. ♂ closely resembles ♂ Planalto Slaty Antshrike, though the latter has a whiter belly. The two are best distinguished by their different songs and habitats. Unlike Planalto Slaty, Amazonian *does not habitually wag its tail*. ♀♀ of the two are very different. Usually in pairs, often with small mixed flocks and regularly foraging into midlevel viny tangles. Song an accelerating series of nasal trogon-like notes that starts slowly and fades at end, "kuh, kuh, kuh-kuh-kuh-kuhkuhkuh-kunh." The ♀ sometimes echoes with a higher-pitched version.

PLANALTO SLATY ANTSHRIKE *Thamnophilus pelzelni* 14.5 | 5.5"

Locally fairly common in undergrowth of deciduous and gallery woodland. ♂ gray above with black crown and mixed black-and-gray back; *wings black with bold white markings*; tail and upper tail-coverts black, *feathers tipped white*. Below gray, belly whiter. ♀ brownish above, crown and back more rufescent, with *wings and tail marked as in ♂*. Below buffy brownish, brightest on breast and whiter on belly. Amazonian Antshrike is similar, but occurs only in N of our area. Amazonian's back is blacker, but it is best distinguished by its very different ♀, very different voice, and *not* wagging its tail. Usually found in pairs, sometimes accompanying loose mixed flocks; can often be instantly recognized by habit of *almost constant tail wagging*. Song a fast-paced series of well-enunciated nasal notes that starts slowly, rises in pitch, and ends in a rattle: "anh-anh-anh-anh-ah-ah-ahahahahah."

Bolivian Slaty Antshrike (*T. sticturus*) occurs in deciduous woodland undergrowth near Corumbá. Aside from its grayish irides (both sexes), it closely resembles the Planalto Slaty. Song also similar.

VARIABLE ANTSHRIKE *Thamnophilus caerulescens* 14.5 cm | 5.75"

Locally fairly common in lower growth and borders of forest and woodland in S of our area. ♂ basically gray with black crown, black wings and tail with white markings, and *tawny on flanks and crissum*. ♀ olivaceous brown above, slatier on crown; wings and tail much as in ♂; throat and chest grayish olive, *lower underparts bright tawny*. Both sexes can be known by their tawny belly. Usually in pairs, often quite tame, Variables routinely accompany mixed understory flocks but also forage separately. The tail is occasionally raised and then lowered but is not habitually wagged. ♂'s simple song is a fairly fast repetition of 5-7 "kaw" or "kow" notes. The ♀'s version is higher-pitched.

PLAIN ANTVIREO *Dysithamnus mentalis* 11.5-12 cm | 4.5-4.75"

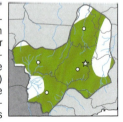

Fairly common and widespread in lower growth of forest and woodland, including N Pantanal. *Chunky*. ♂ gray to olive grayish above, darker on crown and neck, with *dusky auriculars*; wing-coverts edged whitish. Below paler gray, throat and belly whiter. ♀ olive to olive brown above with *contrasting rufous crown, dusky auriculars*, and *white eye-ring*. Below mostly pale olive grayish, whiter on throat and midbelly. In SE of our area (illustrated) both sexes have a pale yellow belly. No other similar antbird shows the contrasting dark auriculars; ♀'s rufous on crown is also distinctive. *Thamnophilus* antshrikes are larger and longer-tailed; *Myrmotherula* antwrens are smaller and tend to be more active. Gray-eyed Greenlet vaguely recalls the ♀ antvireo. Most often in pairs, foraging lethargically, sometimes peering about for protracted periods; regularly accompanies understory flocks. Song a short series of rapidly repeated semimusical notes that accelerate into descending clipped roll. Both sexes give soft "ert" call, sometimes interminably repeated; also has a more abrupt and nasal "nyah."

AMAZONIAN ANTSHRIKE
CHOCA-CANELA

PLANALTO SLATY ANTSHRIKE
CHOCA-DO-PLANALTO

VARIABLE ANTSHRIKE
CHOCA-DA-MATA

PLAIN ANTVIREO
CHOQUINHA-LISA

HERPSILOCHMUS antwrens are small, active, dimorphic antbirds with *fairly long tails*. All of our species are *arboreal* birds that usually range in pairs and habitually accompany mixed flocks.

BLACK-CAPPED ANTWREN *Herpsilochmus atricapillus* 12 cm | 4.75"
Fairly common and widespread in midlevels and subcanopy of *deciduous and gallery woodland*, locally in Pantanal. ♂ has *black crown and postocular line, long white superciliary*; back gray. Wings and tail black, wings with two white bars; tail feathers broadly tipped white. Below whitish. ♀ has *buff forehead, black crown streaked white*, whitish superciliary, and grayish olive back; wings and tail as in ♂. Below whitish, breast tinged buff. Large-billed Antwren can occur with this smaller *Herpsilochmus*; its ♂ has obvious spotting below; ♀ very different. Usually not too hard to observe, descending lower at woodland edge. Song a chippered trill often introduced by a hiccupping note, accelerating into a louder sputter toward end.

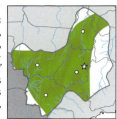

Rufous-winged Antwren (*H. rufimarginatus*) occurs locally in forest in N of our area. It recalls Black-capped Antwren, but both sexes have obvious rufous edging in flight feathers and are pale yellow below.

LARGE-BILLED ANTWREN *Herpsilochmus longirostris* 13 cm | 5.25"
Locally fairly common in midlevels and subcanopy of gallery and deciduous forest and woodland, including Pantanal. Rather long bill. ♂ has black crown, long white superciliary, and black postocular line. Back mottled gray, black, and white. Wings and tail black; wings with two bold white bars, tail feathers broadly tipped white. Below white, *throat and breast with gray "streaky" spots*. Pretty ♀ has *bright orange-rufous head and neck*; gray back; wings and tail as in ♂. Below bright cinnamon-buff. Black-capped Antwren ♂ is smaller, has no spotting below. Song a series of ca. 15 even chippered notes, slowing markedly toward end: "wh-chchchchchchchchchchch-chu-chu."

MYRMOTHERULA are small, *short-tailed* antwrens found in forest and woodland *understory*.

AMAZONIAN STREAKED ANTWREN *Myrmotherula multostriata* 9.5 cm | 3.75"
Fairly common in shrubby growth along margins of lakes, streams, and rivers in NW of our area. ♂ *black above streaked white*, semiconcealed dorsal patch white; wings black with two white bars. *White below streaked black.* ♀ has *crown and nape orange-rufous streaked black*; dorsal patch as in ♂; wings black with two white bars. *Below pale ochraceous with fine black streaking*, belly whiter. Forages low in bushes and low trees close to the water, often where vegetation is partially submerged. Most often not with flocks, though pairs are sometimes joined by other small antbirds. Vocalizations include a dry trill, "dr-r-r-r-r-r," and a distinctive lilting, more musical song: "pur-pur-peé-peé-peé-pur." Also gives a "chee-pu" contact note.

WHITE-FLANKED ANTWREN *Myrmotherula axillaris* 10 cm | 4"
Fairly common in lower and middle growth of forest and woodland in *NW of our area*. ♂ blackish to dark gray above; wing-coverts with two white-spotted bars, tail tipped white. *Below blackish, axillars and long silky plumes on flanks white* (often conspicuous, but sometimes hidden beneath wings). ♀ grayish brown above; wings and tail browner; wings with two faint buff-dotted bars. Ochraceous below with whiter throat and *white plumes on flanks* (sometimes concealed). Unique in our area, though several similar *Myrmotherula* range just to the north, in Amazonia. Frequently with understory flocks. Gleans actively in foliage and vine tangles, often flicking its wings, thereby exposing white flanks. Song a descending series of whistled notes: "pyii, pii, pee, pey, peh, puh, pu;" also gives a faster, run-together descending chipper. Oft-heard call a fast "chee-du" or "chee-doo."

BAND-TAILED ANTBIRD *Hypocnemoides maculicauda* 12 cm | 4.75"
Fairly common in undergrowth of swampy forest, thickets at lake and stream margins, and gallery woodland in NW of our area, including N Pantanal. Iris pale grayish. ♂ gray, somewhat paler below, with *black throat* and *large semiconcealed white dorsal patch*; wing-coverts black with three white bars; *tail black with wide white tip*. ♀ similar above (including dorsal patch); whitish below *mottled grayish especially on breast*. No other semiterrestrial antbird occurs with this species. Mato Grosso Antbird is larger and longer-tailed. Forages singly and in pairs, hopping low in thickets, almost always near water or at water's edge. Usually not shy. The tail is often held slightly cocked. Song a loud series of notes that accelerate and become raspy, "pee-pee-pi-pipipipipipipi-pe-pe-peh-pez-pez-pzz."

BLACK-CAPPED ANTWREN
CHOROZINHO-DE-CHAPÉU-PRETO

LARGE-BILLED ANTWREN
CHOROZINHO-BICUDO

AMAZONIAN STREAKED ANTWREN
CHOQUINHA-ESTRIADA-DA-AMAZÔNIA

WHITE-FLANKED ANTWREN
CHOQUINHA-DE-FLANCO-BRANCO

BAND-TAILED ANTBIRD
SOLTA-ASA

WHITE-BROWED ANTBIRD *Myrmoborus leucophrys* 13.5 cm | 5.25"
Uncommon in undergrowth of forest borders, woodland, and regenerating clearings in *NW of our area*. The *dark bluish gray* ♂ has a *broad snowy white superciliary extending back from forehead*, black face and throat. ♀ has *broad cinnamon-buff superciliary* contrasting with *black mask*; brown above, buff dots on wing-coverts. *Below mainly white*. A chunky, short-tailed antbird with a distinctive bold brow in both sexes. Pairs forage close to ground in dense younger growth, perching on slender saplings. Rarely with mixed flocks. Song a fast, descending stream of loud, ringing notes that increases in volume, then fades away, "pipipipipipipipip'p'p'p'p'rr."
Black-faced Antbird (*M. myotherinus*) inhabits forest undergrowth in NW of our area. ♂ recalls ♂ White-browed but has a less prominent white brow and shows conspicuous white fringing on wing-coverts. ♀ olive brown above with contrasting black mask and buff fringing on wing-coverts; ochraceous below.

FORMICIVORA antwrens are attractive, fairly long-tailed antbirds found in scrubby habitats, *not* forest. Vocal, but hard to see, they skulk in dense low growth. ♂♂ of our three species have a *distinctive white "fringe" extending from the brow down the sides* (less marked in Black-bellied).

SOUTHERN WHITE-FRINGED ANTWREN *Formicivora grisea* 12.5 cm | 5"
Fairly common but local in dense undergrowth of low shrubby woodland and regenerating clearings in *N of our area*. ♂ *grayish brown above* with *white superciliary extending as stripe down neck and sides to flanks* (where especially wide); wings blackish with white bar and spotting on coverts; tail feathers broadly white-tipped, outermost edged white. *Below black*. ♀ similar above, but white superciliary does not extend to underparts. *Below cinnamon-buff*, throat whiter. ♂ Black-bellied Antwren has a less conspicuous "fringe" on lower underparts; its ♀ is very different. Hops methodically and gleans in foliage; often swivels its spread tail side to side. Usually remains in thick cover. Song a repeated sharp, penetrating "chup-chup-chup..." (up to 20 or more such notes), sometimes varied to a "chedep...."

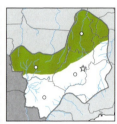

BLACK-BELLIED ANTWREN *Formicivora melanogaster* 13 cm | 5.25"
Uncommon and somewhat local in scrub and lower growth of deciduous woodland. ♂ dark grayish brown above with *broad white superciliary extending as stripe down sides of neck*; wings blackish with bold white bar, spotting on coverts, and *whitish tertial edging*; tail feathers broadly white-tipped, outermost edged white. *Below black*. ♀ similar above, but browner and with *blackish cheeks*. Below whitish. ♂ of similar S. White-fringed Antwren has a narrower white superciliary, wider white stripe on flanks. Its ♀ is very different. Behavior as in S. White-fringed Antwren. Song a rather slow series of "cha" notes, often long-continued.

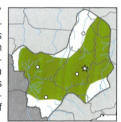

RUSTY-BACKED ANTWREN *Formicivora rufa* 12.5 cm | 5"
Locally fairly common in low shrubby areas, cerrado, and gallery woodland undergrowth. ♂ *rufous brown above* with *white superciliary extending as stripe down neck to sides*; wings blacker, coverts with white dotting; tail feathers broadly tipped white, outermost edged white. Below otherwise black, *flanks buff*. ♀ like ♂ above but below white, *streaked with black on throat and breast*, flanks buff. ♂ S. White-fringed Antwren is much less rufescent above and its white "fringe" incorporates flanks. ♀♀ of all three *Formicivora* are very different below. Behavior as in S. White-fringed. Song a fast gravelly "che-de-de, che-de-de, che-de-de..." repeated up to 12-15 times, sometimes run together almost as a trill.

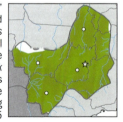

STRIPE-BACKED ANTBIRD *Myrmorchilus strigilatus* 16.5 cm | 6.5"
Fairly common on or near ground in arid woodland and scrub in *extreme SW Mato Grosso do Sul (Corumbá area) and in N Minas Gerais and Tocantins*. ♂ *boldly streaked black and rufous above*, narrow superciliary pale buff; wing-coverts black with two white bars and white dots. Rump and central tail feathers rufous, lateral ones black-tipped and edged white. *Throat and breast black*, lower face and belly contrastingly whitish. NE birds have superciliary white. ♀ similar but buffy whitish below (no black bib), breast streaked dusky. Not likely confused in range and habitat, where there are few other antbirds, *none of them more or less terrestrial*. Found singly or in pairs, hopping on or near ground where, because of its dense habitat, hard to see without tape playback. Distinctive song a loud wheezy "cheem, cheery-gweér" or "chree, chree-cho-weé," with ♀ sometimes following with a descending "pur, cheer-cheer-chur-chur-chr-chr."

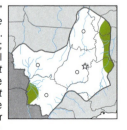

WHITE-BROWED ANTBIRD
FORMIGUEIRO-DE-SOBRANCELHA

SOUTHERN WHITE-FRINGED ANTWREN
FORMIGUEIRO-PARDO

BLACK-BELLIED ANTWREN
FORMIGUEIRO-DE-BARRIGA-PRETA

RUSTY-BACKED ANTWREN
FORMIGUEIRO-VERMELHO

STRIPE-BACKED ANTBIRD
PIU-PIU

BLACK-THROATED ANTBIRD *Myrmeciza atrothorax* 14 cm | 5.5"
Fairly common in undergrowth of *forest borders*, secondary woodland, and riparian growth in *NW of our area*. ♂ umber brown above with semiconcealed white dorsal patch; wing-coverts with white dots, tail blackish. *Throat and breast black; face, sides of neck, and belly gray*. ♀ brown above, dorsal patch as in ♂; wing-coverts with buffyish dots; *tail blackish. Throat white, breast orange-rufous*. Forages close to ground, usually in pairs; adept at keeping out of view due to its active, fidgety behavior. Only rarely with mixed flocks. Incisive song a fast ascending series of high-pitched notes, "chee-ch-chee, chi-chi-chi-chí-chí," sounding like the peeping of baby chicks.

Southern Chestnut-tailed Antbird (*M. hemimelaena*) is also found along NW edge of our area, but is much more a bird of forest interior (not edge). ♂ has gray head and neck, black bib, and rather short rufous-chestnut tail. ♀ also has a rufous-chestnut tail; below mainly orange-rufous. Song a series of ringing notes that desends, "klee-klee-kli-kli-kli-klu."

MATO GROSSO ANTBIRD *Cercomacra melanaria* 16.5 cm | 6.5"
Fairly common in thickets of *gallery and deciduous woodland in Pantanal*. Slim and slender-billed. ♂ *jet black*; white interscapular patch, *wing-coverts prominently fringed white*, tail feathers tipped white. ♀ gray above; wings and tail as in ♂. *Below pale grayish*. ♂ Band-tailed Antbird is smaller and much more terrestrial; its shorter tail has a bolder white tip. Ranges in pairs that hop through dense undergrowth, sometimes moving with small loose flocks. Distinctive and frequently heard song a slow, measured, guttural "ker-cheeeér-chk, ker-cheeeér-chk, ker-cheeeér." Both sexes also give low "churk" call.

BANANAL ANTBIRD *Cercomacra ferdinandi* 16 cm | 6.25"
🇧🇷
VU Uncommon in dense thickets of riparian woodland and river islands in N of our area along Rio Araguaia, mainly on and near Ilha do Bananal. ♂ closely resembles ♂ of Mato Grosso Antbird. ♀ differs from ♀ Mato Grosso in its *darker gray underparts* with *fine white streaking*. Does not overlap with Mato Grosso Antbird. Behavior similar, ranging in pairs in undergrowth, usually not too hard to observe; regularly accompanies small flocks. Song a rather nasal series of jumbled notes, e.g., "caw-cuh-didi-cuhdidi-cúh-chuk-chuk," with many variations. Both sexes also give single "cawh" and "chuk" calls.

PYRIGLENA fire-eyes are quite large, long-tailed antbirds found in forest and woodland undergrowth where they often attend swarms of army ants. They have distinctive *bright red irides*.

WHITE-SHOULDERED FIRE-EYE *Pyriglena leucoptera* 18 cm | 7"
Locally fairly common in undergrowth of forest and woodland in *S of our area*. Iris bright red. ♂ *all glossy black* with semiconcealed white dorsal patch. *White bend of wing and tipping on wing-coverts form two bars*. ♀ rufescent brown above, tail blackish. Throat whitish, *below dingy buffy brownish*. White-shouldered Fire-eye's range comes close to (overlaps?) White-backed's. ♂ White-backed *lacks white in wing*; ♀ has a bold white supraloral. Usually in pairs and often pounds tail downward. Frequently heard song a series of loud, penetrating whistled notes, "peer-peer-peer-peer-peer-peer," less run together than in White-backed.

WHITE-BACKED FIRE-EYE *Pyriglena leuconota* 17.5 cm | 6.75"
Uncommon and local in undergrowth of forest and woodland in *W of our area*, though seeming absent from actual Pantanal (only around its periphery). Iris bright red. ♂ *all glossy black* with semiconcealed white dorsal patch. ♀ rufescent brown above with dorsal patch as in ♂, *white supraloral*, and *dusky lores*; tail blackish. *Throat and breast ochraceous buff*, belly olivaceous brown. ♂ unique in range, and apparently does not occur with White-shouldered Fire-eye. ♀, though dull-plumaged, is easily known by her red eye. Behavior as in White-shouldered. Song a far-carrying, fast series of ringing whistled notes, "peer-peer-peer-peer-peer-peer-peer-peer-peer-peer;" ♀ echoes with softer, higher-pitched version. Calls include a loud "chik" or "chi-djik."

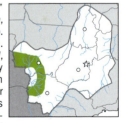

BLACK-THROATED ANTBIRD
FORMIGUEIRO-DE-PEITO-PRETO

MATO GROSSO ANTBIRD
CHORORÓ-DO-PANTANAL

BANANAL ANTBIRD
CHORORÓ-DE-GOIÁS

WHITE-SHOULDERED FIRE-EYE
OLHO-DE-FOGO

WHITE-BACKED FIRE-EYE
PAPA-TAOCA

ANTTHRUSHES (Formicariidae) are shy and solitary birds that *walk on the forest floor like little rails*, their tails cocked up, flicking leaves with their bills as they search for invertebrate food. Both of our species have distinctive far-carrying songs.

BLACK-FACED ANTTHRUSH *Formicarius analis* 17.5 cm | 7"

Uncommon and local on or near ground inside forest and woodland in NW of our area. Brown above with bare ocular area bluish white, white lores. *Lower cheeks and throat black*; below gray, paler on belly, crissum rufous. Juvenile has throat whitish scaled dusky. Cf. Rufous-capped Antthrush. Far-carrying musical song an emphasized note followed by a pause, then a fast series of shorter notes, first ascending and becoming louder, then slowly descending: "tüüü, ti-ti-tí-tí-ti-ti-te-te-tu-tu-tu-tu." Alarm call, given by both sexes, an abrupt "churlew!"

RUFOUS-CAPPED ANTTHRUSH *Formicarius colma* 18 cm | 7"

Uncommon and local on or near ground inside forest along NW edge of our area but generally scarcer than Black-faced Antthrush. Brown above with *bright rufous crown and nape; forehead black.* Sides of head and foreneck sooty black, belly grayish. Some ♀♀ have whitish throat. Behavior as in Black-faced. Distinctive song a fast series of trilled musical notes which falter and drop at the start, then gradually rise: "tr-r-r-rew-u-u-u-u-u-u-u-u-u-u-u-u-u-u?"

White-browed Antpitta (*Hylopezus ochroleucus*) occurs on or near ground in caatinga woodland and scrub in N Minas Gerais and Tocantins. *The only antpitta in our area.* Inconspicuous, aside from voice; very hard to see. Long-ish legs, short tail; *plump and "round."* Olive brown above, crown grayer with white lores, eye-ring, and postocular stripe. White below, breast spotted dusky, flanks buff. Song a ringing, two-part "whu-whú, whu-whú-whu-whu-wheú-wheú-wheú-wheú-wheú-wheú-wheú-wheú."

GNATEATERS (Conopophagidae) are plump, short-tailed, long-legged birds like small antpittas but sexually dimorphic and sporting silvery or gray postocular tufts. They range near the ground in thickets where usually hard to see.

RUFOUS GNATEATER *Conopophaga lineata* 13 cm | 5"

Uncommon and local in undergrowth and borders of forest and woodland in S of our area; can persist even in degraded, fragmented habitat. ♂ rufous brown above with *gray superciliary merging into silvery postocular tuft* (tuft gray in ♀). Throat and breast orange-rufous; midbelly white. Found singly or in pairs, perching on vertical saplings and remaining low but rarely actually on ground. Not particularly shy. Occasionally moves with small mixed flocks, perhaps attracted by the group activity. Distinctive song a series of 8-10 high-pitched whistled notes, hesitating at first, then rising: "tew; tew; tew, tew, tew-tew-tiw-tiw-ti-ti." Also frequently gives a sharp, sneezing "chiff" or "cheff" call.

TAPACULOS (Rhinocryptidae) are semiterrestrial birds that *only rarely fly. Scytalopus* are small, *mouse-like,* short-tailed gray birds; though usually hard to differentiate, our one species is distinctive.

BRASILIA TAPACULO *Scytalopus novacapitalis* 11.5 cm | 4.5"

Uncommon and local in gallery forest and woodland, often in seasonally flooded areas, in S Goiás (mainly Distrito Federal) *and W Minas Gerais (Serra da Canastra; same species?).* Legs yellowish. Bluish slate above and on sides, median underparts pale gray to whitish. Other *Scytalopus* occur in SE Brazil. Creeps about in dense undergrowth, generally in pairs. Very hard to see except after tape playback (and often not easy even then). Song a steady repetition of rather high-pitched "chet" notes with an insect-like quality.

CRESCENTCHESTS (Melanopareiidae), recently separated as a family from the tapaculos, comprise a group of *handsome*, rather long-tailed birds of semiopen terrain that are almost as hard to see.

COLLARED CRESCENTCHEST *Melanopareia torquata* 14.5 cm | 5.75"

Uncommon in grassy cerrado with scattered bushes. Rufous brown above with long narrow white superciliary; *black sides of head, rufous nuchal collar,* large but semiconcealed white dorsal patch. Throat buff, *narrow pectoral band black*; underparts tawny-buff. Unique in our area. Found singly or in pairs, creeping and hopping on or near the ground in tall grass and shrubbery, almost always out of sight. Rarely noted except when singing and even then often only after tape playback has lured a singing bird up into a bush, where it usually remains hidden. Far-carrying song a distinctive, monotonous series of sharp but melodic "tü" notes, continuing for up to 20-30 seconds.

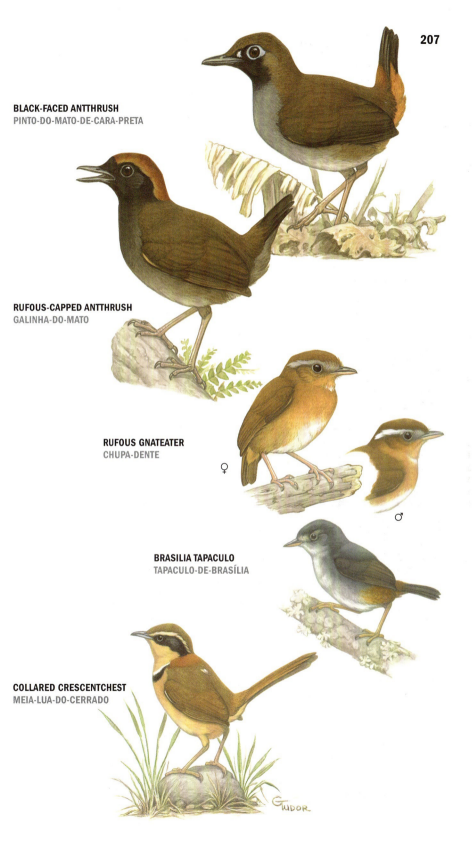

BLACK-FACED ANTTHRUSH
PINTO-DO-MATO-DE-CARA-PRETA

RUFOUS-CAPPED ANTTHRUSH
GALINHA-DO-MATO

RUFOUS GNATEATER
CHUPA-DENTE

BRASILIA TAPACULO
TAPACULO-DE-BRASÍLIA

COLLARED CRESCENTCHEST
MEIA-LUA-DO-CERRADO

TYRANT FLYCATCHERS (Tyrannidae) comprise a huge heterogeneous family of birds found throughout the Neotropics, well represented in our area where they range from the large and spectacular (e.g., Fork-tailed Flycatcher and Streamer-tailed Tyrant) to some of our drabbest and most obscure birds (e.g., various elaenias and tyrannulets); many species remain rare and poorly known. Flycatchers occur in all terrestrial habitats, from the most open to closed-canopy forests, and while a large majority are insectivorous, many also consume much fruit. Identification can be tricky, and in some of the more difficult groups *voice is often the best clue.*

CAMPO SUIRIRI *Suiriri affinis*   16 cm | 6.25"

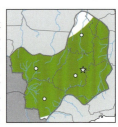

Uncommon in cerrado and campos with scattered shrubs and trees and at edge of gallery woodland. Sometimes considered conspecific with *S. suiriri*, (Chaco Suiriri) found further south; has been called Campo Flycatcher. Olive above with grayer head, *buffy yellowish rump and base of tail*; wings and tail blacker, *wings with yellowish bars and edging*, tail feathers edged yellowish. Throat and chest pale grayish; lower underparts pale yellow. Superficially elaenia-like, the Campo Suiriri differs in having an *all-black bill* (elaenia bills are basally pale). Southern Scrub Flycatcher has a shorter bill and lacks the suiriri's pale rump. Usually in pairs, foraging low, sometimes dropping to ground; often hovers, spreading tail and exposing its pale rump. Pairs give a fast jumbled series of notes as a duet (a common phrase is "pi-chu!"), sometimes while vigorously drooping and flapping their wings.

Chapada Suiriri (*S. islerorum*) is a recently described species found locally in many of the same areas as the Campo Suiriri. It differs in slightly shorter but heavier bill and pale-tipped tail feathers. Its exuberant display is similar to Campo's, but it tends to hold wings above horizontal much more. Voice similar, and also often given as a duet, with phrases often being doubled, e.g., "weer-weer."

ELAENIA elaenias are *dull-plumaged* midsized tyrannids found in semiopen and edge terrain. When not vocalizing, they are usually very difficult to identify; distinguishing their calls requires practice, but *voice generally represents the best way to separate the various species*. All elaenias have *pale lower mandibles*, and some species are *crested with white in center*. They perch quite upright and rarely accompany mixed flocks but regularly congregate with other birds at fruiting trees.

YELLOW-BELLIED ELAENIA *Elaenia flavogaster*   16-16.5 cm | 6.25-6.5"

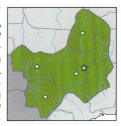

Widespread and generally common in semiopen shrubby areas, clearings, and gardens; more frequent in *disturbed areas* than most elaenias. Usually shows a *conspicuous upstanding bushy crest*; *not* particularly yellow-bellied. Brownish olive above with faint whitish eye-ring; wings duskier with two whitish bars. Throat whitish, breast grayish olive, belly pale yellow. Learn this elaenia well as in many areas it is the most numerous member of a difficult genus. Cf. especially Lesser and Large Elaenias. Though drab in appearance, this elaenia's demeanor is animated and it regularly perches in the open, thus attracting attention. The upright crest is especially noticeable on vocalizing birds. Noisy, this elaenia's most frequent calls, all exuberant with a hoarse quality, include a "breeeyr" and a "wreek-kreeeyuup," the latter often doubled.

LARGE ELAENIA *Elaenia spectabilis*   18 cm | 7"

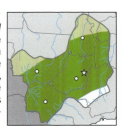

Fairly common and somewhat local summer visitant to forest and woodland borders, and clearings; migrates north during austral winter. Brownish olive above with faint whitish eye-ring; wings duskier with two or three whitish bars. Throat and breast pale gray, belly pale yellow. Resembling the more numerous Yellow-bellied Elaenia, the Large is less conspicuous, less vocal, and larger with a *reduced crest*; its foreneck is a purer gray. Large is more a woodland bird. Cf. Highland Elaenia and Short-crested and Swainson's Flycatchers. Most frequent call an abrupt "p-cheeu" *totally lacking Yellow-bellied's raucous quality.*

HIGHLAND ELAENIA *Elaenia obscura*   18 cm | 7"

Uncommon and local in lower growth and borders of forest and woodland in S of our area. A *large, round-headed* elaenia whose *notably short bill imparts a "snub-nosed" effect.* Dark olive above with *yellowish partial eye-ring* (crescents above and below) and *no crest or crown patch*; wings duskier with two yellowish bars. Throat pale yellowish, breast and flanks dull olive, midbelly clear pale yellow. Large Elaenia is less olive overall and has a markedly grayer foreneck; it usually shows a vague crest. Highland is an inconspicuous elaenia, more forest- or woodland-based than most of its congeners. Call a fast "burrr" or "burrreep."

CAMPO SUIRIRI
SUIRIRI-CINZA

YELLOW-BELLIED ELAENIA
MARIA-É-DIA

LARGE ELAENIA
GUARACAVA-GRANDE

HIGHLAND ELAENIA
TUCÃO

LESSER ELAENIA *Elaenia chiriquensis* 13.5-14 cm | 5.25- 5.5"
Fairly common and widespread in cerrado and shrubby areas. Usually shows a *slight crest, exposing some white.* Grayish olive above with narrow whitish eye-ring; wings duskier with two or three whitish bars; belly pale yellowish. *A difficult elaenia, easily confused with other species*: non-vocalizing birds often cannot be identified with certainty. Plain-crested Elaenia, often found with Lesser, lacks white in crest; but remember that Lesser's white isn't always visible. Yellow-bellied Elaenia is slightly larger, with a fuller crest; it is more likely to be found in disturbed habitats than the Lesser. Cf. also Small-billed Elaenia. Voice is often the best clue. Lesser's distinctive song is a burry, bisyllabic "chíbur" or "jwebu."

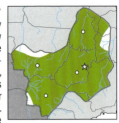

PLAIN-CRESTED ELAENIA *Elaenia cristata* 14.5 cm | 5.75"
Locally not uncommon but *strictly confined to cerrado*; not found in Pantanal. Usually appears *crested*, but *never shows white in crown.* Very dull. Olive brown above, wings duskier with two whitish bars. Breast olive gray; belly pale yellow. Lesser Elaenia often occurs with Plain-crested but has white in its crown (check carefully, as it is often inconspicuous); voice differs. Behavior as in other open-country elaenias. Plain-crested seems closely tied to unmodified habitats, such that it is much less wide-ranging than many of its congeners. It also seems less vocal, with distinctive song a fast gravelly "jer-jéhjeh" or "jer, jujujujuju." Also gives a repeated "wheeu," and a faster series of raspy "dzeeu" notes sometimes given by two birds at once.

SMALL-BILLED ELAENIA *Elaenia parvirostris* 14.5 cm | 5.75"
Uncommon transient and (perhaps) austral winter visitant to woodland, forest borders, and clearings; may breed in S Mato Grosso do Sul, but mostly does so further south. Olive above with usually visible *narrow white crown stripe*. *Lacking a crest, it looks round-headed.* Distinct round white eye-ring; wings duskier with two or three whitish bars. *Throat and breast pale gray,* midbelly whitish. Small-billed's stubby bill is *not* a field character. The similar White-crested Elaenia is dingier (notably so on forehead), has a less obvious eye-ring, and it never seems to have third wing-bar. Olivaceous Elaenia *lacks* white crown stripe, has more uniform underparts. Cf. also Lesser Elaenia. Calls include a sharp "chu" or "cheeu." Distinctive song, apparently given only when breeding, a "weedable-wee" phrase repeated steadily at 3-second intervals.

Olivaceous Elaenia (*E. mesoleuca*) is uncommon and somewhat local in forest patches and woodland in SE of our area, where it seems to be a nonbreeding visitant. Closely similar to Small-billed Elaenia, Olivaceous is drabber and more uniform, with a more olive breast and a less obvious eye-ring. Voice is the best clue, with song a harsh, fast "whik, whikiur" which sometimes is heard here even though it is not known to breed.

WHITE-CRESTED ELAENIA *Elaenia albiceps* 14.5-15 cm | 5.75-6"
Rare to uncommon (perhaps overlooked) austral winter visitant to woodland borders and shrubby areas. Only slightly crested. Dark olive above with white crown stripe and whitish eye-ring; wings duskier with two or three prominent whitish bars. Breast pale gray, belly whitish. The very similar and apparently more numerous Small-billed Elaenia is a cleaner-cut bird that has a *distinct* eye-ring. Many individuals cannot be identified to species, especially as they vocalize so rarely while here. Lesser Elaenia is a more typical open-country, cerrado inhabitant, showing more of a crest and with a very different voice. White-crested's behavior is similar to Small-billed Elaenia. Breeds in S South America. Very vocal when breeding, giving a burry "feeur" or "feeo," but seems generally silent in our area.

SOUTHERN SCRUB FLYCATCHER *Sublegatus modestus* 14 cm | 5.5"
Uncommon in scrub, lighter woodland, and cerrado, seeming to vacate our area during austral winter. *Short all-black bill.* Olive brown above with *narrow whitish supraloral,* wings duskier with *two bold whitish bars.* Throat and breast pale gray, contrasting with pale yellow belly. *Elaenia* elaenias have longer bills with a distinctly pale lower mandible; they also show less contrast on underparts, have no pale supraloral. Quiet and unobtrusive, scrub flycatchers perch upright and usually remain within cover. They sally into the air and to foliage, and also eat small fruits, sometimes accompanying elaenias and other flycatchers while doing so. Song a weak, high-pitched "pseeu" repeated at 1-second intervals.

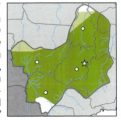

LESSER ELAENIA
CHIBUM

PLAIN-CRESTED ELAENIA
GUARACAVA-DE-PENACHO

SMALL-BILLED ELAENIA
GUARACAVA-DE-BICO-CURTO

WHITE-CRESTED ELAENIA
GUARACAVA-DE-CRISTA-BRANCA

SOUTHERN SCRUB FLYCATCHER
SERTANEJO

MYIOPAGIS elaenias are *very obscure* forest flycatchers that range in woodland and forest. Smaller and more slender than true Elaenia, they usually keep their *crown patches concealed*. They often are best known for their *distinctive voices*.

GRAY ELAENIA *Myiopagis caniceps* 12-12.5 cm | 4.75-5"

Uncommon and local in canopy and borders of more humid forest and woodland. ♂ *gray above* with white crown stripe; wings black with *two bold white bars and edging*. Throat and breast pale gray, belly grayish white. ♀ bright olive above, head grayer with a pale yellowish crown streak; wings black with *two bold pale yellow bars and edging*; below pale greenish yellow, brightest on belly. Young ♂♂ are more olive-backed. Forest Elaenia is duller overall, with blurry breast streaking. ♂'s color pattern recalls certain ♂ becards but the elaenia is smaller and more slender. Usually in pairs, often accompanying mixed flocks; gleans in foliage, holding tail partially cocked. Rarely comes low, even at edges. Rhythmic song a fast, shrill chipper that descends and fades toward the end, introduced by several sharper notes: "swee swee swee wee-ee-ee-ee-ee-ee-ee-ee."

FOREST ELAENIA *Myiopagis gaimardii* 12.5 cm | 5"

Fairly common in canopy and borders of forest and woodland. Head brownish gray, crown darker with indistinct whitish superciliary and *white crown stripe*; brownish olive above, wings duskier with two pale yellowish bars. Throat whitish, breast pale yellow *mottled and streaked olive*, belly pale yellow. *Best known by its distinctive voice*. No vaguely similar flycatcher shows the blurry breast streaking. Cf. Mouse-colored and Southern Beardless Tyrannulets, both with very different voices. Generally ranges high in trees and thus easier to see in secondary woodland and edge, where canopy is lower. Often forages with mixed flocks, gleaning in foliage and slightly cocking tail. Oft-heard song a distinctive sharp and emphatic "ch-veét," sometimes varied to a fast "cheewi chi-chi-chi."

GREENISH ELAENIA *Myiopagis viridicata* 13.5 cm | 5.25"

Fairly common in lower growth, midlevels, and borders of deciduous forest and woodland. Crown grayish with yellow crown stripe; *short superciliary and eye-ring white*; olive above, wings duskier with *dull yellowish edging* (but *no* bars). Throat pale grayish, breast grayish olive, belly pale yellow. Forest Elaenia has obvious wing-bars (not just edging). Pale-bellied Tyrant-Manakin is similar but has a plain facial area, no wing-edging, etc. Greenish Elaenia is a quiet, unobtrusive bird, that often is overlooked unless it is vocalizing. Usually found singly, perching quite upright, generally not with mixed flocks. Most frequent call a high-pitched, strident, and burry "cheerip" or "cheeyree" or "zrreeeeer."

PHYLLOMYIAS tyrannulets are small, *short-billed*, forest-canopy flycatchers that are best located through their *characteristic voices*.

PLANALTO TYRANNULET *Phyllomyias fasciatus* 11-11.5 cm | 4.25-4.5"

Uncommon in canopy and borders of forest and woodland, but absent from actual Pantanal. *Brownish olive above*, darker and grayer on crown, with short whitish superciliary and eye-ring; wings duskier with *two dull olive to whitish bars*. Throat whitish; below yellow, breast clouded olive. Dull-plumaged like so many small tyrannids, the *Planalto is best recognized through its distinctive voice*. Cf. the rarer Reiser's Tyrannulet. Mottle-cheeked Tyrannulet has a longer bill and a longer, often-cocked tail. Planalto is arboreal, remaining high in trees; it regularly forages with mixed flocks. Ultra-distinctive call a soft but far-carrying "pee, puu, puuit?"

White-lored Tyrannulet (*Ornithion inerme*) occurs in forest and borders along NW edge of our area. Very small, with a prominent short white superciliary and eye-ring ("spectacles") and two rows of white spots on wing-coverts.

REISER'S TYRANNULET *Phyllomyias reiseri* 11.5 cm | 4.5"

◢ Uncommon and local in canopy of deciduous and gallery forest. Lower mandible yellowish at base. Quite long-tailed. *Bright olive above* with grayer crown, yellowish short superciliary and eye-ring; wings duskier with *two bold whitish bars* and edging. *Below clear yellow*, breast clouded olive. Planalto Tyrannulet is duller olive above with less well-marked wing-bars; tail shorter; very different voice. Gleans in forest and woodland canopy, regularly with small mixed flocks. Song a rough, descending "bzuu, bzi-bzi-bzi-bze-bzu."

GRAY ELAENIA
MARIA-DA-COPA

FOREST ELAENIA
MARIA-PECHIM

GREENISH ELAENIA
MARIA-DE-CRISTA-DOURADA

PLANALTO TYRANNULET
PIOLHINHO

REISER'S TYRANNULET
PIOLHINHO-DO-GROTÃO

MOUSE-COLORED TYRANNULET *Phaeomyias murina* 12 cm | 4.75"
Fairly common and widespread in cerrado, scrub, light woodland, and gardens; in far N of our area also in riparian growth. *Brownish olive* with a *weak whitish superciliary*; wings duskier with two buffyish bars. Breast dull olive grayish, belly pale yellow. *Drab* and inconspicuous, *usually identified by voice*. Perkier-looking Southern Beardless Tyrannulet is smaller with expressive crest, usually cocked tail, and *no* superciliary. Plain Inezia is smaller with a notably grayer head, whiter wing-bars. Southern Scrub Flycatcher is larger, with more contrast below and an all-black bill. Pairs glean insects in dense foliage, and are rarely in the open for long. Distinctive dry, gravelly song a fast chattering or jumbled "jejejejejéjew" or "jejejejéjéw," sometimes shortened to just "ji-jéw."

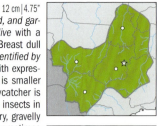

SOUTHERN BEARDLESS TYRANNULET *Camptostoma obsoletum* 9.5 cm | 3.75"
Common and widespread in semiopen and edge habitats. *One of our most frequently seen small tyrannids*. Jaunty, bushy crest, perky mannerisms, and confiding nature are more characteristic than its actual appearance. Bill short, lower mandible pale at least at base. *Olive grayish above*, grayest on crown; wings duskier with *two ochraceous to cinnamon bars*. Throat and breast pale grayish, belly yellowish white. *The short tail is usually held cocked*. Forages in foliage at all levels, singly or in mixed flocks. Very vocal. Most frequent call a slightly husky "freee?" or "weeeé?" and a more musical descending "kleeu, klee--klee-klee-klee-klee."

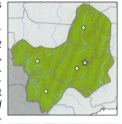

PLAIN INEZIA *Inezia inornata* 10 cm | 4"
Uncommon in woodland, scrub, and forest borders in and near Pantanal; likely only an austral migrant to our area, breeding further south. Formerly called Plain Tyrannulet. *Head gray* with short white superciliary; *base of lower mandible yellowish*; back olive, wings dusky with two white bars. *Below whitish*, belly tinged pale yellow. Similar White-crested Tyrannulet has an all-black bill, crown with white patch and black streaking. Gleans in foliage at varying heights, often partially cocking its tail. Quite vocal. Most frequent calls a fast, high-pitched trill: "pseee, tee-ee-ee-ee-ee," a descending "pseeeeu."
Amazonian Inezia (*I. subflava*) occurs in shrubby riparian growth along Rio Araguaia, e.g., Ilha do Bananal. White superciliary and "spectacles," two prominent white wing-bars, tail feathers tipped and edged whitish.

SERPOPHAGA are small arboreal tyrannulets with *obvious white crown patches*; the genus also includes a drab waterside specialist.

WHITE-CRESTED TYRANNULET *Serpophaga subcristata* 11 cm | 4.25"
Uncommon in forest borders, gallery woodland, shrubbery, and groves; *likely only an austral migrant to our area*, but status uncertain. Grayish olive above, grayer on head and neck with short white superciliary, *long blackish crown feathers* and *semiconcealed white crown patch* (both usually visible); wings and tail blackish, with two bold whitish bars. Breast pale gray, *belly pale yellow*. White-bellied Tyrannulet is grayer above, all whitish below; Plain Inezia lacks the black and white in its smooth gray crown. Forages actively at varying levels, often in open. Song a soft chippering trill introduced by one or more rising notes: "psee? psee psee-ee-ee-ee-ee" or "chit-chit-chit-trrrreeeeu." Call a fast, rhythmic "cheedidireep" or "cheedeedit" or "chit, chu-rít-za."

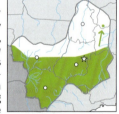

WHITE-BELLIED TYRANNULET *Serpophaga munda* 11 cm | 4.25"
Rare austral winter visitant to deciduous scrub and woodland in S Pantanal. Similar in behavior and voice to more numerous White-crested Tyrannulet, and perhaps thus overlooked. Grayer above; not as olivaceous as White-crested. *All white below (no pale yellow)*. Perhaps conspecific with White-crested.

SOOTY TYRANNULET *Serpophaga nigricans* 12 cm | 4.75"
Uncommon in shrubby and semiopen areas, *strictly near water* (usually where flowing) in S of our area. Unmistakable; *dark and drab*. Brownish gray above, paler and purer gray below; *tail contrastingly black*. Found in pairs that hop at the water's edge, often perching on rocks or branches just above water. The tail is often partially spread or jerked upward. Quiet, but sometimes gives a repeated "ch-víyt, ch-víyt...."
River Tyrannulet (*S. hypoleuca*) occurs in early-succession growth and riparian shrubbery along Rio Araguaia, e.g., Ilha do Bananal. Drab and rather long-tailed; grayish brown above with blackish feathers in crown, whitish below.

MOUSE-COLORED TYRANNULET
BAGAGEIRO

SOUTHERN BEARDLESS TYRANNULET
RISADINHA

PLAIN INEZIA
ALEGRINHO-DO-CHACO

WHITE-CRESTED TYRANNULET
ALEGRINHO

WHITE-BELLIED TYRANNULET
ALEGRINHO-DE-BARRIGA-BRANCA

SOOTY TYRANNULET
JOÃO-POBRE

SHARP-TAILED GRASS TYRANT *Culicivora caudacuta* 10.5 cm | 4.25"

vu Uncommon and local in tall grass of less-disturbed campos and grassy cerrado. Tail very long and narrow, feathers pointed and frayed. Buffy brown above broadly streaked blackish; long and conspicuous superciliary white. Whitish below, sides and flanks cinnamon-buff. Virtually unmistakable, though vaguely recalls a spinetail. Found singly and in pairs, moving through stands of tall grass; once located, often quite tame. Inconspicuous, and likely often overlooked, but seems to require extensive tracts of high-quality grasslands not burned for at least several years. Call a weak and somewhat nasal "wree? wree? wree?" (up to a dozen or more "wree" notes).

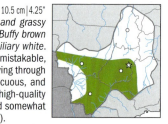

EUSCARTHMUS pygmy-tyrants are small dissimilar *brownish* tyrannids found in shrubby habitats. One is rare, the other numerous.

RUFOUS-SIDED PYGMY-TYRANT *Euscarthmus rufomarginatus* 11 cm | 4.25"

▲ Rare in shrubby cerrado; overall numbers depleted by habitat destruction and disturbance. Brown above; short buffy whitish supraloral, rufous crown patch. Wings and *long slender tail* duskier, *wings with two ochraceous bars* and edging. Throat white, below pale yellow, sides and flanks ochraceous. Similar ♀ Bearded Tachuri favors grassier cerrado, is shorter-tailed and more uniform buffyish below, lacks contrasting white throat and rufescence on flanks. Found singly or in pairs; generally inconspicuous except when vocalizing. Song, given with throat puffed out, a rapidly repeated chattered "cht-cht-cht-chiririréet, cht-cht-cht-chiririréet...."

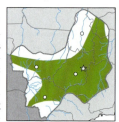

TAWNY-CROWNED PYGMY-TYRANT *Euscarthmus meloryphus* 10 cm | 4"

Fairly common and widespread in scrub, shrubby clearings, and undergrowth and borders of deciduous woodland. Brown above with inconspicuous rufous crown patch; wings with two faint rufescent bars. Whitish below, belly yellower. *A small, plain, brownish tyrannid that resembles no other in its "thickety" habitat*; cf. Fuscous Flycatcher (larger, etc.). An inconspicuous bird, remaining in heavy cover and usually located by its frequent vocalizing. Characteristic, often endlessly repeated call a sharp staccato: "plee-tirik" or "plee-ti-re-tik." Song a fast "tr-tr-tr-tr-tr-tr-treétrrt...."

POLYSTICTUS are small *dissimilar* tyrannids found locally in grasslands. Both are scarce.

GRAY-BACKED TACHURI *Polystictus superciliaris* 9.5 cm | 3.75"

🇧🇷 Rare and very local in shrubby grasslands on a few rocky serras in Minas Gerais, e.g., Serra da Canastra and Serra do Cipó. Brownish gray above, grayer on head with *short white superciliary* and a semiconcealed white crown patch. Two faint pale brownish wing-bars and edging. *Below pinkish buff*, midbelly white. An often tame tyrannid that clings to grass stems and perches in low shrubs. Found singly or in pairs; distinctive in its very limited range and habitat. Call a distinctive, throaty "tchudi" repeated at several-second intervals; apparent song a strange, reeling, buzzy trill: "tzzzzzzzzzzzzzz," lasting about 5 seconds.

BEARDED TACHURI *Polystictus pectoralis* 10 cm | 4"

vu Uncommon and local in tall grass of little-disturbed campos and grassy cerrado. Numbers are now greatly reduced by over-grazing, too-frequent burning, and the loss of so much habitat to agriculture. Tiny and slender. ♂ brown above, crown gray and blackish with white coronal patch; lower face and upper throat ("beard") finely streaked black and white; wings duskier with two cinnamon-buff bars. Below whitish with cinnamon wash on breast and flanks. ♀ has a brown crown and lacks the "beard." Rufous-sided Pygmy-Tyrant differs from ♀ in its longer and narrower tail and more contrasting white throat. Found singly (less often in pairs), occasionally with other grassland birds; clings to tall grass stems, flying only weakly. Generally quiet, but displaying ♂♂ give a plaintive "wheee? whidididrrr" song, sometimes in brief low flight.

CRESTED DORADITO *Pseudocolopteryx sclateri* 11 cm | 4.25"

Rare and local in marshes and adjacent shrubbery in Pantanal. Status in our area uncertain. Olive above mottled dusky on back; face and crest blackish, the latter often parted to show a yellowish white stripe. Wings dusky with whitish bars and edging. Below bright yellow. Generally remains in grassy cover, but can perch in the open soon after dawn. Presence seems erratic and is perhaps dependent on local water levels. Infrequent song a high-pitched "tsit-tsit-tsi-tit," sometimes given while jumping into air.

SHARP-TAILED GRASS TYRANT
PAPA-MOSCA-DO-CAMPO

RUFOUS-SIDED PYGMY-TYRANT
MARIA-CORRUÍRA

TAWNY-CROWNED PYGMY-TYRANT
BARULHENTO

GRAY-BACKED TACHURI
TRICOLINO-DE-DORSO-CINZA

BEARDED TACHURI
TRICOLINO-CANELA

CRESTED DORADITO
TRICOLINO

HEMITRICCUS tody-tyrants are *inconspicuous* small tyrannids found in wooded or scrubby habitats. Like *Todirostrum* they have *rather broad spatulate bills* but they tend to *perch upright*.

PEARLY-VENTED TODY-TYRANT *Hemitriccus margaritaceiventer* 10 cm | 4"
Locally common in scrub and lower growth of deciduous woodland and gallery forest. Iris yellow whitish. Olive above with grayer crown, whitish lores and eye-ring; wings duskier with two bold yellowish bars and edging. White below, *throat and foreneck gray-streaked.* Stripe-necked Tody-Tyrant is bright yellow below, has plain wings. Rusty-fronted Tody-Flycatcher has obvious buff on face. All three species can occur together. A drab but cute little flycatcher that ranges singly or in pairs, sometimes with flocks but at least as often alone. Perches vertically, flitting upward to underside of leaves. Call a staccato "tik" or "stik" in series. Song several notes followed by a descending trill: "tik, tik, tr-r-r-r-r-r-r-r."

STRIPE-NECKED TODY-TYRANT *Hemitriccus striaticollis* 11 cm | 4.25"
Fairly common in shrubbery, low woodland, and cerrado edge. Iris yellow. Olive above, *crown browner; white lores and eye-ring; wings plain.* Throat white sharply streaked blackish; below bright yellow, olive streaking on breast. Spotted Tody-Flycatcher has gray crown contrasting with olive back, *no* eye-ring, prominent wing-bars and edging. Found singly or in pairs; inconspicuous, but vocalizes a lot. Call a simple, inflected "kweep," sometimes doubled. Song a fast "pit-pit-pit-pit, whi-didit."

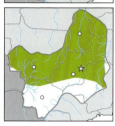

TODIROSTRUM tody-flycatchers also have *long, broad, spatulate bills*, but are *more arboreal* with a *more horizontal posture* than most *Hemitriccus*. They are generally easier to see.

SPOTTED TODY-FLYCATCHER *Todirostrum maculatum* 10 cm | 4"
Locally common in riparian woodland and shrubby areas in N of our area, e.g., Ilha do Bananal. Iris orange-yellow. *Head gray* with whitish supraloral; above olive, wings dusky with *two yellow bars and edging.* Throat white, foreneck with narrow blackish streaks, below yellow. Stripe-necked Tody-Tyrant is duller overall with plain wings, blurrier streaking, less gray on head. Common Tody-Flycatcher shows no streaking, is more often close to ground. Spotted gleans at varying heights and, even though it vocalizes frequently, is a less conspicuous bird than Common. Song a short series of sharp, loud "peek" notes, sometimes a pair's syncopated duet: "pik-peek, pik-peek, pik-peek...."

✓ **COMMON TODY-FLYCATCHER** *Todirostrum cinereum* 9.5 cm | 3.75"
Widespread and often common in a variety of secondary and semiopen habitats as well as gardens. Iris pale yellow. *Forecrown and face black,* shading to gray on nape and olive on back; wings blackish edged yellow; longish, graduated tail black with outer feathers whitish. Below bright yellow. Yellow-lored Tody-Flycatcher occurs only along SE edge of our area, has conspicuous yellow lores. Easy to see, often foraging in the open, from eye-level up into low trees. Behavior animated, often cocking and wagging its tail; sometimes stands high on its legs. Usually in pairs, generally not with mixed flocks. Quite vocal, both sexes giving quick, cricket-like trills and various "tik" notes.

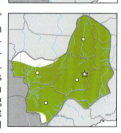

YELLOW-LORED TODY-FLYCATCHER *Todirostrum poliocephalum* 9-9.5 cm | 3.75-4"
🇧🇷 *Fairly common in borders of humid forest and woodland in far SE of our area*, e.g., Serra da Canastra. Iris orange-yellow. Resembles and occurs with more wide-ranging Common Tody-Flycatcher, but has a *conspicuous yellow supraloral spot*, more olive (less gray) back; its olive tail lacks white edging. Common Tody-Flycatcher in SE shows at most a *minute* yellow spot near bill. Behavior similar, though Yellow-lored is a more forest-based bird. Song has a similar quality, though is typically shorter and less run together, a sharp "cheep, chip-chip."

RUSTY-FRONTED TODY-FLYCATCHER *Poecilotriccus latirostris* 9.5 cm | 3.75"
Locally fairly common in dense undergrowth of shrubby clearings, forest and woodland borders, and gallery forest. Crown and neck gray with *prominent buff lores and facial area*; olive above, wings dusky with *two ochraceous wing-bars*, olive edging. *Below dull grayish white.* Pearly-vented Tody-Tyrant has white lores and spectacles, streaking on foreneck. Inconspicuous and generally not with flocks, making upward strikes to leaf undersides. Pairs stay in contact through their soft calling with distinctive call a sharp, low-pitched, rattled "tik, trrrr" or "tik, trrrr, trrrr," sometimes just a "tik."

PEARLY-VENTED TODY-TYRANT
OLHO-DE-OURO

STRIPE-NECKED TODY-TYRANT
SEBINHO-RAJADO

SPOTTED TODY-FLYCATCHER
FERREIRINHO-ESTRIADO

COMMON TODY-FLYCATCHER
RELÓGIO

YELLOW-LORED TODY-FLYCATCHER
TEQUE-TEQUE

RUSTY-FRONTED TODY-FLYCATCHER
FERREIRINHO-DE-CARA-PARDA

YELLOW TYRANNULET *Capsiempis flaveola* 11.5 cm | 4.5"

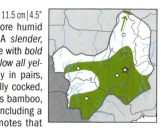

Uncommon and rather local in lower and middle growth of more humid woodland and forest borders; seems absent from Pantanal. A *slender, long-tailed, predominantly yellow tyrannulet*. Yellowish olive above with *bold yellow superciliary*, wings duskier with *two broad yellow bars. Below all yellow*. No other tyrannulet looks so yellow overall. Ranges mainly in pairs, gleaning actively. Usually perches horizontally with its tail partially cocked, but can perch more vertically when at rest. In many regions favors bamboo, but less so in our area. Rather vocal, with a variety of soft calls including a short dry trill: "tr-r-r-r-r." Song a pleasant rollicking series of notes that often lacks discernible pattern, starting slowly but speeding up quickly.

MOTTLE-CHEEKED TYRANNULET *Phylloscartes ventralis* 12 cm | 4.75"

Uncommon and local in canopy and borders of forest and woodland in SE of our area, including Serra da Canastra. Rather long bill. Olive above with *whitish supraloral and partial eye-ring*, dusky line through eye; indistinct mottling on ear-coverts; wings duskier with two pale yellow bars. *Mostly pale yellow below*, throat whiter and chest more olive. Long slender bill, horizontal posture, and long cocked tail aid in identifying this species, numerous elsewhere in its range though less so in our region. Planalto Tyrannulet has a shorter bill; its tail is shorter, and usually held *not* cocked; it has a very different voice. Found singly or in pairs, regularly with mixed flocks. Forages actively, often in the open. Song a somewhat musical "chi-didididit," also has an oft-repeated "whik" note.

🇧🇷 **Minas Gerais Tyrannulet** (*P. roquettei*) is rare and local in canopy and borders of gallery forest in N Minas
CR Gerais. Olive above with a fairly bold yellowish eye-ring; wings with two bold yellowish bars. Below pale clear yellow. Ranges in pairs, perching horizontally with long tail held cocked. Declining due to deforestation.

SOUTHERN BRISTLE TYRANT *Pogonotriccus eximius* 11 cm | 4.75"

Rare and local in lower and middle growth of forest and borders in SE of our area, e.g., Serra da Canastra. Crown gray, *broad superciliary white grizzled with gray, lower face yellow with conspicuous black crescent on ear-coverts*. Above bright olive; wings duskier, olive edged. Throat whitish, breast yellowish olive, belly bright yellow. *The fancy head pattern is unique*. Found singly or in pairs, perching vertically, often motionless for long periods, then sallying to foliage. Occasionally follows mixed flocks. Often lifts a wing up over its back. Call a dry trill that accelerates, then ends abruptly.

SEPIA-CAPPED FLYCATCHER *Leptopogon amaurocephalus* 14 cm | 5.5"

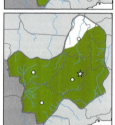

Locally fairly common in lower growth of forest and woodland. *Crown dark brown, facial area dull buff grizzled dusky; blackish patch on ear-coverts*. Olive above, tail more brownish; wings dusky with *two broad buff bars*, yellowish edging. Throat pale grayish, breast dull olive, belly pale yellow. Found singly or in pairs, frequently accompanying mixed flocks of understory birds. Perches erectly, often on open branches and thus easy to see. Sallies out to pick insects from leaf surfaces, less often from twigs or branches. Occasionally lifts a wing up over its back. Call a fast, sputtering chatter that trails off toward the end: "skeúw-k'k'k'k'k'k'kew."

GRAY-HOODED FLYCATCHER *Mionectes rufiventris* 13.5 cm | 5.25"

Uncommon in lower growth of forest and woodland in SE of our area. *Hood gray*, otherwise brownish olive above with mostly plain wings. Breast brownish olive, *belly rich ochraceous*. Unobtrusive and quiet, with quick darting movements; usually ranges singly inside forest and woodland, only rarely coming to edge. Sometimes accompanies mixed flocks or joins aggregations of other small birds at fruiting trees. Frequently flashes a wing up over its back; also ruffles its crown feathers. Displaying ♂'s song, a series of rough nasal notes that starts slowly but accelerates before stopping abruptly.

Ochre-bellied Flycatcher (*M. oleagineus*) occurs in forest borders and woodland in NW of our area (Serra das Araras). Resembles a pallid Gray-hooded Flycatcher but is entirely brownish olive above with some rufous on wings, paler on foreneck, not as rich on belly.

YELLOW TYRANNULET
MARIA-AMARELINHA

MOTTLE-CHEEKED TYRANNULET
BORBOLETINHA-DO-MATO

SOUTHERN BRISTLE TYRANT
BARBUDINHO

SEPIA-CAPPED FLYCATCHER
CABEÇUDO

GRAY-HOODED FLYCATCHER
ABRE-ASA-DE-CABEÇA-CINZA

TOLMOMYIAS flatbills (formerly "flycatchers") are obscure olive and yellow flycatchers of woodland and lighter forest. They have *notably wide flat bills*, but are otherwise confusing.

YELLOW-OLIVE FLATBILL *Tolmomyias sulphurescens*   13.5 cm | 5.5"
Locally common in lower growth and borders of deciduous forest and gallery woodland. *Wide flat bill*. Olive above with *gray crown and nape*, whitish supraloral, dark patch on ear-coverts; two prominent yellowish wing-bars. Throat pale grayish, breast and flanks pale olive, belly pale yellow. Usually perches upright, not cocking tail; sallies briefly to leaves for insects. Found singly or in pairs. Generally easy to see, and often with mixed flocks. Song a series of thin, well-enunciated notes: "swit-swit-swit-swit" or "dzeeyp, dzeeyp, dzeeyp."

Zimmer's Flatbill (*T. assimilis*) occurs in midlevels and canopy of forest along NW edge of our area, e.g., Serra das Araras. Closely resembles Yellow-olive Flatbill but has a darker iris, less gray on crown, and a small pale patch at base of primaries. The two species mostly do not occur together, Yellow-olive favoring drier habitats. Song a very different and shrill "shreeuw... shreeeu... shreee?" (sometimes up to 5-6 notes).

OCHRE-LORED FLATBILL *Tolmomyias flaviventris*   12 cm | 4.75"
Fairly common in open and gallery woodland and gardens in *N of our area*. *Wide flat bill. Yellowish olive above*, wings blackish with two prominent yellowish bars. *Yellow below*. Gives a *very yellow impression overall*. Yellow-olive Flatbill shows more of a facial pattern, has gray on crown. Their behavior is similar but Ochre-lored is more arboreal; it tends to follow flocks less. Song a loud, shrill "shreeeép," usually given at intervals of several seconds or more, sometimes in a rising series of notes.

YELLOW-RUMPED MYIOBIUS *Myiobius mastacalis*   12.5 cm | 5"
Uncommon and local in lower growth of forest and woodland in S Goiás (probably Minas Gerais as well). Often considered conspecific with *M. barbatus* of Amazonia. Unusually long rictal bristles. Above olive (♂ has yellow crown patch, usually concealed); *conspicuous yellow rump and black tail*. Throat whitish, *breast tawny*, belly yellow. Usually remains inside forest where often seen with understory flocks, foraging actively, pirouetting and fanning its tail, and drooping its wings. Quiet, but occasionally gives a sharp "psik."

BLACK-TAILED MYIOBIUS *Myiobius atricaudus*   12.5 cm | 5"
Uncommon and apparently local in lower growth of secondary woodland and forest borders, most often near water, in NE of our area. Similar to Yellow-rumped Myiobius but more uniform yellowish below, with little or no buff tinge. Yellow-rumped has a tawny breast. Behavior similar; ecological relationships of the two species in our area are still somewhat unclear.

WHITE-THROATED SPADEBILL *Platyrinchus mystaceus*   10 cm | 4"
Uncommon in forest and woodland undergrowth; rare in Pantanal. *Very wide flat bill*, lower mandible pale. *Very small and stub-tailed*. Olive brown above with yellow crown patch (usually concealed) and *complex buffy yellowish-and-black facial pattern*. Throat white; buffyish below, midbelly pale yellowish. As the only spadebill in our area, should be easily recognized. Inconspicuous, perching quietly in understory and not joining mixed flocks. Remains in dense growth, with quick abrupt movements that are hard to follow. Most frequent call a sharp "squeep!" or "squik!" sometimes doubled. Distinctive song a fairly musical trill, first descending, then rising, ending with a "whik."

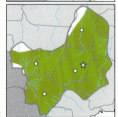

SOUTHERN ANTPIPIT *Corythopis delalandi*   14 cm | 5.5"
Uncommon on or near ground inside forest and woodland; seems absent from actual Pantanal. *Unmistakable*, with *distinctive pipit-like shape*. Brownish olive above, whitish around eye. Throat creamy white, *breast with bold black streaks (sometimes coalescing into an almost solid band)*, belly whitish, grayer on flanks and crissum. *Walks on the ground* with a mincing gait, often nodding head and pumping tail. Regularly perches on fallen logs and when feeding sallies up from the ground, snatching insects from underside of leaves. Also perches on low branches. Generally solitary, less often in pairs; not with mixed flocks. Often first noticed from its frequent and emphatic bill snapping. Distinctive whistled song a shrill "peee, peeur-pi-pi-peépit" ("three cheers for the peépit!").

YELLOW-OLIVE FLATBILL
BICO-CHATO-DE-ORELHA-PRETA

OCHRE-LORED FLATBILL
BICO-CHATO-AMARELO

YELLOW-RUMPED MYIOBIUS
ASSANHADINHO

BLACK-TAILED MYIOBIUS
ASSANHADINHO-DE-RABO-PRETO

WHITE-THROATED SPADEBILL
PATINHO

SOUTHERN ANTPIPIT
ESTALADOR

TROPICAL PEWEE *Contopus cinereus* 14-14.5 cm | 5.5-5.75"

Fairly common at forest edges and in secondary and lighter woodland, clearings, and plantations. Lower mandible yellowish. *Drab.* Grayish olive above with *whitish lores*, crown usually darker; wings dusky with two pale grayish bars. Throat whitish, breast and flanks olive grayish; midbelly whitish to pale yellowish. Usually found singly, *habitually perching upright on exposed branches in the semiopen*. Euler's Flycatcher almost always remains in shady lower growth. Sallies into air after insect prey, sometimes returning repeatedly to the same perch, quivering its tail upon alighting. Most frequent call a fast dry "sree-ip."

EULER'S FLYCATCHER *Lathrotriccus euleri* 13.5 cm | 5.25"

Locally common in lower growth inside forest, woodland, and borders. *Drab.* Lower mandible yellowish. *Olive brown above* with obscure whitish supraloral and obvious eye-ring; *two dull buff wing-bars* and edging; tail olive brown. Throat grayish white, breast brownish olive, belly whitish. Fuscous Flycatcher has an all-black bill, well-marked whitish superciliary. Bran-colored Flycatcher is more rufescent above, streaked below; it favors more open situations. Euler's is an inconspicuous bird, remaining in forest and woodland undergrowth where it perches upright and sallies out to foliage. Usually not with mixed flocks. Song a series of burry notes, "zhweé, zhwee-zhwee-zhwee;" also a simpler "zhwee-buu." Dawn song a less burry, *Elaenia*-like "beeu... beeu-wheéu."

Alder Flycatcher (*Empidonax alnorum*) is a very rare boreal migrant to shrubby areas in Pantanal (may be overlooked?). Resembles Euler's Flycatcher but brownish or grayish olive above (never as brown), and has whitish wing-bars (at most buff-tinged). Most likely to be identified through its song, given occasionally on the wintering grounds, a burry "free-breéo."

FUSCOUS FLYCATCHER *Cnemotriccus fuscatus* 14.5 cm | 5.75"

Fairly common and widespread in undergrowth of gallery and secondary woodland and forest borders. *Drab and inconspicuous*; quite long-tailed. Brown above with *whitish superciliary*; wings dusky with *two broad buff bars* and edging. Below whitish, breast washed brownish, belly pale yellow. Euler's Flycatcher (often with Fuscous) lacks the pale superciliary. Mouse-colored Tyrannulet has a shorter bill and very different behavior and voice. Fuscous favors shady, often swampy undergrowth and is a quiet, retiring bird that avoids open places. Usually in pairs, generally not with mixed flocks. Gives a fast gravelly "wor, jeér-jeér-jeér-jeér-jeér-jeér-jew," and a more piercing "wheeéeu" at long intervals.

BRAN-COLORED FLYCATCHER *Myiophobus fasciatus* 12.5 cm | 5"

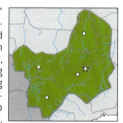

Fairly common and widespread in shrubby clearings, pastures, lighter woodland, and borders. *Reddish brown above* with a weak whitish supraloral and semiconcealed crown patch yellow (faint or absent in ♀); wings dusky with two broad buff bars. Whitish below, *breast and sides with brown streaking*, belly pale yellowish. Neither Euler's nor Fuscous Flycatcher shows streaking below, and neither is as rufescent above. Found singly or in pairs, perching upright on low branches, sometimes in the open but overall not a conspicuous bird (though more so than the Fuscous). Makes short sallies into air and to foliage in pursuit of insects. Does not accompany mixed flocks. Song a "wee-ub" or "wee-eb" note, sometimes given in a protracted series. Frequent call a fast "whee-heeheeheehee."

VERMILION FLYCATCHER *Pyrocephalus rubinus* 14.5-15 cm | 5.75-6"

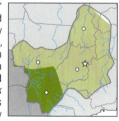

Locally fairly common in semiopen areas with scattered bushes and trees, regularly around houses and farm buildings. In our area probably occurs only during austral winter (but may breed in far S?). Unmistakable, *dazzling ♂* has *crown and underparts brilliant scarlet* with sooty blackish upperparts and narrow mask through eyes. ♀ ashy brown above, paler on forehead and sootier on wings and tail; *wings virtually plain*. Throat and breast white, *breast with dusky streaking*, and with *a little reddish pink on belly and crissum*. Bran-colored Flycatcher and various ♀ *Knipolegus* black tyrants are more rufescent above, have wing-bars, and are unlikely to perch in the open. Vermilion *usually perches in the open*, sallying to the ground and into the air for insects. In spectacular display flight, breeding ♂♂ hover while repeating a musical phrase: "pi-d'd'd'reeít." Displaying ♂♂ often perform during predawn hours.

TROPICAL PEWEE
PIUÍ-CINZA

EULER'S FLYCATCHER
ENFERRUJADO

FUSCOUS FLYCATCHER
GUARACAVUÇU

BRAN-COLORED FLYCATCHER
FILIPE

VERMILION FLYCATCHER
PRÍNCIPE

XOLMIS monjitas are attractive flycatchers that are conspicuous and often quite numerous in open country, and are *simply patterned in white, grays, and black*. They tend to be quiet, and usually perch atop bushes, low trees, fences, and wires, dropping to the ground after insects.

WHITE MONJITA *Xolmis irupero* 17-18 cm | 6.75-7"

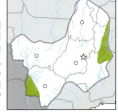

Conspicuous and often common. An *ethereal white* flycatcher of open and semiopen terrain (both natural and agricultural) with scattered bushes and trees in *S Pantanal* and *N Minas Gerais*. Its *pure white plumage* is relieved only by black primaries, primary coverts, and tail tip. ♀ tinged gray on back. E birds are slightly smaller with a wider black tail-band. This unmistakable and lovely bird perches fully in the open, occasionally dropping to the ground in pursuit of prey, sometimes after hovering. Usually quiet, but breeding ♂ gives a soft "preeeyp... tooit... preeeyp... tooit...."

WHITE-RUMPED MONJITA *Xolmis velatus* 19.5 cm | 7.75"

Fairly common in semiopen areas with scattered bushes and small trees. *Head white with pearly gray hindneck*; back gray with *contrasting white rump and basal half of tail*, outer half of tail black. Wings blackish with white edging and *white stripe along inner flight feathers* (latter obvious in flight). Below white. Gray Monjita is larger, grayer on head and breast, lacks white on rump, and has a black malar. Usually in pairs, perching conspicuously and often spreading its tail; quite approachable. Notably quiet.

GRAY MONJITA *Xolmis cinereus* 23 cm | 9"

Fairly common in cerrado and semiopen areas with scattered shrubs and low trees, agricultural areas, and sometimes even around airports. Iris red. *Ashy gray above* with a *broad white supraloral*; wings black with *white patch at base of primaries* (forming a *conspicuous white square in flight*), tail black tipped whitish. Throat white with *black malar streak*, breast gray, belly white. Not likely confused, though vaguely kingbird-like. Chalk-browed Mockingbird has very different shape and behavior. Conspicuous and distinctive, Gray Monjitas remain active even during the heat of day. They sally into the air and also sometimes run on the ground. Flight fast, direct, and graceful. A quiet bird, but breeding ♂♂ give a soft "peee, preeu."

STREAMER-TAILED TYRANT *Gubernetes yetapa* 38-40 cm | 15-16"

Uncommon and local in damp grassland and marshy terrain with shrubbery and low trees near streams. Seems absent from actual Pantanal. A *spectacular* flycatcher with *very long, graduated, deeply forked tail*; bill stout. *Above pale gray*, somewhat streaked forecrown and superciliary whitish; wings blackish with rufous at base of primaries (*showing as a conspicuous stripe in flight*). Throat white, sharply outlined by chestnut *pectoral collar*; breast and sides pale gray. ♀ slightly smaller, duller, and shorter-tailed. Unmistakable, but cf. Fork-tailed Flycatcher (which really is very different). Streamer-taileds are found in sedentary pairs that perch conspicuously atop bushes and trees. They sally out after insects, often for long distances, their long tails whipping around gracefully. Their loud calls attract attention even from long distances, the most notable being a boisterous "whee-irt!" or "wiirt!" Displaying pairs flare and flap their wings exuberantly while vocalizing in tandem.

CATTLE TYRANT *Machetornis rixosa* 19.5 cm | 7.75"

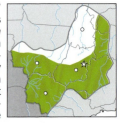

Widespread and generally common in semiopen and agricultural areas and on open ground near buildings. Long legs. Red iris. *Plain olive brown above*, grayer on crown and nape with a usually concealed orange crown patch; *tail narrowly tipped whitish*. Throat whitish, below bright yellow. Superficially kingbird-like, but otherwise very distinctive; *kingbirds are rarely on the ground*. Mainly terrestrial, but also perches on rooftops and in trees. Found in pairs or small loose groups, running about with an erect stance, often attending grazing domestic animals (sometimes also capybaras), even hitching rides on the animals' backs, dropping off to pursue disturbed prey. Often very tame around ranch buildings. Calls include a rising series of squeaky notes, reminiscent of Tropical Kingbird.

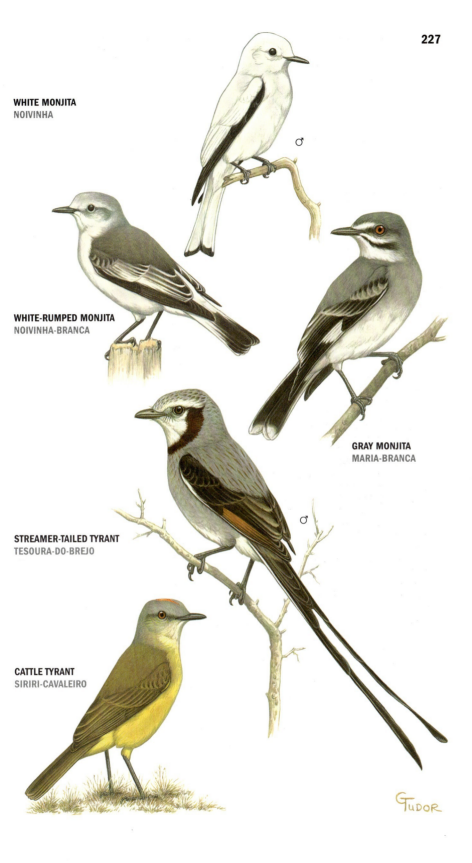

WHITE MONJITA
NOIVINHA

WHITE-RUMPED MONJITA
NOIVINHA-BRANCA

GRAY MONJITA
MARIA-BRANCA

STREAMER-TAILED TYRANT
TESOURA-DO-BREJO

CATTLE TYRANT
SIRIRI-CAVALEIRO

SPECTACLED TYRANT *Hymenops perspicillatus* 15.5 cm | 6"
Rare austral migrant to marshy areas in S Pantanal, breeding in S South America. Both sexes have yellow iris and obvious *yellow wattle around eye* (reduced in nonbreeding birds). Unmistakable ♂ *black* with *mainly white primaries* (tipped black), the white *very* conspicuous on flying birds (and usually visible on perched birds). ♀ streaked brown and dusky above with buff superciliary; wings with two buff bars and *mainly rufous primaries* (visible on perched birds; obvious in flight). Below whitish to pale buff, chest streaked dusky. Vaguely similar ♀ blackbirds, especially Unicolored, lack the ♀ tyrant's diagnostic rufous in wings. A conspicuous bird (especially ♂♂), usually solitary here. It perches low, dropping down to mud or grass in pursuit of insect prey, sometimes even running on the ground.

COCK-TAILED TYRANT *Alectrurus tricolor* 12 cm | 4.75," long-tailed breeding ♂ up to 18 cm | 7"
VU Threatened and now very local in less-disturbed campos and open grassy cerrado (grass must be *tall*, having not burned for several years), but given the right conditions can be locally common. With reduced grazing and less frequent burning, populations can increase. Not in Pantanal, but there are a few on fazendas around its periphery, e.g., Fazenda Rio Negro near Miranda. Unmistakable ♂ is *black above* with gray on rump, white shoulders and secondary edging. *Face and underparts white, with black patch on sides of chest.* Bizarre tail usually held cocked; its *inner pairs of feathers are lengthened, very broad, and held perpendicular to the body*, but usually look frayed. ♀ mottled brown above, buffier on superciliary, shoulders, and scapulars; short tail normally shaped. Whitish below with *brown smudge on sides of chest*. Full-plumaged breeding ♂♂ are spectacular and unique, but all are distinctive enough to be easily recognized. ♀♀ are plump, short-tailed, big-headed, brown-and-buff birds, not likely confused and really quite different from vaguely similar ♀ Bearded Tachuri and ♀ seedeaters. Cock-tailed Tyrants perch in tall grass, gleaning and sallying for insects; they sometimes accompany seedeater flocks. A displaying ♂ launches into air on fast fluttery wingbeats with the tail held so far forward over its back that it almost touches its head, but remains silent. One of the Cerrado's flagship birds.

FLUVICOLA water tyrants are attractively patterned *black-and-white* tyrannids that are *conspicuous near water*. Usually in pairs, they forage on or near the ground and on floating vegetation. Sometimes they flick and fan their slightly cocked tails.

BLACK-BACKED WATER TYRANT *Fluvicola albiventer* 14.5 cm | 5.75"
Uncommon to fairly common in marshes and adjacent shrubby vegetation; most numerous in Pantanal. Head and underparts white, contrasting with black rear crown, back, wings, and tail; rump band and wing-bars (variable in extent) also white. ♀ White-headed Marsh Tyrant is ashy brown above (not black) and lacks white on rump and wings; really not likely to be confused. Not very vocal, but occasionally gives a nasal "zree-zri-zri-zri" and a sharper "treeu!"

MASKED WATER TYRANT *Fluvicola nengeta* 14.5 cm | 5.75"
Fairly common in and around marshes, ponds, and nearby semiopen areas in E of our area. Mostly *white with black stripe through eye*; wings and tail black, the latter broadly tipped white; back pale brownish gray. Black-backed Water Tyrant is mainly black above with a white face, and *lacks* the black eyestripe. This attractive bird is often charmingly tame, and locally is known as the *lavadeira* (washerwoman). Distinctive call a sharp "kirt!" or "kirt-kirt," often given in flight.

WHITE-HEADED MARSH TYRANT *Arundinicola leucocephala* 13 cm | 5"
Fairly common in marshes and damp shrubby grassland; most numerous in Pantanal. Lower mandible yellow basally. Unmistakable ♂ slightly bushy-crested, *black with contrasting white head and throat*. ♀ ashy gray-brown above with white forecrown and blackish tail; whitish below, sides and flanks washed ashy brown. Black-backed Water Tyrant vaguely resembles ♀ but is much more sharply black and white. Conspicuous, perching erectly on grass stems, shrubs, or low trees and sallying for short distances into the air. Unlike the water tyrants, White-headed is only infrequently on the ground and does not fan or cock its tail. Notably quiet.

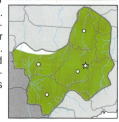

SPECTACLED TYRANT
VIUVINHA-DE-ÓCULOS

COCK-TAILED TYRANT
GALITO

BLACK-BACKED WATER TYRANT
LAVADEIRA-DE-CARA-BRANCA

MASKED WATER TYRANT
LAVADEIRA-MASCARADA

WHITE-HEADED MARSH TYRANT
FREIRINHA

KNIPOLEGUS black tyrants are quiet midsized flycatchers of woodland edges. The more conspicuous ♂ is black or blackish, often with a white wing-band showing in flight. ♀ is usually brownish, with streaking below.

CINEREOUS TYRANT *Knipolegus striaticeps* 13-13.5 cm | 5-5.25"
Rare austral winter visitant to deciduous woodland and borders (S Mato Grosso do Sul); breeds further south in Chaco. ♂ *dark gray with a red iris. Face and foreneck blacker;* wings blackish with two grayish bars, outer tail feathers whitish. *Belly paler gray.* ♀ brownish above, *rufous on crown and nape* and cinnamon on uppertail-coverts; wings dusky with two white bars, tail blackish with rufous inner webs. Whitish below with *fine dusky streaking.* Other *Knipolegus* ♂♂ are black. Compare ♀ to Bran-colored Flycatcher. Found singly, usually in the open; perches erectly, often twitching its tail.

BLUE-BILLED BLACK TYRANT *Knipolegus cyanirostris* 14.5-15 cm | 5.75-6"
Rare austral winter visitant to gallery woodland and borders in S Mato Grosso do Sul; also in N Minas Gerais (status uncertain). ♂ has pale blue bill, *iris red; uniform glossy black with no visible white on wing.* ♀ has iris orangey; rufous brown above, *most rufous on crown and rump;* wings blackish with two buff bars, tail feathers edged cinnamon-rufous. Whitish below *coarsely streaked dark olive brown,* crissum cinnamon. ♀ is so dark and heavily streaked below that confusion unlikely, but cf. Bran-colored Flycatcher. Found singly or in pairs, usually near the ground; inconspicuous and often confiding.

● **Velvety Black Tyrant** (*K. nigerrimus*) is local in woodland borders of NE. ♂ blue-black with white wing-band. ♀ duller with *throat streaked chestnut.*

SÃO FRANCISCO BLACK TYRANT *Knipolegus franciscanus* 16.5 cm | 6.5"
● Rare and local in deciduous woodland, mostly around large limestone outcroppings, in N Minas Gerais and adjacent Goiás. ♂ shiny black with a white band at base of primaries (conspicuous only in flight). ♀ brownish gray above with indistinct *whitish wing-bars;* whitish below with dusky breast streaking. No similar black tyrant in range. Forages in pairs in the subcanopy, sallying to branches and leaves, occasionally into the air. Sometimes accompanies loose mixed flocks. Cf. Bran-colored Flycatcher (very different behavior).

White-winged Black Tyrant (*K. aterrimus*) is a very rare austral winter visitant to extreme S. ♂ like ♂ São Francisco. ♀ *rufous rump and on tail* and is unstreaked buff below.

CRESTED BLACK TYRANT *Knipolegus lophotes* 20.5-21 cm | 8-8.25"
▲ Uncommon in open *grassy and shrubby areas in cerrado and woodland borders;* not in Pantanal. A large, *prominently crested* black tyrant. Bill black. Sexes alike. Glossy blue-black with white band on base of primaries only obvious in flight. The smaller Velvety Black Tyrant has a much smaller crest, bluish bill. Occurs in well-dispersed pairs, perching conspicuously atop shrubs and low trees, sometimes even on wires. Sallies for insects, and eats some fruit. Like the other *Knipolegus,* usually quiet.

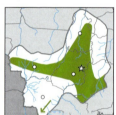

RIVERSIDE TYRANT *Knipolegus orenocensis* 15 cm | 6"
Uncommon and local in shrubby growth on river islands in N of our area (Rio Araguaia). Bill blue-gray tipped black. ♂ *uniform dark slaty.* ♀ very similar but slightly paler and more olive. Amazonian Black Tyrant is smaller and favors undergrowth; ♂ glossier blue-black, ♀ browner above with rufous in tail, definite wing-bars. Pairs generally remain within cover but regularly will perch in the open soon after dawn. Displaying ♂ mounts up into air, returning to its perch accompanied by a mechanical snap.

AMAZONIAN BLACK TYRANT *Knipolegus poecilocercus* 14 cm | 5.5"
Uncommon and local in undergrowth of seasonally flooded woodland in far N of our area (Rio Araguaia). ♂ has blue bill, is *uniform glossy blue-black.* ♀'s bill dusky; olive brown above with *lores and narrow eye-ring whitish;* wings dusky with *two pale buff bars,* tail feathers edged cinnamon-rufous. Whitish below with *heavy olive brown breast streaking.* Riverside Tyrant is larger, ♂ not as glossy black. A quiet and inconspicuous bird, found singly or in pairs, perching at or below eye level and sallying to foliage or water's surface. Displaying ♂ jumps up and returns to perch, sometimes with a soft snapping sound.

Hudson's Black Tyrant (*K. hudsoni*) is a rare austral winter visitant to extreme S. Both sexes resemble Amazonian; ♂ has small whitish area on flanks, ♀ a rufous rump and basal tail.

CINEREOUS TYRANT
MARIA-CINZA

BLUE-BILLED BLACK TYRANT
MARIA-PRETA-DE-BICO-AZUL

SÃO FRANCISCO BLACK TYRANT
MARIA-PRETA-DO-NORDESTE

CRESTED BLACK TYRANT
MARIA-PRETA-DE-TOPETE

RIVERSIDE TYRANT
MARIA-PRETA-RIBEIRINHA

AMAZONIAN BLACK TYRANT
PRETINHO-DO-IGAPÓ

LONG-TAILED TYRANT *Colonia colonus* ♂ 23-25 cm | 9-10" ♀ 18-20 cm | 7-8"
Locally fairly common in forest and woodland, borders and adjacent clearings; mainly in upland areas, and not in Pantanal. Conspicuous and unmistakable, with *long tail*. Stubby bill. *Black* with *white forecrown and superciliary*, grayer on crown and nape; rump patch white; *central tail feathers greatly elongated*. ♀ paler and grayer on belly; tail shorter. Immatures may lack lengthened tail feathers and pale crown and thus can be confusing but for their distinctive behavior and *stubby bill*. Usually in pairs (sometimes family groups) and highly sedentary, not joining mixed flocks. Often perches on dead snags and branches, flicking its long tail and sallying into air after insects. Most frequent call a distinctive soft rising "swee?"

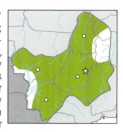

EASTERN SIRYSTES *Sirystes sibilator* 18.5 cm | 7.25"
Locally fairly common in canopy and borders of forest. *Crown black* shading to slaty on sides of head, with *mottled gray back*; wings and tail blackish, coverts and inner flight feathers edged grayish. *Throat and breast pale gray, belly white*. Though *Myiarchus*-like in aspect and behavior, the E. Sirystes has a very different black, gray, and white plumage pattern. Ranges singly or in pairs, frequently accompanying (even leading) mixed flocks. Often leans forward, nodding its head and ruffling crown feathers. Very vocal, with calls including a loud "wheeer-péw" or "wheeer-péwpu," sometimes extended into an excited series "wheeer-pe-pe-pew-pew-péw" or "wheeer-péw-péw-péw."

MYIARCHUS are fairly large flycatchers found in wooded or semiopen habitats. They are *so closely similar that most species are best identified through their characteristic voices*. Nests are placed in tree cavities.

BROWN-CRESTED FLYCATCHER *Myiarchus tyrannulus* 19.5 cm | 7.75"
Widespread and locally fairly common in a variety of semiopen habitats. Bill pinkish at base. Crown dull brown, otherwise dull grayish brown above; wings with *rufous primary edging*; tail feathers' *inner webs broadly edged rufous* (especially evident from below and in flight). Throat and breast pale gray, belly pale yellow. *No other Myiarchus shows as much rufous in wings and tail* (though juveniles of certain species may show a little). Conspicuous and noisy, often perching in the open, peering about and vigorously nodding its head. Most frequent call a sharp "peert!" or "weerp!" and a "hurrip," sometimes repeated.

SWAINSON'S FLYCATCHER *Myiarchus swainsoni* 18-18.5 cm | 7-7.25"
Fairly common in forest and woodland borders and adjacent semiopen areas; at least in S of our area apparently present only during austral summer. *Lower mandible pinkish at least at base*. Rather pale grayish olive above; tail blackish with white outer web. Throat and breast pale gray, belly pale yellow. Short-crested Flycatcher has an all-black bill, and is usually near water. Brown-crested Flycatcher has much more rufous in its wings and tail, and favors less wooded areas. All three have similar behavior and are *best identified by voice*. Most frequent call of Swainson's a soft "whoo."

SHORT-CRESTED FLYCATCHER *Myiarchus ferox* 18-18.5 cm | 7-7.25"
Usually common and widespread in borders of forest, woodland, and adjacent clearings; *most often near water*. Bill all black. Dark olivaceous brown above, sootiest on head; tail uniform dusky. Throat and breast gray, belly clear yellow. Juvenile has dull rufescent wing-bars and edging. Swainson's Flycatcher is paler generally, and its pinkish lower mandible is normally evident. Found singly or in pairs, usually foraging close to the ground. Often heard call a distinctive soft rolling "prrrt" or "dr'r'r'ru."

DUSKY-CAPPED FLYCATCHER *Myiarchus tuberculifer* 16.5 cm | 6.5"
Fairly common in canopy and borders of forest and woodland in *NW of our area*. Bill all black. *Crown sepia brown* and back dark olive; tail dusky. Throat and breast gray, belly clear yellow. *So much smaller* than other *Myiarchus* flycatchers that confusion is unlikely. Behavior similar to the others, but Dusky-capped tends to be more arboreal and is more likely to join mixed flocks. Frequent call a distinctive clear whistled "wheeeuw;" also gives various chatters.

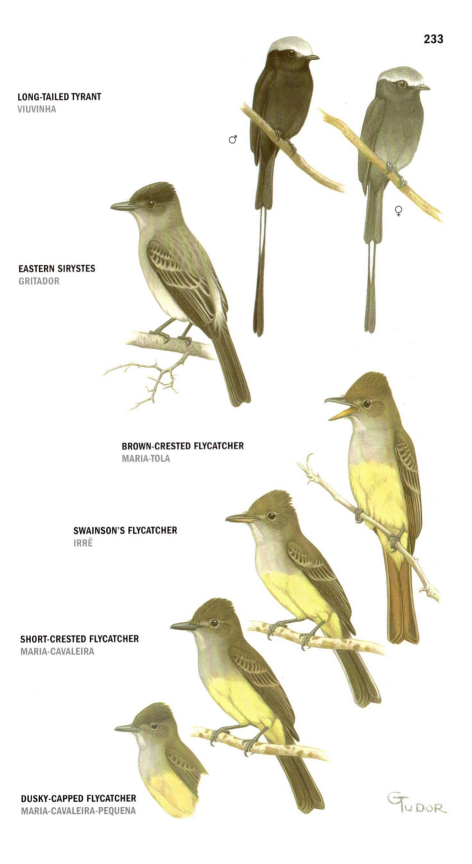

LONG-TAILED TYRANT
VIUVINHA

EASTERN SIRYSTES
GRITADOR

BROWN-CRESTED FLYCATCHER
MARIA-TOLA

SWAINSON'S FLYCATCHER
IRRÊ

SHORT-CRESTED FLYCATCHER
MARIA-CAVALEIRA

DUSKY-CAPPED FLYCATCHER
MARIA-CAVALEIRA-PEQUENA

CLIFF FLYCATCHER *Hirundinea ferruginea* 18.5 cm | 7.25"
Fairly common around cliffs and buildings, locally even in towns and cities. Some austral migrants may occur. Brown above with *conspicuous cinnamon-rufous rump and basal tail*; wings dusky with *most flight feathers rufous*, obvious in flight and evident even at rest. Below mostly cinnamon-rufous. No other flycatcher in our area looks much like it. Found in sedentary pairs or small family groups, usually perching in the open on rock faces or buildings; often seems oblivious to the roar of traffic or proximity of people. Migrants can occur in atypical situations. Sometimes keeps up a near constant chatter of high-pitched calls, including a "wheeeyp!" and "whee, dee-dee-ee-ee-ee" or "wheeuw-d'd'd'r!" and also a continued "wha-deép, wha-deép...."

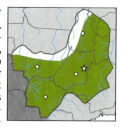

ATTILAS are *rather retiring* forest flycatchers, not often seen except by tracking down their *persistently given songs*. They have *heavy hooked bills*, and are *predominantly rufous*.

WHITE-EYED ATTILA *Attila bolivianus* 19 cm | 7.5"
Uncommon in gallery forest and borders in N Pantanal. Iris yellowish white; bill mainly horn. *Rufous brown above* with grayish crown, bright cinnamon-rufous rump and tail; primaries and greater wing-coverts blackish. Below cinnamon-rufous, belly paler. The *whitish eye is obvious and should preclude confusion*. An arboreal bird that is found singly or in pairs, usually not with flocks. Heard more often than seen, with far-carrying song an ascending series of whistled notes, halting at first: "whup; whup, wheep, wheep, wheep, wheeyp, wheeyp, wheebit, wheeeur," the last note distinctly slurred.

RUFOUS-TAILED ATTILA *Attila phoenicurus* 18 cm | 7"
Very rare transient in canopy and subcanopy of forest and woodland; few records in our area. Breeds in SE Brazil, wintering in Amazonia. Bill blackish. *Head and nape dark gray contrasting with deep rufous upperparts*, primaries blackish. *Below orange-ochraceous*. Crested Becard is less bright overall with a stouter bill, gray on crown only. Rufous Casiornis is more uniformly rufous, with no gray on head. In our area this attila vocalizes little if at all, and thus it usually passes unnoticed. Distinctive song (given on breeding grounds) a far-carrying "whee? whee? whee-bit," repeated over and over. Softer call a repeated "peeur."

The two **CASIORNIS** are *mainly rufous* flycatchers that are *smaller and more slender* than *Myiarchus*. They occur in deciduous forest and are only rarely together.

RUFOUS CASIORNIS *Casiornis rufus* 18 cm | 7"
Fairly common and widespread in deciduous and gallery woodland. Bill basally pinkish. *Uniform rufous above*. Throat and breast cinnamon, belly pale buffy-yellow. No becard is as slender or long-tailed. Cf. rarer Ash-throated Casiornis. Mainly arboreal (at times foraging lower, even in the understory) and usually in pairs, perching erectly and often elevating crest feathers, nodding its head, *Myiarchus*-like. Regularly accompanies mixed flocks. Rather quiet, but does give a weak "pseee," sometimes in a quick series

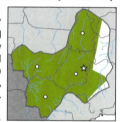

ASH-THROATED CASIORNIS *Casiornis fuscus* 18 cm | 7"
Uncommon and local in deciduous woodland and scrub in N of our area. Bill basally pinkish. Crown rufous, *back dull sandy brown*; tail rufous, *wings dusky broadly edged rufous and buff*. Throat whitish, *breast grayish fawn*, belly creamy yellowish. Limited overlap with Rufous Casiornis, which is more uniform rufous above and brighter cinnamon below. Ash-throated's behavior is similar, though it may tend to remain more in canopy. Call a vigorous "cheeyp, chew-chew;" song (less often heard) a series of about a dozen strong "chew" notes, hesitating at end.

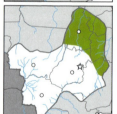

GRAYISH MOURNER *Rhytipterna simplex* 20.5 cm | 8"
Uncommon in midlevels and subcanopy of forest in NW of our area. Iris reddish. *Uniform plain gray*, somewhat paler below with slight yellowish cast. Similarly colored but larger Screaming Piha has a stouter bill, more grayish eye, purer gray plumage (without the yellowish tone below), and a rounder head. Their voices differ markedly; the piha rarely accompanies flocks. Found singly or in pairs, perching erectly and quietly on open branches, peering around in search of insect prey; also eats fruit. A frequent member of mixed flocks. Distinctive call a fast "r-t-t-t-t-t-tchéw!" explosive at end (almost like a sneeze), sometimes with some rising preliminary notes.

CLIFF FLYCATCHER
GIBÃO-DE-COURO

WHITE-EYED ATTILA
BATE-PARA

RUFOUS-TAILED ATTILA
CAPITÃO-CASTANHO

RUFOUS CASIORNIS
CANELEIRO

ASH-THROATED CASIORNIS
CANELEIRO-ENXOFRE

GRAYISH MOURNER
VISSIÁ

BOAT-BILLED FLYCATCHER *Megarynchus pitangua* 23 cm | 9"

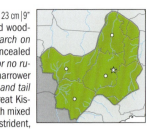

Widespread and fairly common in canopy and borders of forest and woodland, and in clearings. *Bill very heavy and broad with an obvious arch on culmen.* Crown and face black with long white superciliary, semiconcealed yellow crown patch; above brownish olive, *wings and tail with little or no rufous edging.* Throat white, below bright yellow. Great Kiskadee has a narrower bill with straight culmen; it is browner above with *more rufous wing and tail edging.* Though a noisy bird, Boat-billed is less conspicuous than Great Kiskadee, less apt to perch low or in the open. Pairs sometimes range with mixed flocks. Eats mainly large insects, also some fruit. Most frequent call a strident, nasal "kryeeeh-nyeh-nyeh-nyeh," sometimes given as it pumps its head.

GREAT KISKADEE *Pitangus sulphuratus* 23-23.5 cm | 9-9.25"

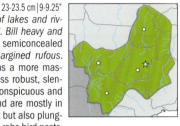

Common and widespread in semiopen areas at margins of lakes and rivers, clearings, and residential areas. Abundant in Pantanal. Bill heavy and straight. Crown and face black with *long white superciliary,* semiconcealed yellow crown patch; olive brown above, *wing feathers margined rufous.* Throat white, below bright yellow. Boat-billed Flycatcher has a more massive bill with arched culmen, more olive upperparts. The less robust, slender-billed Lesser Kiskadee is closely tied to water's edge. Conspicuous and noisy, Great Kiskadees often perch low and in the open, and are mostly in pairs and not with flocks. Consumes mostly insects and fruit but also plunges into water after fish, pursues lizards and snakes, and even robs bird nests. Gives a variety of loud calls, the best known and most frequent a boisterous "kis-ka-dee!" reflected in its English name (*Bentevi* in Portuguese). This can be varied to a "geép geép ga-reér," or a shrill raptor-like "keeeer."

LESSER KISKADEE *Philohydor lictor* 17 cm | 6.75"

Fairly common at margins of lakes, streams, and rivers, sometimes in marshy shrubby pastures. *Bill long and slender.* Resembles Great Kiskadee, but *markedly smaller* with *more slender bill* and *notably different vocalizations.* Rusty-margined Flycatcher has a stubbier bill, more rufescent upperparts, whining voice. Usually in pairs that perch no more than a few meters above the water, sallying to foliage and water's surface. *Much less bold and conspicuous than the better known Great Kiskadee.* Its nasal, raspy calls are distinctive, but are subdued compared to Great Kiskadee's. Most frequent is a "dzreeéy, dzwee" or "dzreeéy, dzwee-dzwee-dzwee."

MYIOZETETES flycatchers resemble the two kiskadees though they have *notably stubbier bills.* In size they are closer to the Lesser.

RUSTY-MARGINED FLYCATCHER *Myiozetetes cayanensis* 17 cm | 6.75"

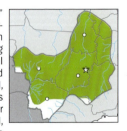

Fairly common and widespread (especially near water) in shrubby clearings and forest and woodland borders. Olive brown above with black crown and face, long white superciliary, semiconcealed *yellow crown patch; wing feathers edged rufous.* Throat white, below bright yellow. Resembles Social Flycatcher, and *often best separated by voice.* Social has an orange-red crown patch (often hidden), duskier (less black) crown and sides of head, pale wing-covert edging, and *lacks rufous primary edging* (juvenile Socials may show some rufous, but they will also have edging on coverts). Cf. Lesser Kiskadee. Rusty-margineds perch, sometimes partially cocking their tail, anywhere from the ground to tall treetops. They are usually in pairs, less often in small groups, but do not move with flocks. They sally to foliage for insects (less often into air), and also eat much fruit. Frequent call a distinctive whining, plaintive "freeeea" or "wheeeeea." Also gives faster, excited-sounding calls similar to Social's.

SOCIAL FLYCATCHER *Myiozetetes similis* 17 cm | 6.75"

Fairly common in shrubby clearings and forest and woodland borders in E and S parts of our area (but nearly absent from N Pantanal and around Chapada dos Guimarães). *Crown dark gray with semiconcealed orange-red crown patch, long white superciliary* and blackish face; olive above, *wing-coverts edged pale grayish or buffyish.* Throat white, below bright yellow. Juveniles can show some rufous wing-edging. Cf. similar Rusty-margined Flycatcher. Lesser Kiskadee has a longer and more slender bill. Behavior much as in Rusty-margined though Socials tend to be noisier. Gives a variety of mostly harsh calls, including a frequent "kreeoouw," chattered "ti-ti-ti-tíchew, chew," and also a single "chew."

YELLOW-BROWED TYRANT *Satrapa icterophrys* 16.5 cm | 6.5"

Widespread but uncommon in semiopen areas with scattered trees, groves, and gallery woodland. Olive above, grayer on crown, with *prominent bright yellow superciliary* and *blackish cheeks*; wings blackish with two pale gray bars, outermost tail feathers whitish. *Below bright yellow.* ♀ slightly duller, with olive breast mottling. Similarly shaped and patterned Rusty-margined and Social Flycatchers have obvious *white* brows. Usually seen singly, less often in pairs, perching erectly and often in the semiopen, mainly feeding by sallying to foliage. Not with flocks. Mainly feeds by sallying to foliage. Very quiet.

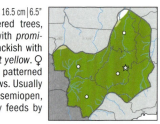

PIRATIC FLYCATCHER *Legatus leucophaius* 14.5 cm | 5.75"

Fairly common in forest and woodland borders and tall trees of adjacent clearings. In S of our area perhaps present only during austral summer. *Stubby black bill. Dark olive brown above with long whitish superciliary*, semiconcealed yellow crown patch, blackish face, whitish malar area, and dusky submalar streak. Below whitish, breast streaked dusky, belly pale yellow. Variegated Flycatcher is larger with a longer bill, more mottled back, pale edging on wings, and rufous on uppertail-coverts and tail. *Persistently vocal when breeding (so learn the voice!)* but otherwise inconspicuous, remaining high in trees where often hard to detect. Eats mostly small fruit. Pairs usurp the pendant nests of various other birds, especially oropendolas and caciques. Singing ♂ gives a whining querulous "wheé-yee," sometimes followed by a fast "pi-ri-ri-ri-ri," repeated tirelessly.

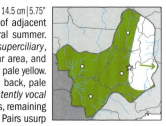

VARIEGATED FLYCATCHER *Empidonomus varius* 18.5 cm | 7.25"

Uncommon in secondary and gallery woodland, borders, and clearings with scattered trees; in S of our area present only during austral summer. Head blackish with semiconcealed yellow crown patch, long whitish superciliary, whitish malar area, and dusky submalar streak; *back olive brown mottled dusky. Wing feathers edged whitish; rump and tail feathers broadly edged rufous.* Throat whitish; below pale yellowish, breast with dusky-brown streaking. The more arboreal Piratic Flycatcher is smaller and has a stubbier bill, *plain* brown back and *lacks* pale wing-edging and rufous in tail. Streaked Flycatcher is larger and has a heavier bill, coarser streaking below. Variegateds are easy to see and are usually in the open, sometimes joining other tyrannids in fruiting trees. They also sally for insects. Quiet, though breeding birds give a high-pitched thin and raspy "pseee."

CROWNED SLATY FLYCATCHER *Griseotyrannus aurantioatrocristatus* 18 cm | 7"

Uncommon in lighter woodland, scrub, and cerrado with scattered trees. Present in S of our area only during austral summer, and in some areas northward may occur only as a transient. *Brownish gray above with contrasting black crown*, semiconcealed yellow crown patch, and gray superciliary. *Smoky gray below*, paler on belly. This obscure flycatcher looks *essentially gray*, and often appears *distinctly flat-crowned.* Cf. various *Knipolegus* black tyrants. The Variegated Flycatcher favors more wooded terrain than the Crowned Slaty. Often a solitary bird, frequently perching fully in the open. Sallies into the air like a pewee or kingbird and also eats considerable fruit, though seemingly less than the Variegated Flycatcher. Breeding birds give a weak "pseee" (recalling Variegated), sometimes in series, but on the whole rather quiet.

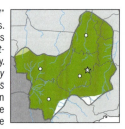

STREAKED FLYCATCHER *Myiodynastes maculatus* 20 cm | 8"

Fairly common in borders of forest, secondary woodland, and clearings. Vacates our area during winter, then migrating to Amazonia. *Large and boldly streaked. Dark brown above broadly streaked blackish* with semiconcealed yellow crown patch, *whitish superciliary*, blackish mask, whitish lower cheeks, and dusky malar streak. Wings with whitish edging; *rump mainly rufous; tail feathers edged rufous laterally.* Throat white; below whitish to pale yellowish *boldly streaked blackish.* Variegated Flycatcher is smaller with a stubbier bill and blackish crown; it is less obviously streaked. Found singly or in pairs, and often noisy and conspicuous. Consumes a variety of insects and also eats much fruit. Has a variety of loud harsh calls, including a repeated "kip!" and a "chup" or "eechup." More musical song, given especially at dawn and dusk, a repeated fast, rhythmic "wheeé-cheederee-wheé" (sometimes without the final "wheé").

SULPHURY FLYCATCHER *Tyrannopsis sulphurea* 19 cm | 7.5"
Rare and local, mainly around groves of Mauritia palms, in NW of our area. Resembles a chunky, short-billed, short-tailed Tropical Kingbird. Head gray, somewhat darker and duskier on face, with semiconcealed yellow crown patch; back brownish olive, brownish dusky on wings and tail. *Foreneck white with a little gray streaking, sides of throat grayer and olive on sides of chest; below bright yellow.* Tropical Kingbird is *much* commoner, larger, and has a longer, distinctly notched tail; its throat is not as contrastingly white. Found in localized pairs, and usually inconspicuous, perching high on palm fronds, though periodically it bursts into frenzied activity with much calling. Vocalizations are high-pitched and piercing, some harsh, some squeaky: a fast "jee-peet! jee-peeteet, jeepeet!" and a "squeezrr-squeezrr-prrr."

TYRANNUS kingbirds and flycatchers are fairly large *conspicuous* tyrannids that range in *open terrain*. Two of our species are yellow-bellied, while the other two are smaller-billed and white below, one with a long tail.

TROPICAL KINGBIRD *Tyrannus melancholicus* 21.5 cm | 8.5"
Very common and conspicuous in open, agricultural, and urban areas. Most individuals move N into Amazonia during austral winter, though it is resident in N of our area. *Head gray* with a darker mask. Grayish olive above; wings and tail brownish dusky, *tail notched*. Throat pale grayish, *chest grayish olive, breast and belly bright yellow*. Use the oft-seen and familiar Tropical Kingbird as the basis of comparison with other scarcer tyrannids such as White-throated Kingbird and Sulphury Flycatcher. Seen singly or in pairs and generally perching in the open, often on wires, and remaining active through the day. Mainly eats insects captured in aerial sallies, but nonbreeders also consume considerable fruit. Has a variety of high-pitched twittering calls: "pee, ee-ee-ee-ee," often accompanied by wing quivering. Dawn song a short series of "pip" or "pee" notes followed by a rising twitter, "piririree?" Frequently one of the first birds to greet the dawn.

WHITE-THROATED KINGBIRD *Tyrannus albogularis* 21 cm | 8.25"
Uncommon and local in shrubby areas and gallery woodland, mostly near water and *favoring stands of Mauritia palms*. A summer breeding visitant here, moving N into Amazonia during austral winter. *Resembles much more numerous Tropical Kingbird*, but White-throated's *paler gray head contrasts more with its blackish mask*; back paler and brighter olive; *pure white throat* contrasts with bright yellow underparts (*chest with at most a tinge of olive*). In strong light Tropical Kingbirds can look pale-headed and white-throated, but they are never as obviously "masked" as White-throated. Behavior and voice similar to Tropical Kingbird.

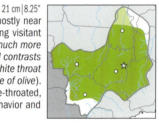

EASTERN KINGBIRD *Tyrannus tyrannus* 20.5 cm | 8"
Rare boreal migrant to forest borders and clearings; few records from our area. Mainly winters further N and W in South America; breeds in North America. *Slaty above, blackest on head*; squared tail black with *white tip. Whitish below.* Birds in worn plumage look faded and show little or no white on tail tip. Juvenile and worn adult Fork-tailed Flycatchers without long tail streamers can resemble this kingbird though at least a tail cleft is invariably apparent. In our area generally seen singly, sometimes with other flycatchers such as Fork-taileds. This kingbird is primarily frugivorous on its wintering ground, and is mainly silent here.

FORK-TAILED FLYCATCHER *Tyrannus savanna* ♂ 38-40 cm | 15-16" ♀ 28-30 cm | 11-12"
Measurement includes the *very long deeply forked tail*. Locally fairly common summer resident in campos and pastures with scattered trees, moving N into Amazonia during winter. Virtually unmistakable. *Head black; back pearly gray*. Wings dusky, *very long forked tail* black (shorter in ♀♀, immatures, and molting birds); outer web of outer tail feathers basally white. *Below white*. Juvenile has a browner cap, lacks extended tail streamers (short-tailed birds vaguely recall the rare Eastern Kingbird). Conspicuous and lovely, Fork-tailed Flycatchers often perch low and make frequent long and graceful sallies in pursuit of insects. Pairs separate out when breeding but at other times often gregarious, migrating in loose flocks. Nonbreeders consume much fruit. Notably quiet, but breeding birds occasionally give weak "tic" notes.

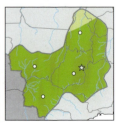

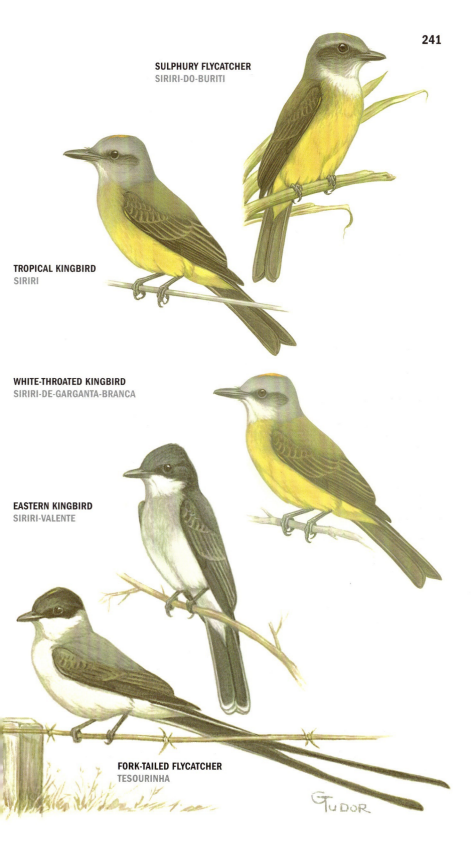

PALE-BELLIED TYRANT-MANAKIN *Neopelma pallescens* 14 cm | 5.5"
Uncommon and local in lower and midlevels of deciduous and gallery woodland. Iris grayish mauve. *Very plain*. Olive above with inconspicuous crown patch bright yellow (visible during display). Below olivaceous gray, *belly creamy whitish*. Greenish Elaenia has a yellower belly, white on face, prominent wing-edging; its vocalizations are very different. An *inconspicuous* bird, though sometimes noted at fruiting trees. Most often seen are displaying ♂♂, which perch on open low branches, periodically giving a soft nasal "wraah, wra-wra," sometimes simultaneously jumping off perch to land facing the other way.

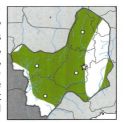

Dwarf Tyrant-Manakin (*Tyranneutes stolzmanni*) occurs locally in lower and midlevels of forest in NW of our area. A tiny, pale-eyed "manakin" (like *Neopelma*, the genus is now thought to be a tyrannid). Dull-plumaged and mainly olive, resembling many ♀ manakins. Best known from ♂'s distinctive song, an endlessly repeated "jew-pit."

WHITE-NAPED XENOPSARIS *Xenopsaris albinucha* 13 cm | 5"
Rare and local in riparian growth and lighter woodland. ♂ has *crown black* with white lores and a pale gray nape; above brownish gray, *brownest on wings*; tail blackish, outer web of outer feathers white. *Below white*. ♀ has browner crown and upperparts, buff-tinged belly; juvenile's crown scaled whitish. Pert and alert-looking, not likely confused, the xenopsaris is found singly or in pairs, not accompanying mixed flocks. Rather quiet, but ♂ does have a thin, high-pitched song, a whistled "tsip, tsiweeé, tseee-ti-ti-ti-ti?"

PACHYRAMPHUS becards are *chunky, large-headed,* and *rather short-tailed* birds; most are sexually dimorphic. Becards are *arboreal* and generally inconspicuous birds, though some are quite vocal. Nests are large globular structures with an entrance on the side or below.

GREEN-BACKED BECARD *Pachyramphus viridis* 14.5 cm | 5.75"
Uncommon in lighter and gallery woodland and forest. ♂ has *glossy black crown* with white lores and *gray sides of head and nuchal collar; bright olive above*, blacker on outer flight feathers. Throat whitish, *pectoral band bright yellow*, belly whitish. ♀ duller, with olive crown, rufous lesser wing-coverts. *The only olive-backed becard*. Not likely confused when seen, but more often only heard. Cf. Rufous-browed Peppershrike. Forages in pairs, gleaning rather sluggishly at varying heights. Sometimes with mixed flocks, but often apart. Pretty song an oft-heard and far-carrying series of musical notes with a slight crescendo: "trididideédeédeédeé?"

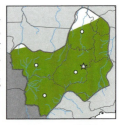

WHITE-WINGED BECARD *Pachyramphus polychopterus* 14.5 cm | 5.75"
Fairly common in secondary, riparian, and gallery woodland. ♂ *blackish above*, crown glossier, with gray nuchal collar and rump; *two bold white wing-bars and scapular edging*; outer tail feathers white-tipped. Below gray. ♀ olive brown above, *crown more brownish*, supraloral and partial eye-ring whitish; wings dusky with *cinnamon-buff scapulars, two bars, and edging*; outer tail feathers tipped cinnamon-buff. Below dull yellowish, breast clouded olive. Forages lethargically in leafy canopy and outer branches; usually in pairs, sometimes accompanying mixed flocks. ♂'s song a series of pretty melodic notes: "teu, teu, tu-tu-tu-tu-tu-tú," always with slow first notes.

Black-capped Becard (*P. marginatus*) occurs locally in forest at NW edge of our area. ♂ resembles ♂ of slightly larger White-winged Becard (which favors more secondary and semiopen areas) but has pale gray lores and a mixed black-and-gray back. ♀ also similar but with a distinctive *rufous-chestnut crown*. Song a shorter "teeu, whee-do-weét."

CHESTNUT-CROWNED BECARD *Pachyramphus castaneus* 14 cm | 5.5"
Uncommon in forest and woodland borders in *SE of our area*. *Cinnamon-rufous above*, darker on crown, with dusky lores, buffy supraloral, and a *gray band that encircles nape*; blackish greater primary-coverts and inner webs to primaries. *Pale cinnamon-buff below*, throat and midbelly whiter. Rufous plumage unlike any of our other becards; not likely confused. Rufous Casiornis lacks the gray nape band. Cf. *Philydor* foliage-gleaners (very different behavior). Usually in pairs that forage lethargically in leafy canopy, generally not with flocks. Easily overlooked unless vocalizing. Soft musical song a melancholy whistled "teeeuw-teeu-teeu-teeu;" also gives a rising "tititíti?"

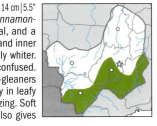

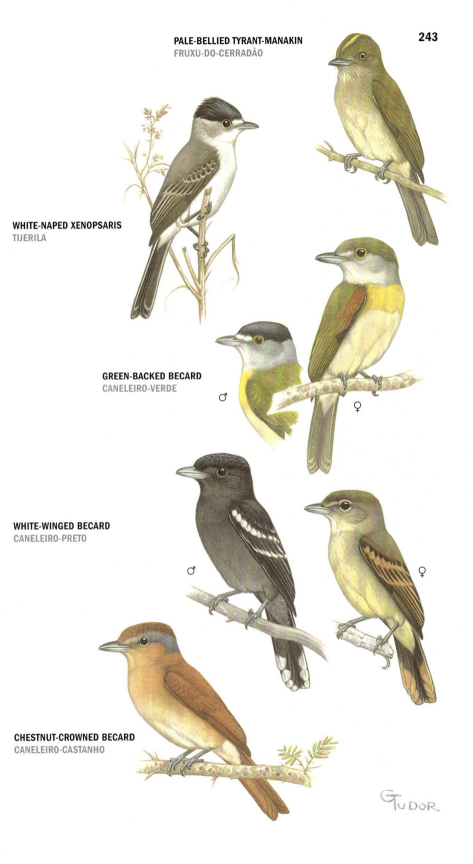

CRESTED BECARD *Pachyramphus validus* 18.5 cm | 7.25"
Widespread but generally uncommon in canopy and borders of forest and woodland, sometimes out into clearings. Heavy bill. ♂ *black above, dark olivaceous gray below*. Often looks all dark. ♀ *rufous above with contrasting sooty crown, buffy lores. Below cinnamon-buff*. The *largest* becard, but *not* especially "crested." Seen singly or in pairs, and usually inconspicuous; sometimes accompanies mixed flocks. Less vocal than many becards but gives various squeaky or twittery notes, e.g., "tsee-eéyk," also up to 6 clearer notes, "sui-sui-sui...."

PINK-THROATED BECARD *Pachyramphus minor* 17 cm | 6.75"
Uncommon in canopy and forest borders in *NW of our area*. ♂ *black above, blackish gray below with small patch of rosy pink on lower throat* (usually inconspicuous). ♀ has *slaty gray crown and back*, rufous wings and tail. Below cinnamon-buff. Some (younger?) ♂♂ have crown and back more olivaceous gray. ♂ White-winged Becard has obvious white in wings and tail. Crested Becard is somewhat larger; ♂ is paler below, lacks pink on throat; ♀ has rufous back (dark only on crown). Usually in pairs, regularly accompanying mixed flocks, often bobbing its head. Easily overlooked as it tends to stay well above ground. Relatively quiet, with ♂'s song a clear, melodic "teeuuuweeet" often followed by twittering notes.

Wing-barred Piprites (*Piprites chloris*) occurs locally in forest at N edge of our area. Plump, with large round head and short tail. Bright olive above with yellow lores and eye-ring, gray nape and sides of neck, broad yellowish wing-bars and tertial tipping. Mainly pale grayish below. Far-carrying song has a distinctive hesitant cadence: "whip, pip-pip, pididip, whip, whip?" Generic relationships uncertain (closest to becards?).

TITYRAS (Tityridae) are nearly unmistakable, *chunky, short-tailed* birds; all three species are *predominantly white*. They are conspicuous at forest and woodland borders. Though their relationships remain uncertain, the tityras are now usually placed in their own family, together with the becards.

MASKED TITYRA *Tityra semifasciata* 21 cm | 8.25"
Uncommon in canopy and borders of forest and woodland in *NW of our area*. Wide orbital area and most of bill rosy red. ♂ has *forecrown and face black, pale pearly gray upperparts, white underparts*. Wings (except for tertials) black, *tail grayish white with wide black subterminal band*. ♀ has *head dusky brownish*, back grayish; below pale grayish. Black-tailed Tityra has an *all-black tail*; ♂ has more black on head than ♂ Masked; ♀ shows streaking on back and breast. Black-crowned Tityra *lacks red on bill, face*. Behavior and voice much as in wider-ranging Black-tailed.

BLACK-TAILED TITYRA *Tityra cayana* 21.5 cm | 8.5"
Fairly common and widespread in canopy and borders of forest and woodland. In most of our area (illustrated) both sexes have *orbital area and most of bill rosy red*. ♂ has *head black, pearly gray upperparts, white underparts*; most of wings and *all of tail black*. ♀ grayer above with black cap; *variable amount of blackish streaking on back and breast*. In S of our area orbital area reddish purple (hence less conspicuous), bill mostly black. ♂ there similar to but whiter than N ♂; ♀ *lacks black cap* and is *more profusely streaked black below*. The *only tityra with obvious streaking*. Black-crowned Tityra shows no red on bill and face. ♂ Masked Tityra has white of back extending to hindcrown and a mainly whitish tail. Often in pairs that perch for long periods on high exposed branches, favoring snags, where they frequently nest. Often aggressive toward other birds. Flight strong and direct. Eats mostly fruit, also some large insects. Unusual call, often uttered in flight, a nasal croak or grunt, "urt," sometimes given in series.

BLACK-CROWNED TITYRA *Tityra inquisitor* 18.5 cm | 7.25"
Widespread but uncommon in canopy and borders of forest and woodland. *Bill blackish, showing no red*. ♂ has *crown and face black*, pale pearly gray upperparts, white underparts. Wings (except tertials) and tail black. ♀ has *buff frontlet and obvious rufous-chestnut on sides of head*; back duller gray, sometimes streaked brown. Other tityras have conspicuous rosy red bills and faces (but beware S Black-taileds with their darker purplish-red). Behavior much as in other tityras, Black-crowned being equally conspicuous though generally less numerous. Calls also similar, likewise often doubled and given in flight; Black-crowned's tend to be drier (less "grunty") and weaker: "zik-zik" or "chet-chet."

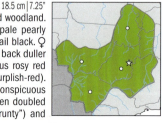

CRESTED BECARD
CANELEIRO-DE-BONÉ

PINK-THROATED BECARD
CANELEIRO-DE-PAPO-ROSA

MASKED TITYRA
ANAMBÉ-MASCARADO

BLACK-TAILED TITYRA
ANAMBÉ-DE-CARA-BRANCA

BLACK-CROWNED TITYRA
ANAMBÉ-DE-BOINA

MANAKINS (Pipridae) are small chunky birds with short bills and tails that range mainly in tropical forests; ♂♂ *are often very colorful*. ♂♂ of most *"true"* manakins engage in *stereotypical displays*, either alone or in leks, to attract ♀♀.

GREENISH SCHIFFORNIS *Schiffornis virescens*   16 cm | 6.25"

Uncommon in undergrowth in forest, woodland, and gallery forest in *SE of our area*. *Round-headed; prominent dark eyes*, narrow but distinct pale eye-ring. Olivaceous; *wings and tail rufescent*. Most likely confused with ♀ Blue Manakin, which is generally greener, lacks rufous in wings, has reddish legs and slightly elongated central tail feathers. The schiffornis is an inconspicuous bird, foraging alone, hopping in undergrowth and clinging to vertical stems. Rarely with mixed flocks. Heard much more often than seen, with simple but musical song a clear "teeeo, toweé?" or "teeeo, to, toweé?" Generic relationships uncertain, apparently not a "true" manakin.

Thrush-like Schiffornis (*S. turdinus*) occurs locally in forest undergrowth in NW of our area. Resembles Greenish Schiffornis but browner above, more grayish below; wings and tail less rufescent. The two species do not occur together; unobtrusive behavior much the same.

FLAME-CRESTED MANAKIN *Heterocercus linteatus*   14 cm | 5.5"

Uncommon and local in undergrowth of riparian forest and woodland in NW of our area. ♂ has *black head* with *red crown stripe* (usually hidden); dark olive above. *Throat silky white, lengthened at sides*; chest sooty olive, breast chestnut, belly rufous. ♀ dark olive above. *Throat gray; below dull cinnamon-buff*. The contrasting throat is distinctive in both sexes. Usually seen singly, occasionally at fruiting trees. Singing ♂♂ perch on regularly used low open branches where they remain quiet and motionless before, at very long intervals, emitting an explosive "skeeeeéeeuw, skeeeu" (with a curassow-like quality) sometimes preceded by a "ts-ik" note, all without moving. Displaying ♂♂ reportedly flare their gorgets and chase each other.

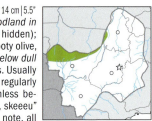

BLUE-BACKED MANAKIN *Chiroxiphia pareola*   12.5 cm | 5"

Uncommon and local in lower growth of forest and woodland on N edge of our area, e.g., Rio Araguaia drainage. ♂'s legs orange; ♀'s orange-yellow. Nearly unmistakable ♂ black with *deep red crown patch, pale azure blue mantle*. ♀ olive above with indistinct paler eye-ring. Paler olive below, belly yellower. ♀ White-bearded Manakin is smaller and shorter-tailed with brighter orange legs; it inhabits younger, more secondary growth. Shy and inconspicuous, most often seen at fruiting trees or around leks where ♂♂ (usually two) perform together, cartwheeling around each other when a ♀ finally arrives. Most frequent call a rich throaty "che-wurrr" or "te-turrr." Displaying ♂♂ give a more nasal "wr-r-r-aang, wr-r-r-aang…." Both sexes also give a clear "tooo-eee," sometimes doubled.

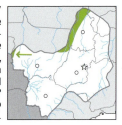

BLUE MANAKIN *Chiroxiphia caudata*   14.5-15.5 cm | 5.75-6"

Locally fairly common in lower growth of forest and woodland in SE of our area. Legs dull reddish. Beautiful ♂ *(by far the bluest manakin) rich cerulean blue with scarlet crown* and black head, wings, and outer tail feathers (*central tail feathers elongated*). ♀ olive, slightly paler below; central tail feathers only slightly elongated; a few (older birds?) have orange on crown. ♀ resembles Greenish Schiffornis though that has dark legs, a "normal" tail, rufescent wings. Behavior much as in Blue-backed Manakin, but Blue tends to be much easier to see and often ranges out to borders and into shrubby clearings. Blue's display is similar, though typically three ♂♂ perform together. Gives a throaty "qua-a-a-a-a" call during display; at other times both sexes frequently emit a "chorreeo, cho-cho-cho."

HELMETED MANAKIN *Antilophia galeata*   14.5 cm | 5.75"

Locally fairly common in lower and middle growth of gallery forest and woodland, including Pantanal. Unmistakable ♂ *black with red frontal crest, crown, nape, and midback*. ♀ has *miniature frontal crest*; it is olive, somewhat paler and grayer below. ♀'s frontal crest is unique, but can be surprisingly hard to see. Helmeted occurs with few other manakins, but cf. ♀ Band-tailed (smaller, with pale eye). Ranges higher above ground than many manakins and thus relatively easy to observe; not particularly shy. ♂'s distinctive, rollicking song a fast, rich, musical phrase, "whip-dip, whih-deh-deh-dédidi?" sometimes given as ♂♂ chase each other through trees. They seem to have no organized display. Both sexes give a throaty "wreee" or "wreee? pur."

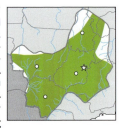

GREENISH SCHIFFORNIS
FLAUTIM

FLAME-CRESTED MANAKIN
COROA-DE-FOGO

BLUE-BACKED MANAKIN
TANGARÁ-FALSO

BLUE MANAKIN
TANGARÁ

HELMETED MANAKIN
SOLDADINHO

FIERY-CAPPED MANAKIN *Machaeropterus pyrocephalus*　　　9 cm | 3.5"
Uncommon in lower and midlevels in forest and woodland in *NW of our area* (including Chapada dos Guimarães). Unmistakable ♂ has *crown and nape golden with red median stripe; back and rump rosy rufous*, wings olive with whitish inner flight feathers. *Below whitish streaked rosy rufous.* ♀ olive above; *pale yellowish below with blurry olive streaking on breast and belly.* No other ♀ manakin is streaked below. Except for calling ♂♂, a very inconspicuous bird, most apt to be found during its quick visits to fruiting trees. ♂'s distinctive song a high-pitched bell-like "pling" or "cling," given at intervals of 10 seconds or more, hard to track to its source and easily passed over as a frog or insect. Displaying ♂♂ reportedly hang upside down and twist rapidly from side to side while emitting a protracted buzzy sound.

PIN-TAILED MANAKIN *Ilicura militaris*　　　11-12.5 cm | 4-5"
Uncommon in lower and midlevels of forest and woodland in *SE of our area*, e.g., Serra da Canastra. *Iris orange.* Unmistakable ♂ *black above with red forehead patch, lower back, and rump*, and olive flight feathers. Sides of head pale gray, below grayish white. *Elongated central tail feathers.* ♀ bright olive above, *sides of head and throat gray*; whitish below. Tail less elongated than ♂'s and *more wedge-shaped.* ♀ Blue and White-bearded Manakins lack gray on sides of face. Usually seen singly; Pin-tailed seems to follow mixed flocks more than most other manakins. Distinctive and oft-heard call a simple, descending series of 3-4 high-pitched notes. Displaying ♂♂ give a longer version of this call.

WHITE-BEARDED MANAKIN *Manacus manacus*　　　11 cm | 4.25"
Locally fairly common in dense undergrowth of secondary woodland and forest borders. *Legs orange.* Unmistakable ♂ *black above* with gray rump. *Throat (puffed out into a "beard" by displaying birds) and nuchal collar white*, underparts grayish. ♀ olive above, paler and more grayish olive below, belly yellower. Most other small ♀ manakins have *dark legs.* Most similar is ♀ of larger and longer-tailed Blue-backed Manakin; its legs are yellower. Inconspicuous away from their leks, then usually encountered singly, often at fruiting trees. At leks each ♂ has a small "court" cleared of leaves; they display on certain low branches, flying to and fro while flaring their throat feathers, giving a variety of sounds including whistled "peeur" calls and also a loud firecracker-like snap produced by the wings. Both sexes utter an occasional "chee-pu" call.

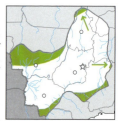

RED-HEADED MANAKIN *Pipra rubrocapilla*　　　10 cm | 4"
Locally fairly common in lower and midlevels of forest and woodland in *NW of our area.* Bill pinkish; iris pale hazel. Unmistakable ♂ *black with scarlet head*; thighs red and white. ♀ olive above, paler and more grayish olive below, belly pale dull yellowish. Most likely confused with ♀ Band-tailed Manakin, which is brighter olive generally with a yellower belly and bold white iris. Red-headeds are unobtrusive, generally quiet manakins that tend to perch motionless, hunched, and then abruptly dart away. They are most apt to be seen at fruiting trees or where ♂♂ are displaying. Groups of up to ten birds gather at leks that may be used for years, there flying back and forth between favored perches, sidling along branches, and turning back and forth. ♂'s calls include a sharp "dzeek, dzeeuw" and "drree-dit, dree-dee-dew."

BAND-TAILED MANAKIN *Pipra fasciicauda*　　　11 cm | 4.25"
Fairly common in undergrowth of riparian and gallery forest and woodland, including Pantanal. *Iris white.* Beautiful ♂ has *crown and nape bright red*; black above with *basal tail yellowish white; white on inner wing obvious in flight.* Bright yellow below, *breast variably stained scarlet.* ♀ olive, brighter above, somewhat paler below, yellow on belly. ♀ is best known by her staring white eye and bright olive and yellow coloration. An inconspicuous manakin that tends to remain closer to the ground than most others; most apt to be seen at fruiting trees and when ♂♂ are displaying at their small leks. Most frequent call a nasal, descending "eeeeuw."

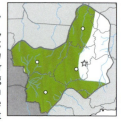

FIERY-CAPPED MANAKIN
UIRAPURU-CIGARRA

PIN-TAILED MANAKIN
TANGARAZINHO

WHITE-BEARDED MANAKIN
RENDEIRA

RED-HEADED MANAKIN
CABEÇA-ENCARNADA

BAND-TAILED MANAKIN
UIRAPURU-LARANJA

SHARPBILL *Oxyruncus cristatus* 17 cm | 6.75"
Rare and local in canopy and borders of forest in S of our area and at Serra das Araras. Sometimes considered a cotinga, but now usually as a separate family. *Pointed bill;* iris reddish to orange. Olive above with *black crown concealing an orange to red crest, head and neck whitish with blackish scaling. Throat white, scaled blackish; below pale yellowish spotted black.* Scaled and spotted plumage in conjunction with pointed bill distinctive. Rather inconspicuous birds, Sharpbills glean along branches and amongst leaves, sometimes even hanging upside-down, probing into moss and curled-up leaves; they also eat some fruit. They can be stolid, perching quietly for long periods, but also accompany mixed flocks. ♂'s song at Serra das Araras is rather different from songs typical of the species (a drawn-out buzzy trill that fades away, "zheeeeeeee-uuu'u'u'u'ur"), consisting of merely a short, rather buzzy "bz-z-z-z-z-z-r" (B. Carlos recording).

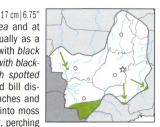

COTINGAS (Cotingidae) are arboreal birds found only in the Neotropics, mainly in lowland forests. Only a few species occur in our area, and none is particularly noticeable here. Some cotingas are among the most colorful, and even bizarre, of Neotropical birds. Most are primarily frugivorous.

SWALLOW-TAILED COTINGA *Phibalura flavirostris* 21.5 cm | 8.5"
Rare and local in forest and woodland borders, adjacent clearings, and gardens in *S Goiás, apparently only with old records. Bill yellow.* Unmistakable, with a *unique long, slender, deeply forked tail.* ♂ has *mostly black crown, yellowish olive upperparts barred black;* wings and tail bluish black. Throat bright yellow, breast white barred black, belly pale yellow with sparse black streaking. ♀ similar but shorter-tailed, with grayer crown, more olive in wings and tail, and duller yellowish underparts with denser black spotting and scaling. In flight the tail is held closed in a point. Perches at varying heights in clearings and at edge; not a true forest bird and often quite conspicuous where it occurs. Apparently feeds mainly on mistletoe berries. Seems very quiet.

Pompadour Cotinga (*Xipholena punicea*) occurs in small numbers in forest canopy along NW edge of our area, e.g., Serra das Araras. Iris white. Stunning ♂ *mostly shiny crimson-purple* with *highly contrasting, mainly white wings.* ♀ ashy gray with *wing-coverts and inner flight feathers edged white.*

SCREAMING PIHA *Lipaugus vociferans* 24.5-25.5 cm | 9.5-10"
Locally fairly common in lower and midlevels of forest in *NW of our area. Very vocal,* but otherwise *surprisingly inconspicuous. Plain gray,* wings and tail duskier, belly somewhat paler. Juvenile has browner wings. Grayish Mourner is smaller and browner with a paler iris, more slender bill, and yellowish cast to belly. Piha's shape is vaguely thrush-like (but no thrush in piha's range is uniform gray). Seen singly as it perches erect, periodically sallying in pursuit of a large insect or fruit. Sometimes moves with mixed flocks but much more often alone. ♂♂ gather in loose leks inside forest, in ample earshot of each other; here they perch 5-8 m above ground, remaining nearly motionless. The classic call is a very loud and ringing "weee weee-ah" (sometimes two initial "weee" notes), but they often warm up with softer and more guttural notes. Also gives a querulous, less loud "kweeeah" and a series of loud "kweeeo" notes. Even vocalizing birds can be very hard to spot because they move so little.

BARE-THROATED BELLBIRD *Procnias nudicollis* 26.5-28 cm | 10.5-11"
Rare and local in canopy and forest borders at S edge of our area; perhaps only wanderers occur now, trapping for the cagebird market having greatly reduced its overall numbers. Unmistakable ♂ *pure white* with *ocular area and throat patch bare bright greenish blue.* ♀ olive above with *dusky-olive crown and sides of head.* Throat blackish with fine pale streaking; below coarsely streaked olive and pale yellow. Subadult ♂♂ require several years to acquire white plumage; at first they resemble ♀♀, acquiring blue on throat early on. Inconspicuous, remaining in leafy canopy, though sometimes one will be spotted at a fruiting tree. Even singing ♂♂ are very hard to locate as they tend to remain in dense foliage; compounding the problem, their calls are surprisingly ventriloquial. There are two vocalizations: the *very loud "bock" or "bonk" call,* given with gape wide open (as illustrated), and a series of 6-8 less loud metallic "clink" or "tonk" calls, given with bill closed (though the throat visibly pulsates).

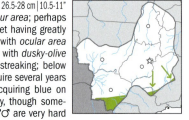

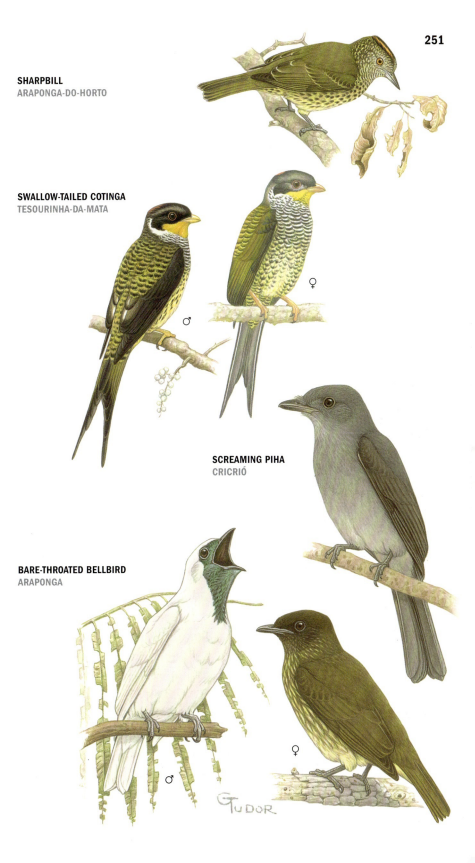

SHARPBILL
ARAPONGA-DO-HORTO

SWALLOW-TAILED COTINGA
TESOURINHA-DA-MATA

SCREAMING PIHA
CRICRIÓ

BARE-THROATED BELLBIRD
ARAPONGA

BARE-NECKED FRUITCROW *Gymnoderus foetidus* ♂ 38 cm | 15" ♀ 33 cm | 13"
Uncommon to locally fairly common at borders of forest, *mainly near water*, primarily in *NW of our area* and certainly more numerous there. Bill mostly pale bluish. ♂ *mainly black* with *contrasting silvery gray wings*; head feathers short and plush-like; *sides of neck and throat unfeathered and pale blue, the skin wrinkled and folded.* ♀ slatier gray generally, wings concolor with back, paler on belly and undertail; bare skin on neck less extensive. Juvenile like ♀ but scaled whitish below, in some very extensively. Found singly and in small groups, often flying high above ground, flapping with distinctive deep, rowing wingstrokes. In flight the silvery on ♂'s wings flashes very conspicuously. Perched birds hop from branch to branch, toucan-like, and often are in the open. Eats both fruit and large insects. Usually silent, but displaying ♂♂ occasionally give a very low-pitched booming sound.

PURPLE-THROATED FRUITCROW *Querula purpurata* ♂ 28 cm | 11" ♀ 25.5 cm | 10"
Uncommon in midlevels and subcanopy of forest along *NW edge of our area*. Bill leaden blue. *Entirely black*, ♂ with *large throat patch shiny reddish purple* (feathers lanceolate). No similar black bird shares the ♂'s purple throat patch. Can be confused with various mainly black caciques, but note the fruitcrow's chunky, short-tailed shape and stocky bill. In flight the fruitcrow shows broad rounded wings, quite different from any icterid. Ranges through forest in noisy bands of 3-6 birds in which purple-throated (adult) ♂♂ are a minority. Often they accompany flocks of other birds such as caciques or nunbirds, but they also troop about on their own. Flight is swooping, often punctuated by glides. They eat both fruit and large insects. Give a variety of loud and mellow calls, including a drawn-out "ooo-waáh" or a "kwih-oo, ooo-waáh," and in flight often a "wah-wah-wheéawoo." Calling birds often flick or spread their tail. Imitating their calls can sometimes lure them in close, the ♂♂ flaring their gorgets and spreading their tails.

RED-RUFFED FRUITCROW *Pyroderus scutatus* ♂ 38-43 cm | 15-17" ♀ 35.5-39.5 cm | 14-15.5"
Rare to uncommon and local in midlevels and subcanopy of forest in *S of our area*. Bill bluish gray (♂) or dusky (♀). *Black* with a *large shiny flame-red bib on throat and chest* (with somewhat crinkled effect). This *large*, spectacular cotinga can hardly be confused. A mostly solitary and inconspicuous bird, this fruitcrow tends to remain inside forest and rarely perches in the open. At times it can seem almost curious, even approaching the observer. ♂♂ gather in small leks (no more than a couple of birds) where they display on branches close to the ground, leaning forward and extending their red foreneck feathers, which hang away from the body like a ruff. They then emit a very deep booming call, "ooom-ooom-ooom." Away from their leks these fruitcrows are mainly silent, but occasionally a single "ooom" is given.

BARE-NECKED FRUITCROW
ANAMBÉ-POMBO

PURPLE-THROATED FRUITCROW
ANAMBEÚNA

RED-RUFFED FRUITCROW
PAVÓ

JAYS & CROWS (Corvidae) are *fairly large, noisy*, and *highly social* birds that are conspicuous and hard to miss in woodland and semiopen areas. Only jays occur in South America. In all of our species *some shade of blue predominates*.

CURL-CRESTED JAY *Cyanocorax cristatellus* 34.5 cm | 13.5"

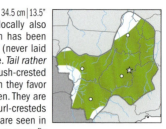

Fairly common in cerrado and adjacent gallery woodland, locally also in agricultural terrain where at least a little native vegetation has been left standing. Iris dark. *Conspicuous upstanding frontal crest* (never laid flat). Head and foreneck sooty black, otherwise dull blue above. *Tail rather short, its terminal half white.* Lower underparts white. Both Plush-crested and White-naped Jays can occur with the Curl-crested, though they favor woodland and are much less likely to range as far into the open. They are yellow-eyed, lack the crest, and have pale facial markings. Curl-cresteds are bold jays that generally range in small groups; they often are seen in straggling, slow flight across semiopen country, and sometimes even fly quite high. Their most frequent vocalization is a loud and harsh "kyaar!" often repeated several times.

WHITE-NAPED JAY *Cyanocorax cyanopogon* 34.5 cm | 13.5"

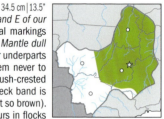

Uncommon and local in deciduous and gallery woodland in *N and E of our area*. Iris yellow. Head black with stiff forecrown feathers, facial markings and a short moustache blue; *broad area of white on hindneck*. Mantle dull brownish; tail broadly tipped white. Throat and chest black, lower underparts white. Plush-crested Jay is similar; it and the White-naped seem never to have been found together, though their ranges come close (Plush-crested replacing White-naped to W and S). Plush-crested's pale hindneck band is considerably smaller and bluer, and its mantle is much bluer (not so brown). White-naped also lacks Plush-crested's rear-crown "bulge." Occurs in flocks of up to 8-12 birds that forage at all levels in trees. Though sometimes bold, they usually do not range much into the open. The most frequent of a wide variety of calls is a loud ringing "cho-cho-cho."

PLUSH-CRESTED JAY *Cyanocorax chrysops* 35.5 cm | 14"

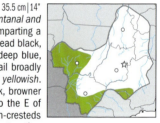

Locally uncommon to fairly common in woodland and forest in *Pantanal and adjacent areas*. Iris yellow. *Crown feathers stiff and plush-like*, imparting a flat-topped look and a *somewhat crested effect on rear crown*. Head black, with spots above and below eye silvery blue, short moustache deep blue, and *milky bluish white hindneck*. Otherwise violet-blue above, tail broadly tipped white. Throat and chest black; *lower underparts creamy yellowish*. White-naped Jay has an obvious and extensive white hindneck, browner upperparts, and lacks the "bulge" on rear crown. It is found to the E of Plush-crested's range, and apparently never occurs with it. Plush-cresteds range in flocks of up to about a dozen birds, sometimes moving together with Purplish Jays. General behavior and voice much as in White-naped; in some areas they can become quite tame.

PURPLISH JAY *Cyanocorax cyanomelas* 37 cm | 14.5"

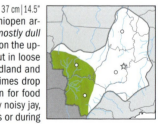

Common in deciduous and gallery woodland and adjacent semiopen areas, *mainly in and around Pantanal*. A *large, relatively drab jay, mostly dull purplish mauve*, blackish on head and foreneck, with violet only on the upperside of its tail. *Our only all-dark jay*. Purplish Jays troop about in loose straggling flocks of up to 8-10 birds. They mainly forage in woodland and edges but also regularly come out into open areas, and sometimes drop to the ground. Conspicuous and often seen, they even come in for food put out for birds at certain ecolodges and ranches. A particularly noisy jay, giving loud raucous "craa-craa-craa..." calls as they perch in trees or during their rather labored flight.

CURL-CRESTED JAY
GRALHA-DO-CAMPO

WHITE-NAPED JAY
GRALHA-CANCÃ

PLUSH-CRESTED JAY
GRALHA-DO-MATO

PURPLISH JAY
GRALHA-DO-PANTANAL

SWALLOWS & MARTINS (Hirundinidae) are familiar *aerial* birds with *long pointed wings*; unlike the superficially similar swifts, they are *often seen perching on branches and wires*. Highly insectivorous, swallows capture their prey in flight.

PROGNE martins are *large, broad-winged* swallows with forked or notched tails. They have semimusical, gurgling calls and are notably gregarious in semiopen areas.

PURPLE MARTIN *Progne subis* 18-19 cm | 7-7.5"
Locally common boreal migrant (Sep-Mar) to semiopen terrain, often near water; recorded mainly from Pantanal. ♂ *glossy blue-black*; tail forked. ♀ brown above with a blue gloss (often faint); *whitish or pale grayish forecrown and nuchal collar*. Below grayish brown with some dark scaling, belly paler, sometimes with dark streaking. *All-dark* ♂ distinctive. Gray-breasted Martin differs from ♀ Purple in *lacking* pale forecrown and nuchal collar. Roosting aggregations can be immense. Breeds in North America.

GRAY-BREASTED MARTIN *Progne chalybea* 19 cm | 7.5"
Common summer resident (Aug-Apr) to semiopen and inhabited areas (to N some are resident year-round). *The most widespread martin here.* Steely blue above (♂ glossier); throat and breast grayish brown; *belly whitish* sometimes with dark streaking. Cf. Purple Martin. Brown-chested is brown-backed, has a chest band. Other swallows are smaller. Often flies high, with a leisurely, languid flight style. Nests are usually placed under the eaves of buildings or other structures; somewhat colonial.

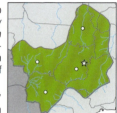

BROWN-CHESTED MARTIN *Progne tapera* 18 cm | 7"
Locally common summer breeding resident (Aug-Apr) in semiopen areas; in N some are probably year-round residents. Favors areas near water. Breeding birds (**A**) *dull grayish brown above*. White below with a *prominent brownish chest band*. Austral migrants (**B**; few records) have brown spots on median breast. Sand Martin has a similar color pattern but is smaller with different flight style. Other martins are never brown-backed. Brown-chested swoops around mainly quite low, often with wings bowed and exposing the silky white on sides of tail. Gregarious, though pairs separate out when breeding. They nest in old hornero nests or burrows dug in banks, occasionally in holes in snags.

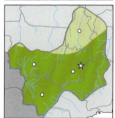

SAND MARTIN *Riparia riparia* 12 cm | 4.75"
Uncommon boreal migrant (Sep-Apr) to open areas and near water. Often called Bank Swallow. A *small* swallow, *uniform brown above*, somewhat darker on wings. White below with *distinct brown chest band* (sometimes hard to see in flight), usually extending down as a "spike" in the middle. Brown-chested Martin has a similar color pattern, but is larger and has a different flight style. Cf. S. Rough-winged Swallow. Often in groups, Sand Martins frequently associate with Barn and American Cliff Swallows. Their flight is fast, erratic, and "fluttery." Call, often given in flight, a short gravelly "chirt." Breeds in North America.

BARN SWALLOW *Hirundo rustica* 14-16.5 cm | 5.5-6.5"
Common boreal migrant (Aug-May), widespread in semiopen terrain, favoring areas near water. *Tail long and deeply forked* (tail streamers shorter in juveniles and molting adults); *some fork is always evident*, and tail has *white spots on inner webs*. Steely blue above, chestnut forehead. Throat chestnut to buff, *cinnamon-buff to buffy whitish below* (deepest in breeding-plumage ♂). Juveniles and worn-plumaged adults duller above, paler below; these are the most apt to be confused with other swallows, especially the stockier American Cliff. Feeding birds often swoop rapidly low over water or nearby. Locally large numbers roost in sugarcane fields. Usually quiet when here, but sometimes gives a rising "veeet?" Breeds in North America.

AMERICAN CLIFF SWALLOW *Petrochelidon pyrrhonota* 13-13.5 cm | 5-5.25"
Fairly common boreal migrant (Sep-Apr) to open and semiopen areas, especially near water but also over cerrado or agricultural areas. Can be abundant during Oct at Emas NP, especially just after termites have emerged with the first rains. Tail square. Mostly steely blue-black above with *prominent buffy whitish forehead* (chestnut in at least one race), grayish nape, and *contrasting cinnamon-rufous rump*. Sides of head and throat mostly chestnut, below grayish white. No other swallow shares the conspicuous tawny rump. Often associates with Barn Swallows. Sometimes gives a rough "drrt" call, especially in flight. Breeds in North America.

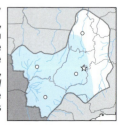

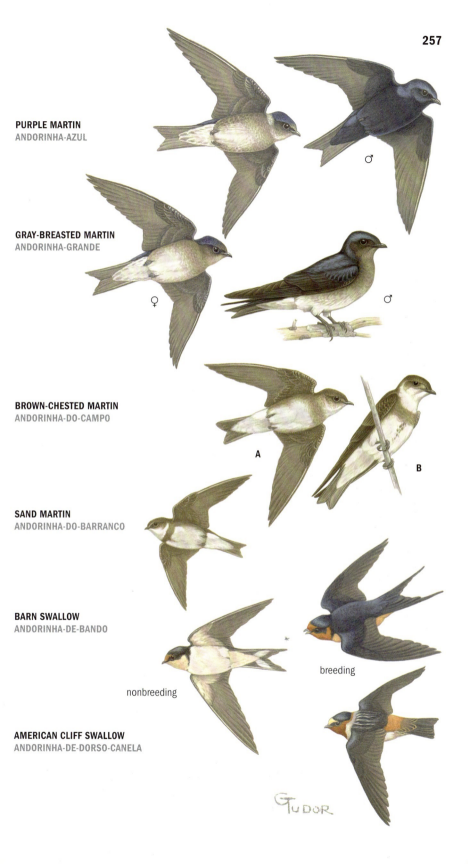

257

PURPLE MARTIN
ANDORINHA-AZUL

GRAY-BREASTED MARTIN
ANDORINHA-GRANDE

BROWN-CHESTED MARTIN
ANDORINHA-DO-CAMPO

SAND MARTIN
ANDORINHA-DO-BARRANCO

BARN SWALLOW
ANDORINHA-DE-BANDO

AMERICAN CLIFF SWALLOW
ANDORINHA-DE-DORSO-CANELA

The swallows on this page, classified in several genera, are attractive and often familiar birds that range in *open and semiopen areas*. Some species favor areas near water. Like the others, they are *adept aerial feeders*.

WHITE-WINGED SWALLOW *Tachycineta albiventer* 13.5 cm | 5.25"
Locally common in open areas near rivers, lakes, and ponds. Unique among our swallows, the White-winged has a *large, usually conspicuous white patch on its inner flight feathers and upperwing-coverts* (but in worn plumage the white can be abraded and not so obvious). Rump also white. Otherwise glossy bluish green above, pure white below. Immature grayer above. Conspicuous and often confiding, this lovely swallow is often seen perching on snags or rocks just above the water, from there flying out gracefully to skim the surface. It rarely strays very far from water and isn't particularly gregarious. Common call a rising "tree-eet?"

White-banded Swallow (*Atticora fasciata*) is a local resident along the Rio Paraguay near Cáceres and around Serra das Araras, where it flies low over forest-bordered rivers. A beautiful swallow with a long, deeply forked tail; all steely blue-black with a conspicuous white chest band.

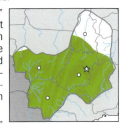

WHITE-RUMPED SWALLOW *Tachycineta leucorrhoa* 13.5 cm | 5.25"
Fairly common and widespread in semiopen areas and woodland edges, but scarce in Pantanal. Dark glossy blue above (greener in fresh plumage) with a *white rump* and *narrow but distinct white supraloral streak*. Pure white below. Immature brown above. Most likely confused with White-winged Swallow, though that has white in the wings that is never seen in White-rumped. White-rumped is generally a less familiar swallow than the White-winged, and is much less strictly tied to the vicinity of water; generally in scattered pairs and small groups. Their calls are similar.

BLUE-AND-WHITE SWALLOW *Notiochelidon cyanoleuca* 12-12.5 cm | 4.75-5"
Locally fairly common in semiopen areas, mainly at higher elevations and scarce or lacking in Pantanal. *Glossy steel blue above*. White below with a *prominent black crissum*. Immature browner above. Austral migrants (which breed in Patagonia) are slightly larger with a grayer underwing, and have black on crissum more restricted to its sides. White-winged and White-rumped Swallows are larger and both have white rumps. White-winged also shows obvious white on wing. Blue-and-white is a familiar swallow that in some areas can be seen around towns and buildings, nesting under eaves and in crevices of rock faces; it also nests in roadcuts. It habitually rests in groups on wires. Frequent call a long, thin, rising "treeeeeelee" or a buzzier "tz-tz-tz-z-z-z-zee?" In flight gives a "tzee."

TAWNY-HEADED SWALLOW *Alopochelidon fucata* 12 cm | 4.75"
Uncommon and local in semiopen areas, especially in campos and grassy cerrado, apparently only in S and E of our area. An attractive, *small swallow whose status is still not well understood here*; it may be somewhat migratory. *Cinnamon superciliary, face, and nuchal collar merge into buff throat and upper chest*. Otherwise grayish brown above; white below. Neither the larger Southern Rough-winged Swallow nor the Sand Martin shows any of this species' cinnamon and tawny on the face (though beware, these colors are sometimes surprisingly hard to see). Tawny-headeds occur in pairs or small flocks, and generally do not associate with other swallows; only infrequently do they perch on wires. They forage while flying low over grasslands and nest in holes in banks, sometimes in isolated buildings.

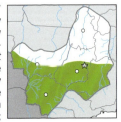

SOUTHERN ROUGH-WINGED SWALLOW *Stelgidopteryx ruficollis* 13 cm | 5"
Fairly common to common and widespread in partially forested or wooded regions and semiopen areas, *almost always near water*. Tail notched. *Above grayish brown*, darkest on forecrown, wings, and tail, somewhat paler on rump. *Throat cinnamon-buff*; below pale grayish brown, midbelly tinged creamy yellowish. Tawny-headed Swallow is smaller, has cinnamon and buff on face; Brown-chested Martin is considerably larger, has a dark chest band. Found in small groups that often rest on dead branches or wires, Rough-wingeds sometimes fly with other swallows but usually are on their own. They are especially numerous along forested or wooded rivers and streams, there nesting in holes in banks, either alone or in small groups. Most frequent call a rough, upslurred "djreeet."

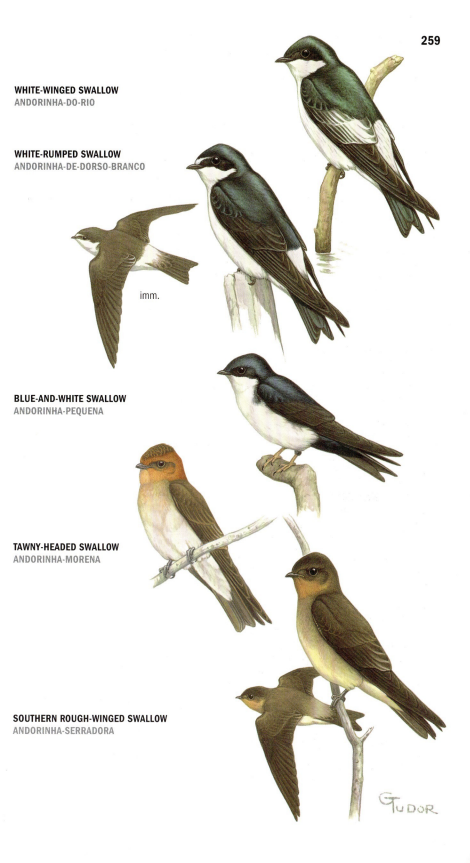

WHITE-WINGED SWALLOW
ANDORINHA-DO-RIO

WHITE-RUMPED SWALLOW
ANDORINHA-DE-DORSO-BRANCO

BLUE-AND-WHITE SWALLOW
ANDORINHA-PEQUENA

TAWNY-HEADED SWALLOW
ANDORINHA-MORENA

SOUTHERN ROUGH-WINGED SWALLOW
ANDORINHA-SERRADORA

MOCKINGBIRDS (Mimidae) are *slender, long-tailed* birds that are *conspicuous in open country*. They are named for their mimicry of other birds, though neither of our species is particularly adept at this.

CHALK-BROWED MOCKINGBIRD Mimus saturninus 26 cm | 10.25"

Common and conspicuous in cerrado, agricultural areas, gardens, and around buildings. Tame and easily seen. Grayish brown above *mottled darker* with a *broad "chalky" white superciliary* and blackish eyestripe; tail long, with *broad white tail corners*. Whitish below, but in the dry season many birds can be tinged rufescent from dust. White-banded Mockingbird (a scarce austral migrant) shows more white in wings, and has all-white outer tail feathers. Bold and hard to miss (it is active even in the heat of mid-day), Chalk-broweds walk and run on the ground with their tail raised, usually in the open. The tail is often thrown forward as the bird alights. They range in pairs or small family groups, never in flocks. Song, usually delivered from atop a shrub or low tree, is variable in form and content and has little pattern; actual mimicry is limited. Frequently heard call a sharp "chert."

WHITE-BANDED MOCKINGBIRD Mimus triurus 23.5 cm | 9.25"

Uncommon austral migrant (May-Sep) to shrubby and semiopen areas in Pantanal and Mato Grosso do Sul; most numerous southward. A handsome mockingbird whose *bold white on inner flight feathers* is especially prominent in flight, showing as a block on the rear wing. Above grayish brown with long white superciliary and *rufescent rump*; tail black with *outer feathers entirely white* (most evident in flight). Mostly whitish below. Much more numerous Chalk-browed Mockingbird has *much less white on wings and tail*. Behavior much as in Chalk-browed, and White-banded's demeanor can be equally bold. Song, heard especially just prior to southward migration, a long continued but rather leisurely stream of pure melodic notes mixed with imitations of the songs of other birds. Breeds mainly in Argentina.

BLACK-CAPPED DONACOBIUS Donacobius atricapilla 22 cm | 8.5"

Fairly common in marshes, lake margins, and wet pastures with tall grass; numerous in the Pantanal. *Sleek* and unmistakable, with a bold yellow iris. *Chocolate brown above* with *black head*; small white wing patch; rather long tail with *broad white tips* (obvious in flight). *Below rich buff*. Usually conspicuous, the donacobius usually is spotted as it perches in pairs in marsh grass and atop low shrubs. Its flight is direct and rather slow (despite its fast wingbeats). A very vocal bird, giving a variety of loud whistled calls and churrs, often repeated in series, e.g., a liquid "whoit, whoit, whoit.... " Pairs often duet, perching next to each other and bobbing their heads while swivelling their fanned tails; they sometimes also inflate an orange pouch on the sides of the neck. Formerly considered a wren, but now classified in its own family.

CHALK-BROWED MOCKINGBIRD
SABIÁ-DO-CAMPO

WHITE-BANDED MOCKINGBIRD
CALHANDRA-DE-TRÊS-RABOS

BLACK-CAPPED DONACOBIUS
JAPACANIM

WRENS (Troglodytidae) are essentially American in distribution; only one species reaches the Old World. They are small, compact birds, basically brownish, often with dark barring on wings and tail. Some are notably skulking, others much more familiar. All are highly vocal. Diversity is greatest in Middle America, with only a few species occurring in our area.

SOUTHERN HOUSE WREN *Troglodytes musculus* 11.5 cm | 4.5"
Common and widespread in open and semiopen habitats. True to its name, *most numerous around buildings and in populated areas,* avoiding extensively wooded regions. *Nearly uniform brown and buffyish,* more whitish below and more rufescent on rump, with a weak whitish superciliary and *indistinct dusky barring on wings and tail* (latter usually cocked). S. House Wrens are almost always in pairs and are confiding, hopping in shrubbery, along walls, or on houses. Their gurgling, warbled song, which usually ends with an accented note, can be repeated more or less constantly, often from a prominent perch. A nasal "jeeyah" scold is frequent. Often called simply House Wren (*T. aedon*), this when considered conspecific with North American birds.

THRYOTHORUS wrens are *midsized* members of their family. Because of their *skulking habits* they are not particularly familiar birds (though their loud vocalizations, sometimes given antiphonally, do often attract attention). All our species are mainly rufous above, buff to grayish below.

BUFF-BREASTED WREN *Thryothorus leucotis* 14-14.5 cm | 5.5-5.75"
Fairly common in woodland thickets and forest borders, almost always near water. Rufescent brown above with white superciliary and dark-streaked cheeks; wings and tail barred black. Throat white, *buff below deepening to rufous on belly*. In Pantanal, cf. the similar Fawn-breasted Wren. Moustached Wren has gray on nape and crown, a bold whisker, grayer underparts, and *unbarred wings*. Buff-breasted ranges in pairs, hopping in dense shrubbery near water. Even when singing it can be hard to see,. The loud vigorous song is a series of fast staccato phrases, each repeated a number of times before switching to another. Oft-heard call a "cheet cho-cho," sometimes just the "cho-cho."

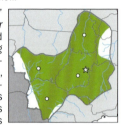

FAWN-BREASTED WREN *Thryothorus guarayanus* 13.5 cm | 5.25"
Fairly common in woodland undergrowth, usually near water, in S Pantanal (N to Porto Jofre area). Resembles Buff-breasted Wren; slightly smaller, duller, and paler than the Buff-breasted race found just to the N. *In the field it's usually best simply to go by range*: in N Pantanal closer to Poconé only Buff-breasted occurs, with Fawn-breasted entirely replacing it by the time you reach Porto Jofre. There is a narrow overlap zone. Behavior and song much as in Buff-breasted, but Fawn-breasted also often gives distinctive shorter and much slower vocalizations; "sweeyr, cheer-chowee, chowee." Also gives a notably slow "chee! cheeu, cheeu," a "cheeyp, cheeu cheeu-cheeu," and a "cheeyp, chee-pu, chee-puu," quite different from Buff-breasted vocalizations.

MOUSTACHED WREN *Thryothorus genibarbis* 15.5 cm | 6"
Uncommon in dense lower growth of deciduous and gallery woodland and forest borders, mainly in NW of our area. Crown grayish brown, gray on nape, rufous-brown above; sides of head boldly streaked black and white, and with a thin, *well-defined black moustachial streak*; tail boldly barred black. *Pale grayish below,* buffier on flanks. Buff-breasted Wren is notably buffier below and lacks the moustachial streak and gray on head. Usually ranges in pairs, tending not to accompany flocks. Skulking, and heard much more often than seen. Song a series of fast rollicking phrases, some ending with a distinctive "chochocho." Also heard are querulous "jeeyr" calls, these often interspersed with various scolds.

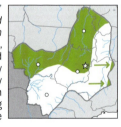

THRUSH-LIKE WREN *Campylorhynchus turdinus* 20.5 cm | 8"
Fairly common (certainly frequently heard) in deciduous and gallery woodland and forest borders in Pantanal and nearby. By far our largest wren, thus nearly unmistakable. Brownish gray above with dark scaling and a whitish superciliary. *Below whitish,* sometimes with vague grayish spots. No other wren is nearly as large; cf. Chalk-browed Mockingbird. Usually found in family groups of 6-8 birds that hop and clamber on larger limbs at varying levels. They are often bold, sometimes coming to the trees and shrubbery around Pantanal ecolodges and farm buildings. Its explosive, far-carrying song, often given as a duet, is rather musical and often includes the phrase "chookada-doh, choh, choh" or "chooka-chook-chook."

SOUTHERN HOUSE WREN
CORRUÍRA

BUFF-BREASTED WREN
GARRINCHÃO-DE-BARRIGA-VERMELHA

FAWN-BREASTED WREN
GARRINCHÃO-DO-OESTE

MOUSTACHED WREN
GARRINCHÃO-PAI-AVÔ

THRUSH-LIKE WREN
CATATAU

GRASS WREN *Cistothorus platensis* 9.5-10 cm | 3.75-4"
Uncommon and local in less-disturbed campos and grassy cerrado. A small buffy brown wren with a *streaked back* and barred wings; crown blackish streaked white; underparts whitish. *Tail coarsely barred.* S. House Wren, *a far more numerous bird*, is much plainer overall, with only indistinct wing and tail barring, a less prominent brow, no back streaking. Grass Wrens are usually hard to see as they habitually creep about, almost mouse-like, on or near the ground in dense vegetation. When breeding the ♂'s complex musical song (repetitions of various gurgles and trills) does reveal its presence, especially as these songs often are delivered from an exposed perch where the bird is much easier to see.

PIPITS (Motacillidae) are *inconspicuous slim, brownish, streaky* birds that are *terrestrial in open, grassy country*. They are most in evidence during the nesting season when ♂♂ engage in frequent flight displays.

YELLOWISH PIPIT *Anthus lutescens* 13 cm | 5"
Locally fairly common in campos and agricultural areas, most often near water and where grass is fairly short. Rather long pinkish legs. Above brown streaked buff and blackish; tail dusky, *outer feathers white* (most evident in flight). *Buffy yellowish below* (whiter in worn plumage) with dark streaking on chest (often faint). Cf. the much rarer Ochre-breasted Pipit; otherwise not likely confused though it can recall various "streaky" finches and sparrows (these have very different behavior, etc). An essentially terrestrial bird that *walks* instead of—like most finches—hopping. Singing ♂♂ give a series of "tzit" or "tizit" notes as they ascend some 10-20 m into the air, then a long slurred "dzeeeeeeeeeeeu" while they glide back to earth.

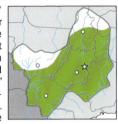

OCHRE-BREASTED PIPIT *Anthus nattereri* 14.5 cm | 5.75"
VU *Rare and local in less-disturbed campos*, usually where the grass is fairly long; in our area known to occur only at *Serra da Canastra NP* but possibly more wide-ranging. *Above boldly streaked blackish and golden ochre*; tail dusky with outer feathers white. Throat whitish; *breast rich golden ochre with heavy black streaking*; belly whitish. Yellowish Pipit is much more common, smaller, and much less richly colored. Ochre-breasted generally occurs in well-dispersed pairs; still not very well known. ♂'s display involves mounting some 20-25 m into the air while giving a complex and long-continued musical song that ends with a slurred "eeeeeeeur" as it drops back to the ground.

GNATCATCHERS & GNATWRENS (Polioptilidae) are small, *long-tailed* birds with animated behavior that range in woodland and borders.

LONG-BILLED GNATWREN *Ramphocaenus melanurus* 12 cm | 4.75"
Uncommon in lower growth of deciduous and gallery forest and woodland. *Very long slender bill; long slim tail.* Brown above, *more rufescent on face*; tail black, *all but central feathers tipped white*. Whitish to pale grayish below. No wren has such a long bill, and no wren's tail is white tipped. An inconspicuous bird that favors viny tangles at borders and in tree-fall openings, the Long-billed Gnatwren occurs in pairs that hop actively, flipping and wriggling their tails around; they sometimes join mixed flocks. Heard more often than seen, with song a distinctive clear musical trill, often preceded by a few "cht" notes.

MASKED GNATCATCHER *Polioptila dumicola* 12.5 cm | 5"
Fairly common in gallery and lighter woodland, sometimes ranging out into adjacent cerrado and scrub. ♂ in most of our area (**A**) *gray above with a narrow black mask*; wings and *long narrow tail* black; inner flight feathers edged white and *outer tail feathers white*. Below paler gray, belly whitish. In Mato Grosso do Sul (**B**) both sexes are darker and more bluish gray above, *more grayish below*. ♂ here has a wider black mask bordered below by a white line; ♀ has *only a narrow black stripe on ear-coverts*. Masked is the only gnatcatcher here, so it's hard to confuse. Usually forages in lower growth, generally in pairs and often accompanying small mixed flocks. Song sweet and musical, typically consisting of a few short notes, sometimes several repetitions of the same lilting phrase, then switching to another, and so on.

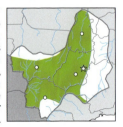

GRASS WREN
CORRUÍRA-DO-CAMPO

YELLOWISH PIPIT
CAMINHEIRO-ZUMBIDOR

OCHRE-BREASTED PIPIT
CAMINHEIRO-GRANDE

LONG-BILLED GNATWREN
BALANÇA-RABO-BICUDO

MASKED GNATCATCHER
BALANÇA-RABO-MASCARADO

THRUSHES (Turdidae) are well-known but *rather dull-plumaged* midsized songbirds that range widely in semiopen as well as wooded and forested habitats; some are quite familiar and numerous. Many species habitually hop on the ground, and are often in the open. Most have notably *melodic songs*.

CREAMY-BELLIED THRUSH *Turdus amaurochalinus*   23 cm | 9"
Fairly common resident in lighter woodland and clearings, but only an uncommon austral winter visitant in N of our area. ♂'s *bill bright yellow*, sometimes with dusky tip, mostly dusky in ♀. Olive brown above with *blackish lores*. Below pale brownish gray, midbelly whiter, with *sharp black streaking on throat*. No similar thrush has the black on its lores. Pale-breasted Thrush has a grayer head, weaker throat streaking. *Creamy-bellied shivers its tail upon alighting*; some other thrushes do this occasionally, but not so habitually. Otherwise behavior similar to other non-forest thrushes. Song a typical *Turdus* caroling, but less forceful than many others. Call a sharp "pok."

PALE-BREASTED THRUSH *Turdus leucomelas*   23 cm | 9"
Locally common in forest borders, secondary woodland, clearings, and gardens. Bill olive yellowish. *Contrasting gray head and nape*; rufescent brown above and mostly pale grayish buff below with dusky throat streaking. Most likely confused with Creamy-bellied Thrush (the two are regularly together), though Pale-breasted lacks Creamy-bellied's dark lores and bold throat streaking; Creamy-bellied's head does *not* contrast. Regularly feeds while hopping on the ground, often in the semiopen and sometimes quite bold. Song a series of melodic phrases, usually delivered from a hidden perch and often hard to distinguish from its congeners. More distinctive is its harsh, usually trebled call, a fast "wert-wert-wert."

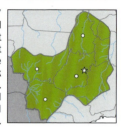

EASTERN SLATY THRUSH *Turdus subalaris*   21.5 cm | 8.5"
Uncommon austral winter visitant (mostly Apr-Oct) to forest and gallery woodland; status here still somewhat unclear. *Yellow bill, eye-ring, and legs*. ♂ gray above, sometimes washed with olive. *Prominent white crescent across upper chest*, below the black throat streaking; below grayish, midbelly white. ♀ has yellowish brown bill and *white crescent across chest*; it is grayish brown above, mostly pale brownish gray below. Rather inconspicuous on its wintering grounds, when mostly arboreal. Though it apparently does not breed here, Eastern Slaty's rather weak song is nonetheless sometimes heard; this consists of a series of brief bursts of high-pitched notes with a "tinny," somewhat squeaky quality. Breeds mainly in southern Brazil.

White-necked Thrush (*T. albicollis*) occurs in forest and woodland lower growth in Mato Grosso. It generally resembles Eastern Slaty Thrush but is browner above and has a *blackish bill and legs*. Mainly a forest interior thrush, often hopping on ground, White-necked has a melodic but rather slow and monotonous song.

COCOA THRUSH *Turdus fumigatus*   23 cm | 9"
Fairly common in gallery and riparian woodland and forest in NW of our area; favors areas near water. *Uniform warm rufescent brown* with lower belly and crissum paler and buffier; (but variable, some birds almost uniform below). Throat lightly streaked dusky. *No other thrush is so uniformly rufescent*. Generally a shy and retiring bird, less familiar than many of our other thrushes. Seen singly or in pairs, foraging both on the ground and in trees, for fruit. Song a long and musical caroling comprised of numerous slurred phrases, usually given from a hidden perch quite high above the ground.

RUFOUS-BELLIED THRUSH *Turdus rufiventris*   24.5 cm | 9.75"
Widespread and often common in semiopen terrain, woodland, and gardens. Bill olive yellowish. A mostly brown thrush with contrasting *orange-rufous belly and crissum* (somewhat paler in ♀); throat white with dusky streaking. *The rufous below is unique among our thrushes*. Though it regularly hops on the ground and often occurs around houses, only infrequently does the Rufous-bellied Thrush emerge very far from cover. Its attractive melodic song is a fast rich caroling, usually given from a hidden perch, often at first light or even before. In many areas often kept (*too often*) as a cagebird.

Veery (*Catharus fuscescens*) is a rare boreal migrant to undergrowth of forest and woodland, with most records from the Oct-Nov period. *The smallest thrush occurring in our area* (18 cm / 7"). Above uniform rufescent brown. Below whitish with *a few brownish spots on chest*. Solitary and inconspicuous on its wintering grounds, usually seen hopping on or near the ground in dense cover. Call a distinctive downslurred "pheeuw," also a more nasal "waaa-a-a-a." Breeds in North America.

CREAMY-BELLIED THRUSH
SABIÁ-POCA

PALE-BREASTED THRUSH
SABIÁ-BARRANCO

EASTERN SLATY THRUSH
SABIÁ-FERREIRO

COCOA THRUSH
SABIÁ-DA-MATA

RUFOUS-BELLIED THRUSH
SABIÁ-LARANJEIRA

VIREOS (Vireonidae) are *rather plain* arboreal birds found only in the Americas. All are best known from their *frequent and repetitive vocalizing*.

RUFOUS-BROWED PEPPERSHRIKE Cyclarhis gujanensis   15 cm | 6"
Widespread and often common in woodland, clearings, and forest borders, sometimes even in towns. *Heavy hooked bill pale horn*; legs pinkish. Olive above with gray head and neck; *obvious rufous superciliary*. Whitish below with *broad olive-yellow chest band*. Cf. Red-eyed Vireo (less robust) and Green-backed Becard (slighter bill; lacks the rufous brow, ♂ with black crown). Peppershrikes glean sluggishly in foliage at varying heights. They regularly join mixed flocks but also forage alone or in pairs. Though often numerous, they are *always heard far more often than seen*. Song, frequently given even at mid-day, a series of brief, melodious phrases, each repeated many times before switching to a new one. A slurred call, a progressively lower-pitched "dreeu" repeated 3-7 times, is also regularly given.

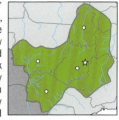

RED-EYED VIREO Vireo olivaceus   14.5 cm | 5.75"
Locally fairly common to common in a variety of wooded and forested habitats; in S of our area apparently only a summer resident. *Crown gray with prominent white superciliary bordered above and below by blackish lines*; olive above. Whitish below with some greenish yellow on sides, flanks, and crissum. Greenlets are smaller and shorter-tailed. Arboreal, gleaning for insects at varying levels in trees; often moves with mixed flocks. Also eats much fruit, especially when not breeding. Song a series of short, clipped, somewhat musical phrases with brief intervening pauses, given only when breeding.

GREENLETS are *small* vireos with pointed bills. Aside from their oft-given vocalizations, they are rather *inconspicuous arboreal* birds of forest and woodland.

BUFF-CHEEKED GREENLET Hylophilus muscicapinus   11.5 cm | 4.5"
Uncommon and local in canopy and borders of forest in *NW of our area*. *Face bright buff; buff wash on breast*; otherwise olive above with gray crown and nape, whitish below. *No other greenlet shows buff on face or breast*, and this is the only greenlet ranging in our area normally found in forest. ♀ Rose-breasted Chat has a vaguely similar face pattern. An inveterate member of canopy flocks comprised of tanagers and other species; often hard to see as it usually remains well above ground. Recorded most often from its spritely call, the fast phrase "itsochuwéet" being typical and most frequent.

ASHY-HEADED GREENLET Hylophilus pectoralis   12 cm | 4.75"
Fairly common in deciduous and gallery woodland and shrubby clearings; numerous in Pantanal. Iris reddish brown. *Head and nape gray*; olive above. Whitish below with *breast contrastingly yellow*. Red-eyed Vireo is larger and lacks the yellow across breast. A rather confiding greenlet that frequently forages quite low and in the open. Usually occurs in pairs, gleaning in foliage, often accompanying mixed flocks. Regularly heard song a fast, musical "pooor pooor pooor pri i i i i i," with a distinctive terminal trill.

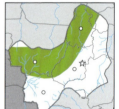

GRAY-EYED GREENLET Hylophilus amaurocephalus   12.5 cm | 5"
Rare to uncommon in deciduous woodland and scrub in *E of our area and in SW Mato Grosso do Sul* (so far only in the Corumbá region). Iris usually gray (can be dark?). Longish tail often held partially cocked. *Crown bright rufous with mottled blackish patch on ear-coverts*; bright olive above. Below buffy yellowish. ♀ Plain Antvireo, also rufous-crowned, is chunkier and shorter-tailed with a heavier, more hooked (not so pointed) bill. Often forages in pairs, sometimes low and in the open; regularly accompanies mixed flocks. Song a fairly loud series of 3-5 "chwee-erter" or "chee-wee" phrases sometimes interspersed with scolds and chatters.

TROPICAL PARULA Parula pitiayumi   11 cm | 4.25"
Widespread and locally common in a variety of wooded and forested habitats, but largely absent from actual Pantanal. ♂ *dull blue above* with *black foreface, olive patch on midback*, and *bold white wing-bars*. Below mostly bright yellow, *throat and breast washed with ochraceous or orange*. ♀ slightly duller. This small, pretty warbler is not likely to be confused. It is basically an arboreal bird, foraging in pairs and often accompanying mixed flocks, usually remaining high when in taller forest. Frequently given song a distinctive (though variable) accelerating and rising buzzy trill preceded by several thin notes: "tzip-tzip-tzip-tzip-tzrrrrrrrrip."

RUFOUS-BROWED PEPPERSHRIKE
PITIGUARI

RED-EYED VIREO
JURUVIARA

BUFF-CHEEKED GREENLET
VERDINHO-CAMURÇA

ASHY-HEADED GREENLET
VERDINHO-DE-CABEÇA-CINZA

GRAY-EYED GREENLET
VERDINHO-DE-OLHO-CINZA

TROPICAL PARULA
MARIQUITA

WOOD WARBLERS (Parulidae) comprise a varied group of small insectivorous birds found mainly in North and Middle America. They are closely related to the tanagers; note that the Tropical Parula (p. 269) is also a warbler. *Basileuterus* warblers are *rather dull-plumaged*. They mainly range in lower growth inside woodland and forest.

GOLDEN-CROWNED WARBLER *Basileuterus culicivorus* 12.5 cm | 5"
Locally fairly common in lower growth of forest and woodland. *Crown streak orange-rufous, superciliary whitish to pale gray*; otherwise olive above, *yellow below*. White-bellied Warbler (sympatric with Golden-crowned at a few localities) has whitish (not yellow) underparts. Forages actively, moving about constantly, flicking its wings and jerking its often cocked tail; easy to see and regularly with mixed flocks. Song a spritely series of 6-8 semimusical "swee" notes, with the next to last usually higher and accented.

WHITE-BELLIED WARBLER *Basileuterus hypoleucus* 12.5 cm | 5"
Fairly common in lower growth of deciduous woodland, gallery forest, and borders. *Crown streak orange-rufous*, superciliary whitish. Otherwise olive above, *whitish below*. Resembles Golden-crowned Warbler, and locally occurs with it, sometimes even in the same understory flock; Golden-crowned is *yellow below*. White-bellied favors drier regions. White-striped Warbler is larger with *no* rufous in crown, bolder white eyestripe. Behavior and vocalizations much as in Golden-crowned; they have been considered conspecific.

WHITE-RIMMED WARBLER *Basileuterus leucoblepharus* 14.5 cm | 5.75"
Fairly common on or near ground in forest and woodland, generally near water, in SE of our area. Head mainly gray with *white supraloral and prominent partial eye-ring*. Otherwise olive above, white below, *breast mottled gray*. White-striped Warbler has a prominent long white superciliary; it more or less replaces this species in central Brazil. Usually in pairs that walk and hop through thick undergrowth with tail frequently fanned, sometimes wagged or swiveled sideways; shy and generally hard to see. Oft-heard and enchanting song a series of tinkling notes, at first high-pitched but then gradually dropping.

WHITE-STRIPED WARBLER *Basileuterus leucophrys* 14.5 cm | 5.75"
Uncommon and local on or near ground inside swampy gallery forest from S Mato Grosso to Goiás and W Minas Gerais; not in Pantanal. Crown slaty with *striking broad white superciliary* and *black stripe through eye*. Otherwise bronzy olive above and white below, breast mottled gray. White-bellied Warbler is smaller with a less stout bill, orange-rufous crown stripe, different behavior. White-rimmed Warbler shows a much more prominent eye-ring and a much less bold brow. Usually in pairs, moving through thick, low undergrowth and hard to see well; not with flocks. The tail is frequently raised, spread, and swiveled sideways. Beautiful song a loud, ringing cascade of pure melodic notes; other shorter versions, e.g., "t-r-r-r-r, ti-i-i-i, tr-r-r," may be repeated several times in succession.

FLAVESCENT WARBLER *Basileuterus flaveolus* 14.5 cm | 5.75"
Fairly common *on or near ground* inside deciduous and gallery woodland. *Simply patterned*. Bright olive above and *bright yellow below*; short *superciliary also yellow*. Legs orange-yellow. Though it favors edge much more, Southern Yellowthroat can occur with the Flavescent; the yellowthroat is much less terrestrial, has gray on crown, no yellow superciliary. Usually in pairs that walk and hop on ground, often swiveling its fanned tail. Generally not with flocks. Heard much more often than seen, with song a loud, ringing, musical series of fast notes, "titi, teetee, teetee, chéw-chéw-chéw-chéw."

SOUTHERN YELLOWTHROAT *Geothlypis velata* 14 cm | 5.5"
Widespread and fairly common in shrubby areas and clearings, most often near water. ♂ with *black mask broadly outlined above with gray*; otherwise olive green above. Below bright yellow. ♀ has a *vague yellow supraloral and narrow eye-ring, crown and ear-coverts at least tinged gray, no mask*. Usually in pairs that tend to skulk in thickets or tall grass. More conspicuous when ♂♂ are giving their sweet, warbled song that resembles a seedfinch's or seedeater's, then often perching in the open. Call a fast chatter which distinctively drops in pitch and strength. Often considered conspecific with Masked Yellowthroat (*G. aequinoctialis*) of northern South America.

ROSE-BREASTED CHAT *Granatellus pelzelni* 12.5 cm | 5"
Rare and local in subcanopy and borders of forest and woodland in NW of our area. Unmistakable ♂ has head black with a *prominent white postocular stripe*; otherwise blue-gray above, tail black. Below mostly rosy red with white throat, white on flanks. ♀ *blue-gray above with cinnamon-buff forehead, face, and underparts; crissum pink*. ♀ Chestnut-vented Conebill lacks ♀ chat's buff on face and its pink crissum, has shorter tail (never cocked). The chat is an arboreal bird that usually forages in pairs and often accompanies mixed flocks; it gleans restlessly in foliage, with a *strong preference for vine tangles*. It tends to perch horizontally with tail cocked and fanned, wings often drooped. Song a greenlet-like series of 5-6 clear sweet notes: "sweet-sweet-tuwee-tuwee-tuwee-tuwee." Also gives a sharp "jrrt" call, sometimes in series.

TANAGERS (Thraupidae) comprise an *exceptionally variable* group of small to midsized passerines found only in the Americas. There is little agreement as to the "boundaries" between the tanagers, wood warblers, and emberizid and even fringillid finches. Tanagers are both insectivorous and frugivorous, and many are colorful though others are quite dull. Only a few are very memorable vocally. Several are among the most omnipresent of our birds, while others are rare and local, unlikely to be seen except by the most dedicated of birders.

ORANGE-HEADED TANAGER *Thlypopsis sordida* 13.5 cm | 5.25"
Fairly common and widespread in lighter and gallery woodland and cerrado. *Crown and sides of head orange-rufous*, becoming *bright yellow on face and throat; above otherwise gray*. Below dingy buffyish. ♀ duller, with face and foreneck yellowish. Ranges in pairs, gleaning actively in foliage and dead leaves at varying heights, usually not too high. Sometimes with mixed flocks. Not very vocal but has a high-pitched stuttering song, usually preceded by several longer and even higher-pitched notes.

HEMITHRAUPIS tanagers are small and arboreal. ♂♂ are colorful and easy to identify though ♀♀ are duller, essentially olive and yellow.

RUFOUS-HEADED TANAGER *Hemithraupis ruficapilla* 13 cm | 5"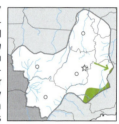
Locally fairly common in canopy and borders of forest and woodland, adjacent clearings, and gardens in *Minas Gerais*, e.g., Serra da Canastra. Bill mostly yellowish. Pretty ♂ has a *deep rufous head, large yellow patch on sides of neck*, and *orange-rufous throat and breast*. Bright olive above with an orange-rufous rump; below pale olive grayish. ♀ probably not distinguishable from ♀ Guira Tanager; fortunately they are not known to occur together (furthermore, both species usually range in pairs, and the ♂♂ are readily separated). Guira Tanager favors drier, more deciduous woodland. Usually in pairs, less often small groups, that glean actively in foliage at varying heights above ground. They often accompany mixed flocks.

GUIRA TANAGER *Hemithraupis guira* 13 cm | 5"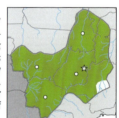
Locally fairly common in canopy and borders of deciduous forest and woodland and gallery woodland. Bill mostly yellowish. Distinctive ♂ has *black face and throat outlined by yellow* and *bordered below by orange-rufous breast*. Mainly olive above, lower back orange-rufous, rump yellow. Belly grayish. ♀ duller, olive above with *vague yellow superciliary and eye-ring*. *Throat and chest pale yellow*, below pale grayish. Rufous-headed Tanager replaces this species further E and S, seemingly with no overlap. Behavior of the two species is much the same.

HOODED TANAGER *Nemosia pileata* 13 cm | 5"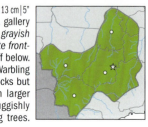
Fairly common and widespread in semiopen areas, forest borders, gallery forest, and cerrado; frequent in Pantanal. *Iris and legs yellow*. ♂ grayish blue above with *contrasting black head and sides of neck* and *white frontlet*. White below. ♀ similar but lacks the black cap and is tinged buff below. Not likely confused, but in S Mato Grosso do Sul cf. Black-capped Warbling Finch. Usually in small groups, sometimes accompanying mixed flocks but more often not. Hooded Tanagers frequently perch and hop on larger horizontal branches, peering at bark; they also glean rather sluggishly in foliage. They eat little fruit, but do sometimes visit flowering trees. Infrequently heard song a short series of "ti-chéw" phrases.

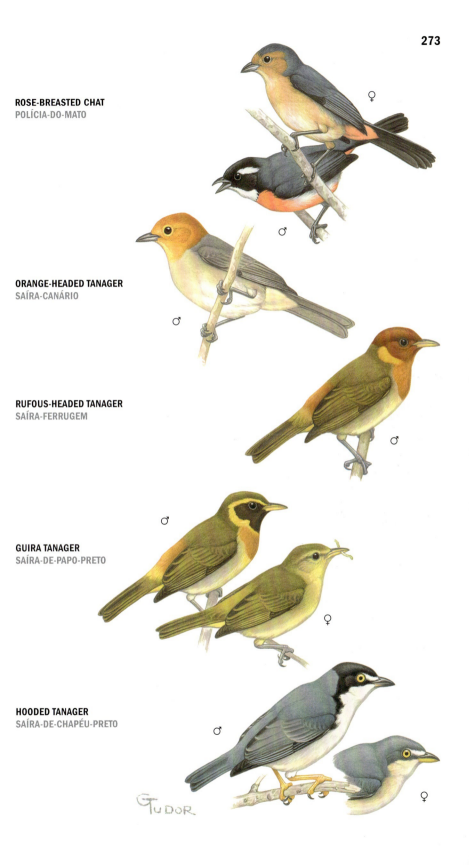

ROSE-BREASTED CHAT
POLÍCIA-DO-MATO

ORANGE-HEADED TANAGER
SAÍRA-CANÁRIO

RUFOUS-HEADED TANAGER
SAÍRA-FERRUGEM

GUIRA TANAGER
SAÍRA-DE-PAPO-PRETO

HOODED TANAGER
SAÍRA-DE-CHAPÉU-PRETO

BANANAQUIT *Coereba flaveola* 10.5-11 cm | 4-4.25"
Fairly common and widespread in a variety of semiopen and disturbed habitats, but seemingly less numerous in our area than it is in many other regions. *Thin, short, and distinctly decurved bill* sometimes shows pink at base. Blackish above with a *long white superciliary* and small white wing-speculum, yellow rump. *Throat gray,* contrasting with *yellow underparts.* Conspicuous birds, Bananaquits are active and seemingly nervous. They forage at all levels and often congregate in flowering trees, there probing for nectar in flowers; they also eat small fruits. The short song is frequently given and consists of a fast and variable series of shrill, buzzy notes. The Bananaquit is often placed in a monotypic family, the Coerebidae.

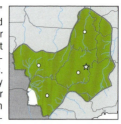

CHESTNUT-VENTED CONEBILL *Conirostrum speciosum* 11 cm | 4.25"
Fairly common and widespread in canopy and borders of woodland and forest. Sharply pointed bill. ♂ *uniform grayish blue,* slightly paler below, with a *chestnut crissum.* Most individuals have a small white wing-speculum. ♀ has *crown and nape bluish gray* contrasting with *fairly bright olive upperparts.* Below grayish white. ♂ is distinctive, but the comparatively dull ♀ can be tricky as it looks vaguely warbler-like; she often is best identified by the presence of her mate. ♀ Blue Dacnis is greener, especially below. The conebill is an arboreal bird that forages energetically in foliage at varying heights, sometimes briefly hanging upside-down. It regularly comes to flowering trees, e.g., *Cecropia* and *Erythrina.* Infrequently heard song a variable, jumbled series of sweet notes showing little pattern.

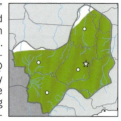

BLUE DACNIS *Dacnis cayana* 12.5 cm | 5"
Fairly common and widespread in a variety of forested and wooded habitats, also sometimes in clearings and gardens; seems largely absent from Pantanal. Pointed bill has pinkish at base; legs pinkish. Pretty ♂ *mostly turquoise blue; throat patch and back black;* wings and tail black edged blue. ♀ *bright green,* paler below, with a *bluish head.* Purple and Red-legged Honeycreepers have obviously decurved bills. The ♀ dacnis is most likely confused with certain immature *Tangara* tanagers, though none of these in our area shows any blue on head. Usually occurs in pairs or small groups that often accompany mixed flocks, foraging at all levels, gleaning for insects, and also eating small fruits. They are notably quiet, only occasionally giving a quick "tsit" call.

CYANERPES honeycreepers are small, arboreal tanagers with distinctive *slender decurved bills* and *brightly colored legs.* The attractive ♂♂ are *mostly blue.*

PURPLE HONEYCREEPER *Cyanerpes caeruleus* 11 cm | 4.25"
Uncommon and local in canopy and borders of forest and woodland in NW of our area, e.g., Serra das Araras. *Slender bill long and decurved; legs bright yellow* in ♂, dusky yellowish in ♀. ♂ *mostly purplish blue;* lores, throat, wings, and tail black. ♀ green above with *buff lores, throat, and face* (the latter streaked), and a *blue malar streak.* Below streaked green and yellowish. Likely only confused with Red-legged Honeycreeper though both sexes of that species have red or reddish legs. Often in small groups that accompany other tanagers and which regularly congregate at fruiting and flowering trees. They can drop quite low at borders, and sometimes come to feeders. Call a high thin "zree," but on the whole rather quiet.

RED-LEGGED HONEYCREEPER *Cyanerpes cyaneus* 11.5-12 cm | 4.5-4.75"
Locally fairly common in canopy and borders of forest and woodland in NW of our area. Bill long and decurved; *legs bright red* in ♂, duller and only reddish in ♀. ♂ *mostly bright purplish blue* with back, wings, and tail black; *crown contrasting pale turquoise.* ♀ olive green above with *vague whitish superciliary.* Dull yellowish below with *blurry olive breast streaking.* Nonbreeding ♂ postbreeding "eclipse" plumage resembles ♀ except that it retains the black. Purple Honeycreeper has yellow or yellowish legs; ♀ Purple has buff throat and blue malar streak. Behavior much as in Purple Honeycreeper though Red-leggeds are less forest-based and sometimes range in monospecific groups. They regularly perch on high exposed branches. Calls include an oft-uttered "tsip" note (sometimes in series) and an ascending "zhreee."

275

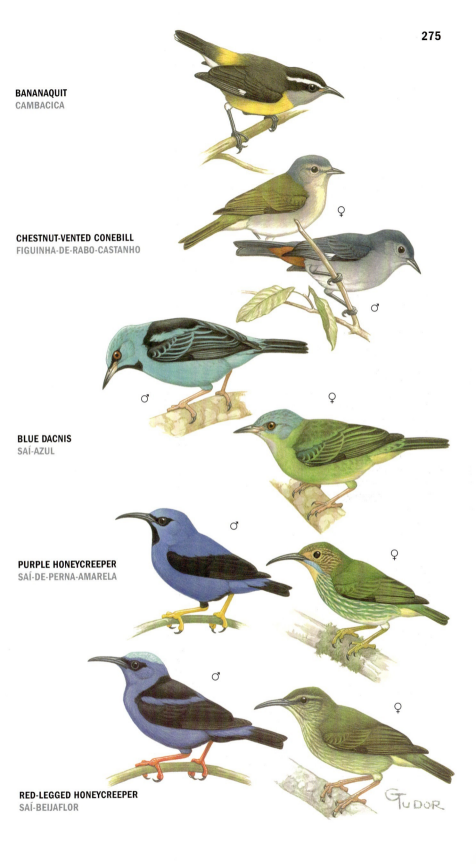

BANANAQUIT
CAMBACICA

CHESTNUT-VENTED CONEBILL
FIGUINHA-DE-RABO-CASTANHO

BLUE DACNIS
SAÍ-AZUL

PURPLE HONEYCREEPER
SAÍ-DE-PERNA-AMARELA

RED-LEGGED HONEYCREEPER
SAÍ-BEIJAFLOR

TANGARA are the "classic" tanagers, with many species being well known for their *bright colors and patterns*. As they are mainly birds of extensively forested regions, there are not that many in our area; only one (the attractive Burnished-buff) is at all widespread. They eat both fruit and insects and are not particularly vocal. Sexes alike, ♀♀ often duller.

BAY-HEADED TANAGER Tangara gyrola   13.5 cm | 5.25"

Uncommon in canopy and borders of forest in *W Mato Grosso*. A distinctive *Tangara* with *brick red head*. ♂ grass-green above with a *broad golden yellow nuchal band* and brick red shoulders. *Below mainly turquoise blue*, green on sides. ♀ duller but with a similar pattern; juvenile can look quite uniform green, showing little or no rufous on head. Often in pairs or small groups, frequently joining mixed flocks and regularly with various birds at fruiting trees. Searches for insects mainly by hopping along larger limbs, pausing to peer underneath.

BURNISHED-BUFF TANAGER Tangara cayana   13.5 cm | 5.25"

Common and widespread in cerrado, gallery woodland, and forest borders. ♂ *mostly shiny ochraceous buff above*, wings and tail mainly bluish. *Broad area from face and throat down to median belly black to blackish*, contrasting with *ochraceous sides*. ♀ duller and *lacking the black below; mask dusky* and back more greenish. Not likely confused in its open-country habitat where it sometimes flies long distances between surprisingly isolated trees and shrubs. Usually found in pairs. Frequently found at fruiting trees; even there most often not together with other tanagers, an occasional exception being the Sayaca. Sometimes comes to feeders.

MASKED TANAGER Tangara nigrocincta   13 cm | 5"

Uncommon in canopy and borders of forest in *W Mato Grosso*. *Head pale lavender blue*, with a small black mask and greenish cheeks, contrasting with *black back and broad pectoral band*; wings and tail black with greenish blue edging, shoulders blue; *median belly white*. ♀ duller, with breast band dusky. Blue-necked Tanager has head a much more intense blue, coppery wing-coverts, and all-dark underparts (with no white). Turquoise Tanager lacks the pale head, has a yellow belly. Behavior as in other Tangara, and regularly with them, but usually in pairs or at most small groups. Often perches on bare branches in the open; eats mainly small fruits.

Paradise Tanager (*T. chilensis*) also occurs in the forests of N Mato Grosso. Paradise is *the gaudiest Tangara of them all*, unmistakable with an *apple green head, bright turquoise blue underparts*, and a *red rump*.

TURQUOISE TANAGER Tangara mexicana   13.5 cm | 5.25"

Uncommon in canopy and borders of forest and woodland in *NW of our area*. Mostly black above with *mainly cobalt blue face and underparts*, black bases to the feathers showing through especially on flanks. *Belly rich yellow*. Should be readily recognized (though it shows no "turquoise!"), at a distance looking very dark with a contrasting pale belly. Usually occurs in groups of 3-6 birds, sometimes accompanying mixed flocks (especially when at fruiting trees) but more often moving independently of them. Eats primarily fruit; forages for insects by inspecting limbs, often dead branches.

BLUE-NECKED TANAGER Tangara cyanicollis   13 cm | 5"

Uncommon in forest borders and woodland in *NW of our area*. *Hood bright turquoise blue contrasting with black back and underparts*. Wing-coverts and rump contrastingly glistening straw to yellowish green. The less intensely colored Masked Tanager has a white belly; otherwise the beautiful Blue-necked will surely not be confused. Generally does not occur in continuous forest, instead favoring edge and clearings. Forages mainly for fruit at all levels but usually not too high above ground; often ranges in small groups.

🇧🇷 **Gilt-edged Tanager** (*T. cyanoventris*) occurs in forest and borders in S Minas Gerais, e.g., Serra da Canastra. A beautiful tanager: *mostly golden yellow above*, brightest on face and streaked black on back; *mostly turquoise blue below*.

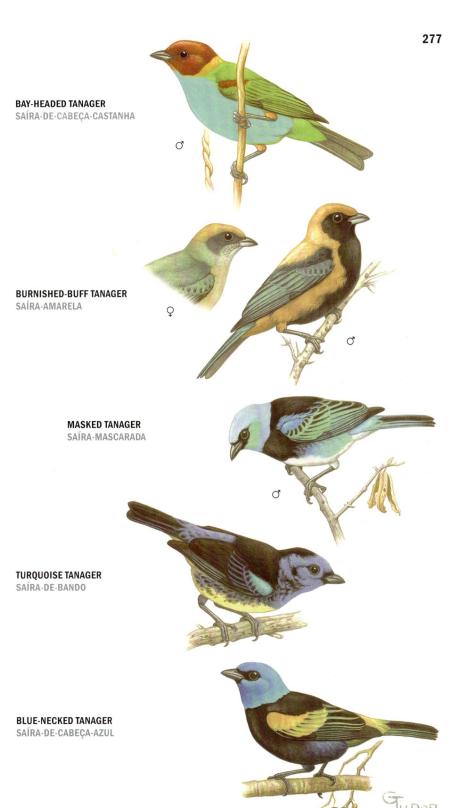

BAY-HEADED TANAGER
SAÍRA-DE-CABEÇA-CASTANHA

BURNISHED-BUFF TANAGER
SAÍRA-AMARELA

MASKED TANAGER
SAÍRA-MASCARADA

TURQUOISE TANAGER
SAÍRA-DE-BANDO

BLUE-NECKED TANAGER
SAÍRA-DE-CABEÇA-AZUL

EUPHONIAS are distinctive *small, short-tailed* birds with *stubby bills* that range in a variety of wooded or forest habitats. Long associated with the tanagers, they are now believed to be more closely related to the siskins (*Carduelis* spp., Fringillidae).

GOLDEN-RUMPED EUPHONIA *Euphonia cyanocephala* 11 cm | 4.5"
Rare and local (possibly seasonal?) in woodland borders in SE of our area and SW Mato Grosso do Sul (only one old record, from near Corumbá). Both sexes have *bright turquoise blue crown and nape*. ♂ glossy blue-black above with *yellow rump*; throat black, below orange-yellow. ♀ olive above and yellowish below with a *tawny frontlet*. Nearly unmistakable, but cf. Fawn-breasted Tanager. Occurs in small groups that perch high but usually are quiet and inconspicuous. Rarely associates with other euphonias, and seems even more fond of mistletoe berries than the others. Song a fast stream of twittering notes intermixed with more distinctive low-pitched "tueer" and "chuk" notes.

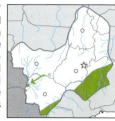

PURPLE-THROATED EUPHONIA *Euphonia chlorotica* 10 cm | 4"
Fairly common and widespread in forest and woodland borders, gallery woodland, and gardens. *The most numerous euphonia in our area*. ♂ steely blue-black above with yellow forecrown. *Throat black*, below yellow. ♀ olive above, *grayish white below* with *yellowish confined to sides and crissum*. The only ♂ euphonia with a dark throat in most of our area; in ♂ Violaceous and Thick-billed Euphonias underparts are all yellow. ♀♀ of those two species are more extensively yellow below. Purple-throated is smaller than either. Usually in pairs or small groups that feed on a variety of small fruits (including mistletoe). Though often with tanagers and other euphonias at fruiting trees, usually does not accompany mixed flocks. Both sexes frequently give a far-carrying clear "bee-beem" call.

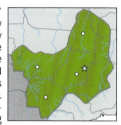

VIOLACEOUS EUPHONIA *Euphonia violacea* 11.5 cm | 4.5"
Locally fairly common in forest and woodland borders, gallery woodland, and clearings. Like ♂ Thick-billed Euphonia, ♂ Violaceous has *all-yellow underparts* (*no* dark throat). Violaceous differs from Thick-billed in having a *smaller yellow forecrown patch* (it reaches back only to just above eye) and a *richer more ochraceous yellow throat and breast*. ♀ olive above, olive-clouded yellow below. Immature ♂♂, frequently seen, resemble ♀♀ but early on acquire the yellow forecrown and a blue-black face. Usually in pairs or small groups, foraging at all levels, sometimes with other euphonias. Song an often long-continued series of short phrases, some musical, others harsher, these interspersed with trills and often excellent imitations of a wide variety of other bird species.

THICK-BILLED EUPHONIA *Euphonia laniirostris* 11.5 cm | 4.5"
Uncommon and local in forest borders, secondary woodland, and clearings in *NW of our area*. As in Violaceous Euphonia, bill slightly thicker than in other euphonias (but note that this difference is hard to see in field). ♂ glossy steel blue above with a *relatively large yellow patch on forecrown*. *Below all yellow*. ♀ olive above, olive-clouded yellow below. ♂ Violaceous Euphonia is more ochraceous orange below and has a smaller crown patch. ♀ indistinguishable from ♀ Violaceous; note that in our area the two species are not known to occur together. Behavior as in Violaceous Euphonia. Rather vocal, both sexes often giving a variety of calls including a musical "chweet." Song a variable and often long-continued series of short phrases that includes brief imitations of the calls of other bird species.

CHESTNUT-BELLIED EUPHONIA *Euphonia pectoralis* 11.5 cm | 4.5"
Uncommon and local in canopy and borders of forest and woodland. ♂ *entirely glossy steel-blue above* (with *no* crown patch) and on throat and chest. *Lower underparts chestnut* with a *golden patch showing near bend of wing*. ♀ olive above with gray-tinged nape. *Gray below* with *sides and flanks olive-yellow* and a *rufous crissum*. Nothing really similar in range. An arboreal euphonia, usually in pairs, often joining mixed flocks. Regularly twitches tail sideways. Frequently heard call a harsh, gravelly "drrt-drrt-drrt-drrt."

Rufous-bellied Euphonia (*E. rufiventris*) occurs in forest canopy and borders in Mato Grosso. ♂ resembles ♂ Chestnut-bellied Euphonia but its underparts are paler (tawny instead of chestnut). ♀ similar, but crissum also tawny (not rufous). The two species do not occur together.

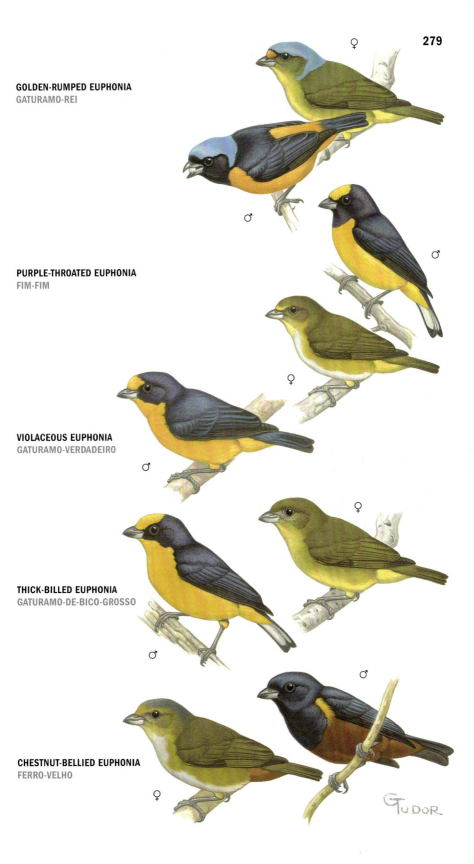

GOLDEN-RUMPED EUPHONIA
GATURAMO-REI

PURPLE-THROATED EUPHONIA
FIM-FIM

VIOLACEOUS EUPHONIA
GATURAMO-VERDADEIRO

THICK-BILLED EUPHONIA
GATURAMO-DE-BICO-GROSSO

CHESTNUT-BELLIED EUPHONIA
FERRO-VELHO

SWALLOW TANAGER *Tersina viridis* 14.5-15 cm | 5.75-6"
Widespread but nomadic (can be seasonally common) in semiopen areas and borders of forest and woodland; absent or nearly so from Pantanal. Bill broad and flat; wings rather long and somewhat swallow-like. Unmistakable ♂ *mostly bright turquoise blue* with *contrasting black face and throat*; midbelly white, bold black barring on flanks. ♀ *mostly bright green* with *pale yellow median breast and belly, prominent dusky-olive barring on flanks* (the latter unique among other vaguely similar greenish tanagers). Immature ♂ resembles ♀ but irregular patches of blue gradually become more extensive. Swallow Tanagers are gregarious and conspicuous birds that frequently perch on high bare branches (sometimes even wires) and often sally into the air after insects. They also consume much fruit. Unique among the tanagers or their close relatives, Swallow Tanagers nest in holes that they dig into banks. They can be semicolonial. Call a sharp unmusical "tzeep," surprisingly distinctive and often given in flight.

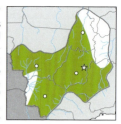

FAWN-BREASTED TANAGER *Pipraeidea melanonota* 14 cm | 5.5"
Rare and local in forest and woodland borders and adjacent clearings; recorded mainly from *Mato Grosso do Sul,* occasionally elsewhere. ♂ has *crown and nape bright pale blue* contrasting with a *broad black mask.* Otherwise dusky-blue above, *uniform fawn-buff below.* ♀ has same pattern but is duller. ♂ Golden-rumped Euphonia (also blue-crowned) is smaller with a black throat, yellow rump. Usually found in pairs, foraging quietly at varying levels in trees, often perching or hopping on larger limbs; most often does not accompany mixed flocks. Song an infrequently given series of simple high-pitched unmusical "see" notes.

THRAUPIS are among the most *widespread* and *familiar* tanagers. Their plumages are relatively dull, and bills quite heavy. These tanagers are arboreal in many semiopen and edge habitats.

SAYACA TANAGER *Thraupis sayaca* 16.5 cm | 6.5"
Common and widespread in forest and woodland borders, gallery woodland, semiopen areas, and around houses. *Mostly dull bluish gray;* wings with flight feathers edged greenish turquoise blue. Not likely confused, but compare to the drabber and more olivaceous Palm Tanager (in poor light the two can look surprisingly similar; in some lights the Sayaca can look surprisingly greenish). The Sayaca is a conspicuous and well-known tanager in most areas and is often tame around ranch buildings, ecolodges, and in urban areas, regularly even dropping to the ground. It forages in a variety of situations, and subsists mostly on fruit; frequently accompanies mixed flocks. Song a jumbled, fast series of rather squeaky notes; calls have a similar quality.

PALM TANAGER *Thraupis palmarum* 17-18 cm | 6.75-7"
Fairly common and widespread in semiopen terrain, woodland, and forest borders. *Primarily grayish olive* with a *glossy sheen* on back and underparts; forecrown yellowish olive. *Mainly black flight feathers contrast with olive wing-coverts, so that the closed wing looks essentially bicolored* (in poor light often this dull-plumaged tanager's best field mark). Sayaca Tanager often occurs with the Palm and, though they are colored very differently, in poor light the two can be confused. Palm Tanager is a conspicuous bird, though it is nowhere near as familiar to most people as the Sayaca; it sometimes ranges in small groups, and often accompanies mixed flocks. True to its name the species shows a predilection for palms, often foraging among their fronds, even hanging upside-down from their tips. Squeaky song and calls recall Sayaca's.

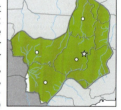

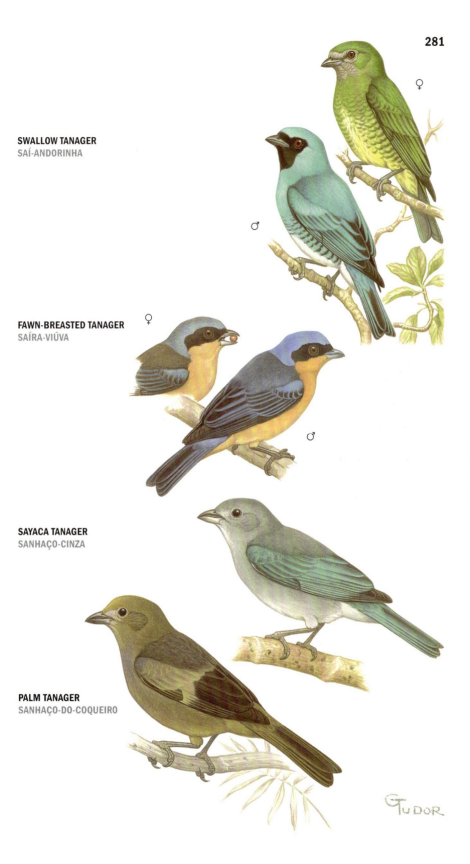

SWALLOW TANAGER
SAÍ-ANDORINHA

FAWN-BREASTED TANAGER
SAÍRA-VIÚVA

SAYACA TANAGER
SANHAÇO-CINZA

PALM TANAGER
SANHAÇO-DO-COQUEIRO

LOWLAND HEPATIC TANAGER *Piranga flava* 18.5-19 cm | 7.25-7.5"
Widespread and locally fairly common in cerrado, cerradão, and gallery and deciduous woodland. Bill dusky above, horn below. Unmistakable ♂ *all red*, paler and brighter below and on forecrown; lores dusky, wings and tail also somewhat duskier. ♀ yellowish olive above with *dusky lores; rather bright yellow below and on forecrown*. ♀ is the only uniformly olive and yellow bird of its size in our area. Hepatics are arboreal birds, usually found in pairs that forage in foliage and often seem sluggish. They sometimes accompany mixed flocks but at least as often move independently of them. Song a fast series of rich and melodic phrases, not all that often heard. More frequent is its call, a soft "chef," sometimes doubled. The genus is now believed to be more closely allied to the cardinalid grosbeaks than to the tanagers.

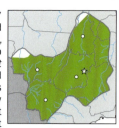

GRAY-HEADED TANAGER *Eucometis penicillata* 18 cm | 7"
Uncommon and somewhat local in lower growth of gallery forest and woodland. Bill pinkish. *Head pale grayish with a bushy crest* (feathers showing white bases); otherwise olive above. *Throat buffy whitish, underparts yellow*. A distinctive tanager of the understory, infrequently in the open and not often seen unless specifically searched out. ♀ White-shouldered Tanager (a much more arboreal bird) is similarly patterned and colored but is much smaller and shows no crest. Gray-headeds are active and excitable birds with an expressive crest that is raised when the bird is nervous or agitated. It moves about in pairs, sometimes accompanying small mixed flocks. Sharp chattered calls are frequent while the song, less often heard, is a jumbled series of high-pitched sputtered notes.

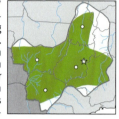

BLACK-GOGGLED TANAGER *Trichothraupis melanops* 16.5 cm | 6.5"
Uncommon in lower growth of forest and woodland in *S of our area*. ♂ has *forehead and ocular area black* (the "goggles") and a *golden yellow coronal patch* (often hidden); otherwise dusky olive above with *contrasting black wings and tail*. Buff below. A white pectoral patch sometimes protrudes from behind wing, and in flight shows a *conspicuous white stripe along base of primaries*. ♀ lacks the black on face and yellow in crest; its wings and tail are duskier, but still contrast. ♂ distinctive, while the duller ♀ can still be recognized by her *contrastingly darker wings*. Forages in pairs or small groups that often accompany mixed understory flocks; in some areas regular at ant swarms. Generally bold and easy to observe. Song a variable series of phrases with a mixture of sweet and sharper notes, often repeated several times.

SILVER-BEAKED TANAGER *Ramphocelus carbo* 18 cm | 7"
Common and widespread in shrubby, non-forest habitats, often around water. ♂ has a *swollen, silvery white lower mandible* and its plumage has a *beautiful velvety sheen. Mainly blackish with a maroon tinge*, but with a *dark crimson throat and chest*. ♀ bill duller, with lower mandible silvery grayish. Dark reddish brown above *and on throat*; rump and underparts a paler reddish. ♂ should be easily recognized, especially from its conspicuous silvery on bill. ♀, however, is duller and more easily confused, though its quite contrasting red lower underparts are distinctive. Conspicuous and familiar birds, Silver-beakeds move through their favored semiopen terrain in lively groups of up to 6-8 birds; usually there is only one adult ♂ per group. They forage at all levels, mostly for fruit, but generally remain low; they are often around houses, and regularly come to feeders. Loud "chink" or "chank" call notes frequently announce their presence but less often heard is the monotonous song, a repetitive series of short simple musical phrases, given mainly at and soon after dawn.

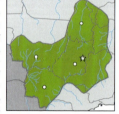

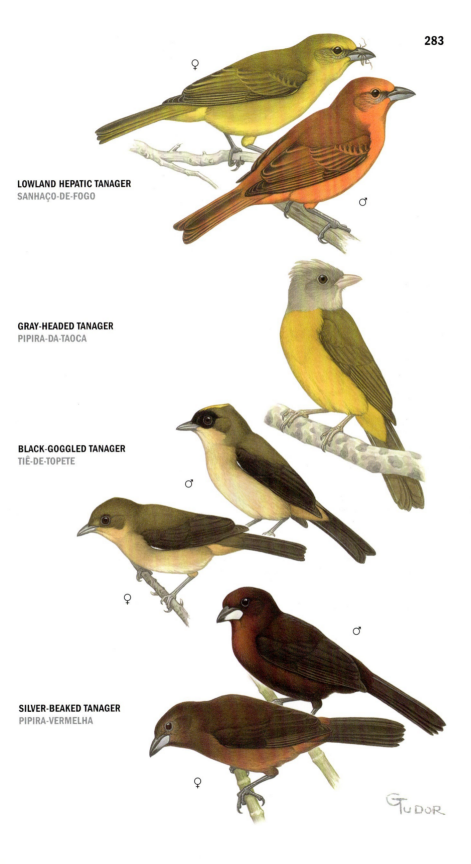

LOWLAND HEPATIC TANAGER
SANHAÇO-DE-FOGO

GRAY-HEADED TANAGER
PIPIRA-DA-TAOCA

BLACK-GOGGLED TANAGER
TIÊ-DE-TOPETE

SILVER-BEAKED TANAGER
PIPIRA-VERMELHA

TACHYPHONUS tanagers have *bluish on their lower mandible or on the base of the entire bill*. ♂♂ are *mainly black*, while ♀♀ are dull, brown or olive. In our region there are two arboreal *Tachyphonus*, both occurring only in the north, and two others that are more widespread and range in lower growth and edge.

WHITE-SHOULDERED TANAGER *Tachyphonus luctuosus* 13-13.5 cm | 5-5.25"

Uncommon in canopy and borders of forest and woodland in *NW our area*. Bill bluish gray. Unmistakable ♂ black with a *large white patch on wing-coverts*. ♀ olive above with *contrasting gray head*. Throat grayish white, *below bright yellow*. ♂ of larger White-lined Tanager shows little or no white on its closed wing. ♀ White-shouldered recalls a miniature Gray-headed Tanager in color and pattern, but lacks that species' bushy-crested effect; Gray-headed is a bird of undergrowth and is not arboreal. Forages actively in foliage and vine tangles and easy to see, often feeding in the open and regularly quite low in second growth and at edge. Occurs most often in pairs and regularly accompanies mixed flocks. Most frequent calls a piercing "tseeyt,tsee-tsee" or sharp "tsit," often repeated.

FLAME-CRESTED TANAGER *Tachyphonus cristatus* 15-16.5 cm | 6-6.5"

Uncommon in canopy and borders of forest in NW of our area. Base of bill bluish gray. ♂ black with *broad flat crest orange to scarlet*, golden buff rump, and *ochraceous buff midthroat* (the latter often not too conspicuous). ♀ rufescent brown above. *Throat whitish, below ochraceous*. ♂'s crest is always evident, and is the mark for this forest-inhabiting tanager. The dull ♀ is often best distinguished by her accompanying mate. Ranges in pairs, generally foraging well above the ground, gleaning actively in foliage and typically with mixed flocks of other tanagers, furnariids, etc. It seems to eat little fruit. Not very vocal; gives thin "tseep" notes while foraging. Natterer's Tanager (*T. nattereri*) is a taxon of uncertain status; known only from a single ♂ specimen taken in the 19th century in Mato Grosso, it resembles Flame-crested but apparently lacks the throat patch and the pale area on rump is smaller.

White-winged Shrike-Tanager (*Lanio versicolor*) occurs locally in canopy of forest in Mato Grosso. ♂ is a stunning, vaguely oriole-like tanager, mainly ochraceous yellow and black with a large white patch on its wings. ♀ much duller, resembling ♀ Flame-crested but more ochraceous below with a yellower midbelly, no whitish on throat. Frequent call a loud "tchew!"

WHITE-LINED TANAGER *Tachyphonus rufus* 18-19 cm | 7-7.5"

Fairly common and widespread in shrubby clearings, gardens, and woodland borders. Bill bluish gray. ♂ *entirely glossy black*. White underwing-coverts flash in flight and as bird flicks its wings; a little white on bend of wing and scapulars also sometimes shows. ♀ *uniform rufous, slightly paler below*. The very similar ♂ Ruby-crowned Tanager is slightly bluer black overall with a red crown streak (often hard to see); ♀ Ruby-crowned differs more, being much less uniformly rufous with a grayer head and dusky breast streaking. ♂ White-shouldered Tanager shows much more white on its wing; its ♀ is very different. Forages in pairs, usually within cover and not terribly conspicuous, though regularly seen flying across a road or some other opening. Generally not with flocks. Not very vocal, with infrequently heard song a series of simple, somewhat musical phrases: "cheép-chooi...cheép-chooi...."

RUBY-CROWNED TANAGER *Tachyphonus coronatus* 18 cm | 7"

Fairly common in low growth and borders of forest and woodland in *S of our area*. Bill bluish gray. ♂ *lustrous blue-black* with a *narrow scarlet streak on rear crown* (usually hidden); white underwing-coverts flash in flight, and also has some white on scapulars (usually hidden). ♀ rufous brown above with *grayer head. Ochraceous below with dusky flammulation across breast*. ♂ resembles ♂ White-lined Tanager, though the latter is more a matte black and lacks the "ruby" in crown. The two are best distinguished by their respective ♀♀, ♀ White-lined being a very different *uniform rufous*. White-lined is much less of a forest bird. Cf. also vaguely similar Screaming and ♂ Shiny Cowbirds. Forages in pairs, generally remaining independent of mixed flocks. Variable but usually monotonous song a series of notes, "ch-chweek... ch-chweek...," most often heard around dawn. Call a steadily repeated "chef."

WHITE-SHOULDERED TANAGER
PIPIRA-DE-ENCONTRO-BRANCO

FLAME-CRESTED TANAGER
TIÊ-GALO

WHITE-LINED TANAGER
PIPIRA-PRETA

RUBY-CROWNED TANAGER
TIÊ-PRETO

285

SCHISTOCHLAMYS tanagers are fairly large, very *dissimilar-looking*, but *equally svelte* tanagers that range in semiopen terrain.

CINNAMON TANAGER *Schistochlamys ruficapillus* 18-18.5 cm | 7-7.25"

Uncommon and local in semiopen terrain with scattered trees and bushes, patches of secondary woodland, and clearings in *N and E of our area*. Rather heavy bill bluish gray tipped black. *Foreface black* contrasting with *cinnamon-buff sides of head, throat, and breast*. Otherwise mainly bluish gray above. Belly grayish and white, crissum cinnamon. Immature similar but duller, with duskier foreface and paler foreneck. Unlikely to be confused. Found singly or in pairs, often perching in the open atop shrubs or low trees. Though only infrequently heard, its short musical song can be repeated over and over and is typically given from an exposed perch.

BLACK-FACED TANAGER *Schistochlamys melanopis* 18-18.5 cm | 7-7.25"

Locally fairly common in semiopen areas, cerrado, patches of low woodland; largely absent from Pantanal. Rather heavy bill bluish gray tipped black. *Mostly gray*, somewhat paler below and becoming whitish on midbelly; *forehead, face, throat, and midchest contrastingly black to brownish black*. Immature entirely different: olive above with a *yellow eye-ring*; yellowish olive below. Subadults begin to show the black face and bib when they are still mostly olive. The gray and black adult is unmistakable, but the dull and olive immatures are potentially confusing, as they somewhat resemble various saltators (though they are plainer than any). Behavior much as in Cinnamon Tanager, though Black-faced seems a more numerous bird. Its song, a rich and melodic grosbeak-like phrase, seems to be given more often.

SHRIKE-LIKE TANAGER *Neothraupis fasciata* 16 cm | 6.25"

Fairly common in cerrado and cerradão. This distinctive tanager has often been called the White-banded Tanager. *Above gray* with a *prominent black mask extending from lores back over ear-coverts*; wing-coverts black with a white band on lesser coverts. Throat whitish; below pale gray. ♀ slightly duller. Immature browner with reduced mask and a yellowish tinge below. Virtually unmistakable in its limited habitat; many foreign observers will be struck by its resemblance to various shrikes (*Lanius* spp.). Forages in small groups, pairs when breeding, regularly accompanying loose flocks. Though basically arboreal, sometimes drops to the ground and sometimes sallies to capture insect prey; also consumes some fruit. Call a chipping note; less often heard is the song, delivered mainly at dawn, a loud, complex, whistled phrase that sometimes is given as a duet.

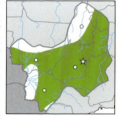

WHITE-RUMPED TANAGER *Cypsnagra hirundinacea* 16-16.5 cm | 6.25-6.5"

Fairly common in cerrado and cerradão; absent from actual Pantanal, though approaching it in various areas. A handsome, *mostly black and white* tanager. *Black above* with *conspicuous white rump*, band across primaries, and patch in primaries. Throat rufous, buffier on chest; below creamy whitish. In SW Mato Grosso throat and chest paler, only washed with ochraceous buff. Immature browner above, uniform creamy buff below. Though often found with Shrike-like Tanager, White-rumped Tanager favors more open terrain with fewer trees. Ranges in small groups of 4-6 birds, foraging in foliage and along gnarled branches, only occasionally dropping to the ground. Periodically (most often soon after dawn) pairs burst into a loud, rollicking duet: the ♀ gives a continuous low churring while the ♂ joins in with a vigorous, melodic phrase, "cheedoocheechoo, cheedeereeyou-chee-choo," the whole repeated numerous times and one of the classic bird sounds of the Cerrado.

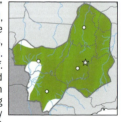

CONE-BILLED TANAGER *Rhynchothraupis mesoleuca* 16 cm | 6.25"

Long known only from a ♂ collected in 1938 in Mato Grosso, *this species was rediscovered in 2004* in Emas NP, where a pair was found in an area of gallery woodland with periodic standing water; since then others have been found as well as an additional population further west. Heavy bill whitish in ♂, blackish in ♀. ♂ *black above* with a small white wing speculum. Throat and chest black contrasting with *white median breast and belly*. ♀ dull brown with a *whitish midbelly*. Has been observed foraging in woodland undergrowth, sometimes accompanying pairs of White-striped Warblers. ♂'s canary-like song is a three-part "jew-jew-jew-jew, trrrrrrrr, triiiii" (the final trill high-pitched). Long classified in the genus *Conothraupis*, but we revert to the genus in which it was described.

CINNAMON TANAGER
BICO-DE-VELUDO

BLACK-FACED TANAGER
SANHAÇO-DE-COLEIRA

SHRIKE-LIKE TANAGER
CIGARRA-DO-CAMPO

WHITE-RUMPED TANAGER
BANDOLETA

CONE-BILLED TANAGER
TIÊ-BICUDO

SCARLET-THROATED TANAGER *Compsothraupis loricata* 21.5 cm | 8.5"

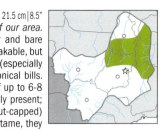

🇧🇷 Uncommon in gallery woodland and trees in clearings in NE of our area. Mostly glossy blue-black, ♂ with *scarlet throat and midchest* and bare black skin around and behind eye. Striking adult ♂♂ are unmistakable, but ♀♀ and young ♂♂—lacking any red—resemble various icterids (especially Chopi Blackbird), though they differ in their heavier and less conical bills. *General behavior also very icterid-like.* Ranges in small flocks of up to 6-8 birds, red-throated ♂♂ a minority but with at least one normally present; sometimes they are accompanied by blackbirds (e.g., Chestnut-capped) and Shiny Cowbirds. Conspicuous birds, often sluggish or even tame, they sometimes perch on exposed branches and snags for extended periods. They have a loud, blackbird-like "chirt" or "kyuh" call that is often repeated and regularly given in flight. A Brazilian endemic whose taxonomic affinities remain uncertain (possibly an icterid?).

MAGPIE TANAGER *Cissopis leverianus* 25.5-29 cm | 10-11.25"

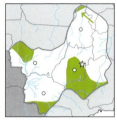

Uncommon and local in borders of forest and adjacent clearings with large trees. *Large* and *very long-tailed*, with a conspicuous *golden yellow iris*. Glossy blue-black above and on foreneck, extending to a point on breast; *white below and on rump*; wings and *strongly graduated tail* mostly black, tail feathers broadly tipped white. Northwestern birds are somewhat smaller than those of the southeast (though even they are still pretty large!), and have more white on back. A conspicuous and ultra-distinctive tanager, vaguely jay-like, that ranges in pairs or small groups. Generally not with mixed flocks. Often perches in the open and usually easy to observe. Gives a variety of arresting metallic calls, the most frequent being a distinctive loud "tchenk," sometimes in series. Its soft, disjointed song is heard much less often.

CARDINALS, GROSBEAKS, & ALLIES (Cardinalidae) are fairly large "finches" that tend to have stout bills but otherwise are quite variable in appearance. There is considerable confusion as to actual family limits.

PAROARIA cardinals are *conspicuous* cardinalid "finches" that are often *found near water or around houses*. They are among the most attractive of our birds, all three species having *bright red on the head*.

YELLOW-BILLED CARDINAL *Paroaria capitata* 16.5 cm | 6.5"

Common and conspicuous around marshes and along the shores of lakes and rivers in *Pantanal*. Bill bright pinkish yellow. *Head bright red* with a *black bib*. Above black with partial white nuchal collar. Below white. Juvenile has head brown, bib buff, and upperparts duskier. To be confused only with Red-crested Cardinal, which is *larger* with a *crest* and *gray upperparts*. Does not occur with Crimson-fronted Cardinal. The lovely Yellow-billed Cardinal is one of the Pantanal's characteristic birds and can occur in large flocks. It regularly congregates around houses to feed on rice and food scraps, and large numbers visit many ecolodges. Also often forages close to the water's edge, sometimes venturing out onto floating vegetation. Musical song a simple repeated "tsit-whit-tu-cheu," given mainly in the early morning.

CRIMSON-FRONTED CARDINAL *Paroaria baeri* 16.5 cm | 6.5"

🇧🇷 Rare and local in shrubby riparian growth in N of our area, along the middle and upper Rio Araguaia. Head black with *crimson forecrown and throat*. Otherwise glossy blue-black above, white below. Behavior much as in Yellow-billed Cardinal, though the Crimson-fronted is nowhere near as numerous or familiar.

RED-CRESTED CARDINAL *Paroaria coronata* 19 cm | 7.5"

Uncommon in semiopen areas, lighter woodland, and around houses in *Pantanal*. Bill mostly whitish. *Head, long crest, and pointed bib bright scarlet*. Above mostly gray, with nuchal collar and underparts white. Juvenile has a brown head and bib, duskier upperparts. This beautiful bird is virtually unmistakable. Yellow-billed Cardinal is smaller, *not* crested, black above. Usually occurs in pairs or small groups, though flocks can gather during nonbreeding season. Numbers have been reduced in many areas due to its popularity as a cagebird, and sadly this may have happened in the Pantanal. Feeds mainly on the ground on seeds and insects. Attractive song a rhythmic series of melodic notes: "weerit, churit, weer, churit."

SALTATORS are fairly large arboreal "finches" with *heavy "swollen" bills* (brightly colored in some species) and rather subdued coloration. Their attractive, melodic songs are often heard.

BUFF-THROATED SALTATOR *Saltator maximus*   20.5-21 cm | 8-8.25"

Fairly common and widespread in canopy and borders of forest, woodland, and adjacent clearings; not present in Pantanal. *Bright olive above* with gray sides of head and a *short white superciliary. Lower throat buff* bordered by *black malar stripe*; grayish below with cinnamon-buff crissum. Grayish Saltator is obviously gray (not olive) above; it favors *non-forested* areas. Basically arboreal, foraging at all levels (though rarely inside forest), coming lower at edge; frequently with mixed flocks. Consumes a variety of fruits, flowers, and other vegetative items. Oft-heard song a sometimes subdued series of short sweet and warbled phrases, the individual phrases repeated several times.

GRAYISH SALTATOR *Saltator coerulescens*   20.5-21 cm | 8-8.25"

Locally common in scrub, lighter woodland, and clearings; avoids open cerrado. *Gray above* with a *short white superciliary. Throat pale buff* bordered by a black malar stripe; below grayish, becoming buff on lower belly and crissum. Immature more olivaceous generally with flight feathers edged green. Buff-throated Saltator is olive (not gray) above. Generally quite conspicuous and bold, often perching in the open (more so than other saltators), especially when singing. Comes regularly to feeding stations around ecolodges. Feeds mainly on fruit and flowers. Usually not with mixed flocks. Loud song varies, but typically a simple phrase of well-enunciated musical notes, often with a chortling quality: "chew-chew-chew, chu-chu, cheéuw" or "chu-chu-cheéu." Call, also often heard, a ringing metallic "tchink."

GREEN-WINGED SALTATOR *Saltator similis*   20.5-21 cm | 8-8.25"

Uncommon and local in canopy and borders of woodland and forest and adjacent clearings. Mostly gray above with a *long white superciliary*; back tinged olive and *much of wings rather bright olive green*. Throat white bordered by black malar stripe; below dingy buffy grayish, buffiest on midbelly and crissum. Grayish Saltator (also gray-backed) differs in having a *shorter* eyestripe and a *buff-tinged throat*; usually the green on Green-winged's wings is distinctive, but note that Grayish immatures also show green wing-edging. Green-winged is much more a forest bird, and is now thought to be declining due to capture for the cagebird market. Its behavior is much as in Buff-throated Saltator. Loud, oft-heard song a short phrase of clear whistled notes: "chew, chew, cho, chewee."

GOLDEN-BILLED SALTATOR *Saltator aurantiirostris*   19-20.5 cm | 7.5-8"

Uncommon in woodland and scrub in *W Mato Grosso do Sul*. ♂ has *bill mainly or entirely orange*, the orange reduced in ♀♀ and immatures. Handsome ♂ gray above with a *broad white superciliary extending back from eye and black forehead and face extending down as a pectoral band* that encloses a white throat. Buffy grayish below. ♀ similar but more olive above with pectoral band narrower or even absent. Immature like ♀ but with brow yellowish. The only saltator whose superciliary does *not* start at lores. A conspicuous bird that usually occurs in pairs, sometimes moving with mixed flocks. Song a far-carrying, almost explosive phrase, "switch-it, tchweet-a-sweéu" or "ch-ch-chew-chwueeé," variable but with the last note always emphasized; often given from a prominent perch.

BLACK-THROATED SALTATOR *Saltator atricollis*   20.5 cm | 8"

Fairly common in cerrado and shrubby campos. *Bill mostly reddish orange*. Mainly brown above with *obvious black foreface and throat* and gray sides of head. Buffyish below. Immature's foreface and throat ashy brown; it is dark-billed. Simply patterned but distinctive; not likely confused. A conspicuous bird that often perches atop low trees and shrubs, sometimes dropping to the ground to feed. Song differs notably from that of other saltators, a fast and musical jumbled warbling, sometimes given by several birds more or less at once. Call a sharply inflected "wheék... wheék...." Probably not a "true" saltator.

BUFF-THROATED SALTATOR
TEMPERA-VIOLA

GRAYISH SALTATOR
SABIÁ-GONGÁ

GREEN-WINGED SALTATOR
TRINCA-FERRO

♂

imm.

GOLDEN-BILLED SALTATOR
BICO-DURO

BLACK-THROATED SALTATOR
BATUQUEIRO

BLACK-BACKED GROSBEAK *Pheucticus aureoventris* 21.5-22 cm | 8.25-8.5"
Uncommon austral migrant to deciduous woodland and scrub in Pantanal, mainly to W Mato Grosso do Sul. *Massive bill* blackish, paler below. ♂ *mostly black above and on throat and breast*, wings with bold white markings and outer tail feathers white-tipped. Belly bright yellow. ♀ like ♂ but browner and more or less speckled yellow above; *below yellow speckled black*. Hardly to be confused: no other similar bird in our area has such a heavy bill. Found singly or in pairs, often perching in the open, sometimes allowing a close approach; not with flocks. Both sexes often give a metallic "pink" call. ♂'s song, given from an exposed perch, is a rich and melodic caroling; infrequently heard here. Breeds mainly in Andes.

The "**BLUE GROSBEAKS**" are *mainly blue* (♀♀ brown) with *heavy to very heavy bills*. They range in woodland, and usually are hard to see.

ULTRAMARINE GROSBEAK *Cyanocompsa brissonii* 15.5 cm | 6.25"
Rare to uncommon in undergrowth of lighter woodland, scrub, and thickets near water. *Very heavy bill blackish*. ♂ *dark blue with forehead, brow, malar area, and shoulders paler and brighter blue*. ♀ *cocoa brown above, paler and brighter fulvous brown below*. Glaucous-blue Grosbeak is smaller with a stubbier bill and its ♂ shows a grayish suffusion on the blue. ♀ Ultramarine also resembles ♀ Great-billed Seedfinch, but the latter's brown is somewhat less rich and more olivaceous, and its bill even more massive. Ultramarine Grosbeaks occur in pairs and are reclusive, generally remaining inside heavy cover though singing ♂♂ can perch in the open. Song fairly loud and musical, typically starting slowly and ending in a warble. In many areas heavily trapped for the cagebird market.

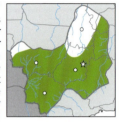

GLAUCOUS-BLUE GROSBEAK *Cyanoloxia glaucocaerulea* 15 cm | 5.5"
Rare austral migrant to lower growth of forest and woodland borders in Mato Grosso do Sul; only a few records from our area. *Bill shorter and stubbier than in Ultramarine Grosbeak*, with a *curved culmen* more like that of *Sporophila*. ♂ *uniform glaucous blue*, slightly brighter on brow. ♀ resembles ♀ Ultramarine Grosbeak though somewhat paler. Ultramarine Grosbeak is larger with a *much more massive bill*; ♂ is a *notably darker blue*. Behavior much as in Ultramarine, tending to be equally or even more shy. Song, likely not often heard here, a fast, jumbled warbling, higher-pitched than the Ultramarine's and with a more even tempo (lacking the Ultramarine's slow start). Breeds in SE Brazil region.

EMBERIZINE FINCHES (Emberizidae) comprise a large and diverse group of small birds found widely in the Neotropics. Most have *deep conical bills* adapted for eating seeds; most inhabit open or lightly wooded terrain.

ORYZOBORUS seedfinches resemble seedeaters but have *much more massive bills*. ♂♂ are black or mostly black. Because of their popularity as cagebirds, both our species have been heavily trapped.

LESSER SEEDFINCH *Oryzoborus angolensis* 12.5 cm | 5"
Now rare in shrubby clearings, woodland, and forest borders. *Very heavy, squared-off black bill*. ♂ *glossy black with chestnut breast and belly*. Small wing speculum and underwing-coverts white. ♀ *brown above, fulvous brown below*; underwing-coverts white, but shows no speculum. The much rarer Great-billed Seedfinch is larger and has a more massive bill, but ♀ especially can be confused. Cf. also ♀ seedeaters, all of which have considerably stubbier and rounder bills. Found singly or in pairs, sometimes associating with seedeaters but in general less tied to grassy areas and much less prone to flock. Song (the reason this bird is so popular a cagebird in Brazil) an attractive and prolonged series of musical whistled notes that gradually becomes more jumbled and twittery.

GREAT-BILLED SEEDFINCH *Oryzoborus maximiliani* 16 cm | 6.25"
CR Now very rare and local in vicinity of marshy areas with luxuriant grass and shrubby clearings; a few have been recently recorded from Emas NP. *Bill enormously thick, chalky whitish in ♂, blackish in ♀*. ♂ *glossy black* with conspicuous wing-speculum and underwing-coverts white. ♀ *brown above, fulvous brown below*; underwing-coverts white. This finch has been so reduced in numbers by trappers for the cagebird market that it remains very poorly known in the wild. Found singly or in pairs, not associating with other finches. Its canary-like song is individually variable but always complex and highly melodic, a lovely series of notes, trills, and rattles.

BLUE FINCH *Porphyrospiza caerulescens* 12.5 cm | 5"
Uncommon and local (seasonal?) in grassy cerrado with scattered bushes and low trees; favors rocky terrain. Slender bill *bright yellow* (blackish culmen in ♀); legs reddish. Unmistakable ♂ *bright cobalt blue*. ♀ *rufous brown above*; buffy whitish below *streaked dusky*. Molting ♂'s feathers are brown-tipped, in some so broadly as to hide the blue. The unique yellow bill should identify the more obscure ♀ and molting ♂. Feeds mainly on ground in or near grassy cover, rarely associating with other finches. Pairs separate out when breeding. ♂'s distinctive song, a series of high-pitched "swee-sweeu" phrases, is given from atop a shrub, low tree, or rock.

GRASSQUITS are small dull finches that resemble seedeaters but have more pointed bills. *Tiaris* build globular (not cup-shaped) nests.

BLUE-BLACK GRASSQUIT *Volatinia jacarina* 10 cm | 4"
Widespread and locally abundant in grassy areas in open and settled regions. *Bill slender and pointed*. ♂ *glossy blue-black* with white axillars. ♀ *dull brown above*; pale buff below, *breast and flanks streaked dusky*. Immature and molting ♂ mottled blackish and brown. Seedeaters have thicker, stubbier bills. Cf. Sooty Grassquit. Occurs in flocks when not breeding; regularly associates with seedeaters, especially Double-collareds. Feeds on seeds, mostly on the ground. Breeding ♂♂ give a buzzy, explosive "dzee-u" from low perches, usually accompanied by a short flutter and a jump into the air (sometimes almost a somersault).

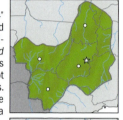

SOOTY GRASSQUIT *Tiaris fuliginosus* 11.5 cm | 4.5"
Rare and local in shrubby clearings, grassy scrub, woodland borders; few records (likely overlooked). ♂ *bill blackish with pink gape*, yellowish in ♀. ♂ *sooty black*, grayer below. ♀ olive brown above, dull buff below, midbelly whiter. ♂ Blue-black Grassquit is a glossier blue-black. ♀ Sooty is hard to identify when alone, but it usually occurs in small groups with some ♂♂ present. Cf. Dull-colored Grassquit. Occurs in pairs or small groups, foraging on or near ground and often with seedeaters. High-pitched song a ringing but rather unmusical "screez-screedelelee."

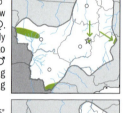

DULL-COLORED GRASSQUIT *Tiaris obscurus* 11 cm | 4.25"
Rare austral migrant to shrubby areas and woodland borders; a few records from Pantanal and Chapada dos Guimarães (Jun-Aug). *Bicolored bill dusky above, yellowish below*. Grayish olive brown above, paler brownish gray below, midbelly whitish. *Nondescript*. Sometimes found with seedeaters; easily mistaken for a ♀. The *bicolored bill* helps (all dark in most seedeaters); most resembles ♀ Double-collared Seedeater though that is buffier overall. ♂'s song, an explosive buzzy "zeetig, zeezeezig" with variations, is distinctive but likely is given only rarely in Brazil. Breeds mainly on Andean slopes.

SPOROPHILA seedeaters are small finches with *stubby bills* that range in *grassy areas*. Many species are rare habitat specialists, some only migratory to our area. ♀♀ are hard to identify, and unless otherwise noted are basically olive brown above, yellowish buff below. Concentrate on the ♂♂!

LINED SEEDEATER *Sporophila lineola* 11 cm | 4.25"
Uncommon to locally fairly common in grassy and shrubby areas. Status here unclear. Bill blackish (♂) or *mostly dull yellowish* (♀). ♂ black above with *white crown stripe* (often not obvious) and *prominent broad malar stripe*. Throat black; white below. Locally numerous and conspicuous when breeding; at other times less often encountered but sometimes with other seedeaters. Feeds primarily by gleaning seeds from tall grass stems. Song a trilled "titititititi-teé;" sometimes sings when not breeding.

YELLOW-BELLIED SEEDEATER *Sporophila nigricollis* 11 cm | 4.25"
Locally common in shrubby and grassy clearings and agricultural areas; *most numerous in Minas Gerais*. Bill pale bluish gray (♂) or dusky (♀). ♂ has *crown, face, throat, and chest black; belly pale yellow*. Above olive, white wing-speculum small or absent. Some (older?) ♂♂ are blacker above; others (in Minas Gerais at least) are *white-bellied* and *gray-backed*. ♂ is the only seedeater with a pale yellow belly (though some birds are faded and pale). Behavior typical of the genus. ♂'s song short and musical, usually ending with two buzzier, emphasized notes: "tsee-tsee-tsee-bseeooo, bzée-bzée."

Hooded Seedeater (*S. melanops*) is known from only one 19th-century specimen, taken in S Goiás. It resembles ♂ Yellow-bellied Seedeater, but black hood extends only over throat.

BLUE FINCH
CAMPAINHA-AZUL

BLUE-BLACK GRASSQUIT
TIZIU

SOOTY GRASSQUIT
CIGARRA-DO-COQUEIRO

DULL-COLORED GRASSQUIT
CIGARRA-PARDA

LINED SEEDEATER
BIGODINHO

YELLOW-BELLIED SEEDEATER
COLEIRO-BAIANO

RUSTY-COLLARED SEEDEATER *Sporophila collaris*   12 cm | 4.75"

Fairly common *in and near marshes*, favoring areas with tall grass and shrubbery. *A boldly patterned, quite large* seedeater. Rather heavy blackish bill. In Pantanal (illustrated) ♂ black above with *prominent white patches around eye* and *tawny-buff nuchal collar*. Throat white, *pectoral band black, cinnamon-buff below*. Toward the north more variable, with much the same pattern but *whiter* where Pantanal birds are buff, this being especially marked on the collar. ♀ *distinctive for a ♀ Sporophila*, echoing ♂'s pattern: brown above with *tawny-buff wing-bars*, a buff speculum, and at least a trace of the nuchal collar. *Throat white; buff below*. No seedeater in its range has the pectoral band or the buff nuchal band (♂) or the bold wing-bars and two-toned underparts (♀). Usually in pairs or small groups, often alone and not associating with its congeners. Song an attractive but rather patternless jumble of musical notes.

DOUBLE-COLLARED SEEDEATER *Sporophila caerulescens*   11 cm | 4.25"

Widespread and generally common in semiopen and shrubby areas and agricultural areas. Bill greenish yellow (♂), *dusky above and dull yellowish below* (♀). ♂ gray above, tinged brownish on mantle (especially in younger birds), sometimes with a white wing-speculum. *Upper throat black, outlined by a wide white malar stripe*, and *white lower throat bordered below by a black pectoral band* (forming the "double collar"). ♂'s face and chest pattern distinctive. Double-collared is an *ecologically unspecialized Sporophila* that occurs primarily in disturbed habitats, and often ranges with Blue-black Grassquits. Its variable song is a fast jumbled series of musical notes.

WHITE-BELLIED SEEDEATER *Sporophila leucoptera*   12 cm | 4.75"

Uncommon and rather local in grassy areas with scattered shrubs and marshy areas; *almost always near water*. Bill dull to pinkish yellow. *Bicolored*. Plainly patterned ♂ *gray above* with an obvious white wing-speculum. *Uniform white below*. ♀'s *relatively large size and pale bill* often help to distinguish her from other ♀ seedeaters. Found singly or in scattered pairs, rarely or never in flocks, and generally not associating with other seedeaters. Distinctive song a repetition of a single clear ringing note, "cleeu, cleeu, cleeu...."

PLUMBEOUS SEEDEATER *Sporophila plumbea*   11 cm | 4.25"

Uncommon to locally or seasonally fairly common in campos, grassy cerrado, and adjacent scrub. *Generally avoids agricultural areas*, and scarce or absent in Pantanal. Bill rather heavy, blackish. ♂ *mostly gray*, paler below, usually with a white subocular crescent and whitish upper throat; wings and tail blackish, wings with a white speculum. ♀ dull, but not as buffy as many other ♀ seedeaters; like the others often best identified by accompanying ♂♂. Regularly occurs in small groups, occasionally with other seedeaters but more often on its own. Song a long series of loud clear phrases, each often repeated several times; imitations of other birds, e.g., Pale-breasted Spinetail, are sometimes incorporated. All or most seedeaters are trapped because of their singing ability and many are declining as a result; this and the next species and all those on p. 299 are the most heavily affected

CAPPED SEEDEATER *Sporophila bouvreuil*   10 cm | 4"

Local -- but can be common -- in campos and grassy cerrado; nomadic when not breeding, but apparently not truly migratory. Bill black (breeding ♂) or yellowish (♀ and nonbreeding ♂). In most of range (**A**) ♂ has *striking black crown* contrasting with brownish gray upperparts; wings and tail black, wings with a white speculum and gray edging. *Below white or whitish*. In N Goiás (also further south?) (**B**) more richly colored: *mostly cinnamon-rufous* with the same *black crown*, wings, and tail. ♀ olive brownish above, wings with a white speculum; ochraceous buff below. *No other seedeater has the black-capped look*; ♀ indistinguishable from many other ♀ seedeaters except by association with ♂♂. Occurs in groups when not breeding, then often with other *Sporophila* in large swirling flocks and usually the most numerous species present. Feeds on grass seeds, mainly while clinging to tall stems as these arch over, sometimes reaching up for the seedheads while remaining on the ground. Variable and pretty song an often long-continued series of melodic phrases and whistled notes.

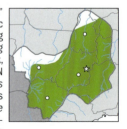

297

RUSTY-COLLARED SEEDEATER
COLEIRO-DO-BREJO

DOUBLE-COLLARED SEEDEATER
COLEIRINHA, COLEIRINHO

WHITE-BELLIED SEEDEATER
CHORÃO

PLUMBEOUS SEEDEATER
PATATIVA

CAPPED SEEDEATER
CABOCLINHO

TAWNY-BELLIED SEEDEATER *Sporophila hypoxantha* 10 cm | 4"

Locally fairly common to common in campos, grassy cerrado, lightly grazed pastures, and roadside shrubbery. Possible migratory status unclear. Bill blackish (♂) or yellowish horn (♀). ♂ brownish gray above; wings and tail duskier, wings with white speculum. *Rump and underparts tawny-rufous*, paler on cheeks and foreneck. Presumed younger ♂ is paler below, more cinnamon-tawny. ♀ like ♀ Capped Seedeater. Cf. similarly patterned but much rarer Rufous-rumped Seedeater. Behavior as in Capped Seedeater. Song a simple clear whistled "cheeu, cheeu, cheweé, chu" with variations.

RUFOUS-RUMPED SEEDEATER *Sporophila hypochroma* 10 cm | 4"

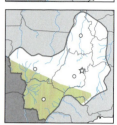

Rare austral migrant (mostly Mar-Oct) to less-disturbed grassy areas of cerrado and Pantanal, favoring moist areas where grass is tall and luxuriant. ♂ gray above, wings and tail blackish with white wing-speculum and gray edging. *Rump and underparts chestnut*, extending up over cheeks. ♀ like ♀ Capped Seedeater. ♂ patterned as in Tawny-bellied Seedeater, but more richly colored (more bluish gray above and deeper chestnut below). Chestnut Seedeater's colors are much the same as Rufous-rumped's, but its pattern differs markedly. Rufous-rumped is mainly seen as the odd individual or small group accompanying a mixed flock of seedeaters. Breeds mainly in NE Argentina.

VU **Black-and-tawny Seedeater** (*S. nigrorufa*) is a rare and still poorly known seedeater of damp grassland in the Pantanal; it is apparently resident (nomadic?). Striking ♂ has a *black crown, hindneck, and back; rump and underparts bright cinnamon-rufous*.

CHESTNUT SEEDEATER *Sporophila cinnamomea* 11 cm | 4"

EN *Rare austral migrant* (mostly May-Oct) to campos and grassy cerrado, mainly in moist areas with taller grass. ♂ *mostly chestnut with a contrasting gray crown*; wings and tail slaty, wings with a white speculum. ♀ like ♀ Capped Seedeater. ♂ Rufous-rumped Seedeater is similarly colored, but the gray of its crown extends down over the back. Behavior much as in the previous seedeater species, small numbers occasionally being found with them in mixed flocks. Breeds mainly in NE Argentina.

MARSH SEEDEATER *Sporophila palustris* 10 cm | 4"

EN *Rare austral migrant* (mostly May-Oct) to grassy cerrado and campos. Like the other migratory seedeaters, Marsh favors damp areas with lush, tall grass; despite its name, *not* particularly associated with marshes. ♂ gray above with rufous-chestnut rump; wings and tail blackish with whitish edging and speculum. *Throat and chest white; rufous-chestnut lower underparts*. ♀ like ♀ Capped Seedeater. With its striking white bib, ♂ is unlikely to be confused. Behavior as in the other grassland seedeaters; this one of the rarest, almost invariably seen as a single individual amongst a large flock. Breeds mainly in NE Argentina.

There is at least one recent sight report of a ♂ **Entre Ríos Seedeater** (*S. zelichi*) from Emas NP, where it was with a large mixed seedeater flock. Breeds in very small numbers in NE Argentina. ♂ resembles Marsh Seedeater but has a *broad white nuchal collar* and *rufous-chestnut back and rump* (not gray). Its taxonomic status remains uncertain; perhaps not a good species.

DARK-THROATED SEEDEATER *Sporophila ruficollis* 10 cm | 4"

Rare to uncommon and local in less-disturbed, often seasonally flooded grasslands in Pantanal; during winter spreads into campos and grassy cerrado, then favoring moister areas with tall grass. ♂ gray above, often browner on back, with a cinnamon-rufous rump; wings and tail dusky, wings with white speculum. *Throat and upper chest blackish to dark brown; lower underparts cinnamon-rufous*. ♀ like ♀ Capped Seedeater. Behavior as in numerous other small seedeaters, though Dark-throated seems more widespread and ecologically tolerant than many.

BLACK-BELLIED SEEDEATER *Sporophila melanogaster* 10 cm | 4"

VU *Rare austral migrant* (Apr-Nov) to grassy cerrado and campos in *SE of our region*; favors damp areas with tall grass. ♂ *gray with striking black throat and median underparts*; wings and tail blackish, wings with gray edging and white speculum. ♀ like ♀ Capped Seedeater. Behavior like other small seedeaters; here on its wintering grounds, Black-bellied flocks with some of them. Breeds in S Brazil.

TAWNY-BELLIED SEEDEATER
CABOCLINHO-DE-BARRIGA-VERMELHA

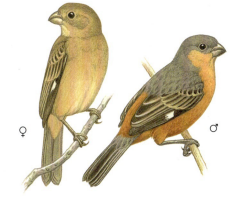

RUFOUS-RUMPED SEEDEATER
CABOCLINHO-DE-SOBRE-FERRUGEM

CHESTNUT SEEDEATER
CABOCLINHO-DE-CHAPÉU-CINZA

MARSH SEEDEATER
CABOCLINHO-DE-PAPO-BRANCO

DARK-THROATED SEEDEATER
CABOCLINHO-DE-PAPO-ESCURO

BLACK-BELLIED SEEDEATER
CABOCLINHO-DE-BARRIGA-PRETA

WARBLING FINCHES are simply patterned, mainly gray emberizid "finches" that are found in scrub and low woodland.

BLACK-CAPPED WARBLING FINCH *Poospiza melanoleuca* 13 cm | 5"
Uncommon in scrub and low woodland in S Mato Grosso do Sul. Iris orange-red. *Head glossy black* contrasting with *uniform gray upperparts* and *snowy white underparts*; outer tail feathers white (flashing in flight). Cinereous Warbling Finch differs in lacking the black head. Cf. very differently shaped Masked Gnatcatcher. Mainly arboreal; when not breeding often in groups of 4-8 birds that accompany mixed flocks. Gleans in foliage, and usually bold and excitable; easy to observe. Song a series of well-enunciated "ch-veet" notes.

CINEREOUS WARBLING FINCH *Poospiza cinerea* 13 cm | 5"
🇧🇷 *Rare and local in cerrado and scrub in E and S of our area*. Iris reddish. *Simply patterned: gray above* and *white below*, lores and cheeks blacker, and throat tinged buff; outer tail feathers white (flashing in flight). Black-capped Warbling Finch has a contrasting black cap; it occurs S of Cinereous's range, with no overlap. Cinereous remains a poorly known bird. It is mainly arboreal, occurring as scattered pairs that occasionally are with mixed flocks.

BLACK-MASKED FINCH *Coryphaspiza melanotis* 13.5 cm | 5.25"
VU *Very local*—though in a few places it can still be common—in *less-disturbed grassy cerrado and campos*. Bill bicolored, mandible yellow. Handsome ♂ has *head black* with *long white superciliary*; back brown with chestnut streaking; shoulders yellow, wing-edging yellow-olive. *Long graduated tail has outer feathers broadly tipped white*, obvious in flight. *Below white*, patch on sides black. Dingier ♀ has a grayish head, more prominent wing-edging, less white in tail. ♂ nearly unmistakable; ♀ best recognized by the company she keeps (the white in her tail helps in flushed birds). Feeds on or near ground, and then inconspicuous, but also sometimes perches in the open, especially early in the morning and when singing. When not breeding can occur in small groups. Song a weak "tsee-slee," easily passed over as an insect and hard to locate.

COAL-CRESTED FINCH *Charitospiza eucosma* 11.5 cm | 4.5"
Uncommon and local in cerrado. Pretty ♂ has *crown and crest black* (crest usually laid flat) and *silvery white face*. Back pale gray, wing-coverts whitish, *basal tail white* (most evident in flight). *Bib on chest black; below cinnamon-buff* (deepest on midbreast). ♀ duller and browner (lacking ♂'s black, *but retaining its crest and the white in tail*), with a weak buff eyestripe. Below dull buff. ♂ unique, but ♀ can be confused with various ♀ seedeaters, though none of these has white in tail, nor a crest. Found in pairs or small groups, mainly feeding on open ground near grassy cover. Seems to favor recently burned areas. Usually quiet, but has a simple warbling finch-like song.

PILEATED FINCHES are small plain birds found in scrubland, one species basically gray, the other red.

GRAY PILEATED FINCH *Coryphospingus pileatus* 13 cm | 5"
Uncommon and rather local in arid scrub, low woodland, and shrubby woodland borders. ♂ has *crown black and a scarlet crown stripe* (the latter often hidden, but can be exposed as a crest, black flaring to sides). *Above pale gray* with white eye-ring. Whitish below, grayish on breast and sides. ♀ lacks black and red in crown, but otherwise like ♂ (including the eye-ring); some grayish streaking below. ♀ seedeaters are browner and have thicker bills. Often in small groups, especially when not breeding; usually forages on the ground. Simple leisurely song a fairly melodic phrase repeated several times, quite similar to Red Pileated Finch's.

RED PILEATED FINCH *Coryphospingus cucullatus* 13.5 cm | 5.25"
Uncommon in scrub, deciduous woodland, and agricultural areas; more numerous southward, in N apparently occurring only as an austral migrant. ♂ has *crown and crest as in Gray Pileated Finch*. Otherwise *mostly dark vinous red above*; eye-ring white. *Below dull crimson*. ♀ like ♂ but lacks black and red on crown, and browner above; *retains the narrow eye-ring*. Below mostly rosy pink. The most uniformly reddish of any of our finches; the eye-ring will help to identify the duller ♀. Forages mostly on ground, gathering in small loose groups when not breeding, then sometimes flocking with other species. Song a simple phrase repeated 3-6 times, usually with short initial and final notes: "chewit, weet-chewit, weet-chewit, weet-chewit...."

301

BLACK-CAPPED WARBLING FINCH
CAPACETINHO

CINEREOUS WARBLING FINCH
CAPACETINHO-CINZA

BLACK-MASKED FINCH
TICO-TICO-MASCARADO

COAL-CRESTED FINCH
MINEIRINHO

GRAY PILEATED FINCH
CRAVINA

RED PILEATED FINCH
TICO-TICO-REI

YELLOW-BROWED SPARROW *Ammodramus aurifrons*   13 cm | 5"
Fairly common in grassy areas, especially along rivers (also in disturbed areas), in N of our area. Above brownish gray streaked dusky and with *conspicuous yellow on foreface*; bend of wing also yellow. Below pale grayish. Grassland Sparrow lacks such prominent yellow on face (though it can show a little), and has more rufous on wing. Yellow-browed hops and feeds on ground, and frequently perches in the open atop shrubs and on fences. Song, given throughout the day, a distinctive buzzy "tic, tzzzz-tzzzz," markedly different from Grassland. Also gives a fast "tseew tseew tsee tsee tsee."

GRASSLAND SPARROW *Ammodramus humeralis*   13 cm | 5"
Widespread and locally common in grassy cerrado, campos, and pastures. Brownish gray above, *grayest on face and neck*, with blackish and chestnut streaking; lores (only) yellow, bend of wing also yellow, and *rufous edging on inner flight feathers*. Whitish below (sometimes stained buff by soil). Cf. Yellow-browed Sparrow. Other streaky sparrows or finches are larger; Grass Wren is very differently shaped. Secretive except when breeding, creeping about on the ground in heavy cover, flushing short distances and then disappearing. Early and late in the day sometimes perches atop a shrub or on a fence to survey its territory; singing birds do so as well. High-pitched and far-carrying song a musical "eee telee, teeeee" with variations.

LONG-TAILED REEDFINCH *Donacospiza albifrons*   15 cm | 6"
Uncommon in damp grassland and shrubbery near water in SE of our area. Rather *long slender tail*. Brown above with *superciliary and crescent under eye white* and *cheeks dark grayish*; back streaked dusky, and shoulders bluish gray. Below buff. ♀ has more streaking above. Wedge-tailed Grassfinch is longer-tailed with more obvious back streaking; it is whitish below. Usually in pairs and, though often inconspicuous, sometimes perches atop grass clumps or on bushes. Fast spritely song a series of paired notes repeated in an almost random fashion.

WEDGE-TAILED GRASSFINCH *Emberizoides herbicola*   19-20 cm | 7.5-8"
Widespread and locally common in less-disturbed, taller grasslands, often with scattered bushes; sometimes in pastures. *Tail very long, graduated, and pointed*. Bill mostly yellow. *Above brownish olive with prominent black streaking*, lores and eye-ring whitish; wings more olive with yellow-olive edging and yellow on bend of wing. Whitish below, sides and flanks tinged buff. Cf. Lesser Grassfinch and Great Pampa Finch. Generally inconspicuous, feeding mostly on the ground and remaining within cover; especially on calm early mornings perches on fences, atop shrubs, or grass clumps. Compared to many grassland birds, ecologically tolerant. Rather musical song a pleasant "jew-lee, jew-lu" or "jew jew-leéu, ju-leéu." Also gives buzzier or more chattered songs, sometimes in flight.

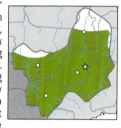

LESSER GRASSFINCH *Emberizoides ypiranganus*   17-18 cm | 6.75-7"
Rare and very local in marshes and damp grassy areas (campos alagados). Resembles much more numerous Wedge-tailed Grassfinch (which favors better-drained grasslands) but slimmer and smaller. More "contrasty" above, with *heavier, blacker streaking on a paler ground color*; cheeks and especially loral area darker and grayer, contrasting with white throat; lower flanks and crissum darker and blackish-streaked. Behavior much as in Wedge-tailed Grassfinch, but song very different: a less musical, fast nasal chattering "ch, ch, ch-ch-ch-ch-ch-ch-ch," somewhat recalling Grass Wren.

PALE-THROATED SERRA FINCH *Embernagra longicauda*   21.5 cm | 8.5"
Uncommon and local in grassy scrub with palms and ground bromeliads atop serras of Minas Gerais, e.g., Serra do Cipó. Bill mostly yellowish orange. *Head gray with white supraloral and lower eyelid*; olive above, flight feathers and *long tail* edged brighter olive. *Throat white* outlined by a gray malar streak; below pale grayish. A rather shy bird, occurring in pairs and usually staying within cover, though ♂♂ typically sing from exposed perches. Song a loud and penetrating "tsi, tsoweeé," repeated steadily at 4-5 second intervals.

🇧🇷 **Great Pampa Finch** (*E. platensis*) occurs in areas with tall grass in S of our area, e.g., at Serra da Canastra and in S Mato Grosso do Sul. Compared with the serra finch, darker on face and foreneck with *no* white facial markings and no white on throat; tail shorter and less graduated.

YELLOW-BROWED SPARROW
CIGARRINHA-DO-CAMPO

GRASSLAND SPARROW
TICO-TICO-DO-CAMPO

LONG-TAILED REEDFINCH
TICO-TICO-DO-BANHADO

WEDGE-TAILED GRASSFINCH
CANÁRIO-DO-CAMPO

LESSER GRASSFINCH
CANÁRIO-DO-BREJO

PALE-THROATED SERRA FINCH
RABO-MOLE

YELLOWFINCHES are compact *predominantly yellow* "finches" (they may turn out to be tanagers) that range in open to semiopen terrain. All are gregarious, conspicuous birds.

SAFFRON FINCH Sicalis flaveola 13.5-14 cm | 5.25-5.5"

Fairly common to common and widespread in semiopen areas, agricultural areas, and around many towns and cities. *Especially numerous in Pantanal*, where flocks gather for food at many ecolodges. ♂ *olive yellow above, yellowest on face*, forehead more orange, and with dusky streaking on back. Yellow below, breast washed olive. ♀ more brownish olive with dusky streaking above; whitish *with fine dusky streaking below*. Grassland Yellowfinch is more olive generally, especially above, and has more marked streaking on back; it ranges primarily in open agricultural terrain, and is infrequent around buildings. Cf. Orange-fronted Yellowfinch (only in N). Saffron Finch forages mainly on the ground (often on lawns), sometimes gathering in large flocks and then often associating with other birds. Vigorous song a variable series of lively, well-enunciated notes and short phrases: "tsip, tsee-tit, tsee, tseeti, tsee, tsee, tseeti."

ORANGE-FRONTED YELLOWFINCH Sicalis columbiana 11.5 cm | 4.5"

Uncommon and local in semiopen areas, riparian growth, and sandbars along rivers *in N of our area*. *Smaller than Saffron Finch*. ♂ *mostly bright yellow with a contrasting orange forecrown* and *dusky lores*. Some birds show an orange breast tinge. ♀ olive brown above; below whitish, sides tinged grayish buff. Ranges in pairs or small flocks, foraging mainly on open ground, but also perching freely in trees and on fences and buildings.

GRASSLAND YELLOWFINCH Sicalis luteola 11.5-12.5 cm | 4.5-5"

Fairly common but erratic austral winter visitant (mostly May-Oct) to campos, agricultural fields, and the edge of marshes. ♂ olive brownish above with dusky streaking on crown and back, *lores and ocular area bright yellow*, rump plain olive; wings and tail brownish. *Below yellow*, breast washed grayish olive. ♀ similar but browner above and buffier below. Unlike the Saffron Finch, this species rarely or never occurs around houses. Stripe-tailed Yellowfinch lacks the obvious facial yellow. Grassland Yellowfinches can occur in large swirling flocks, and in our area they seem nomadic; migrants sometimes fly in compact groups. They feed mainly on the ground, but perch freely in shrubbery and on fences. Call a "tzi-tzit" or "tsip-tseep." Song (not often heard here) a series of buzzy and more musical trilled notes, sometimes given in hovering flight. Birds that occur here breed mainly in N Argentina region.

STRIPE-TAILED YELLOWFINCH Sicalis citrina 12 cm | 4.75"

Uncommon and local in grassy cerrado and campos mainly in E of our area and especially atop serras. ♂ olive above with *forecrown contrastingly citrine yellow* and back streaked dusky; *inner web of outer two tail feathers mostly white on terminal half* (hard to see in the field). Yellow below, breast clouded olive. ♀ above more brownish and *conspicuously streaked dusky*; below pale yellow *extensively streaked dusky; tail much as in* ♂. Grassland Yellowfinch differs in having yellow on face and a streaked crown. ♀ Stripe-tailed is the only yellowfinch having extensive streaking on a *yellowish* ground color. Behavior much as in Grassland Yellowfinch, though Stripe-tailed is less numerous and never seems to gather in sizeable flocks. Musical song a fast sputtering, "switchity, switch-you, switch-you," usually given from an open perch but sometimes during a display flight.

HOODED SISKIN Carduelis magellanica 11-12 cm | 4.25-4.5"

Uncommon and local in semiopen and cultivated areas with scattered trees, gardens, and forest borders; scarce or absent in Pantanal. *Short conical bill.* ♂ has *black hood* contrasting with yellow-olive back and bright yellow underparts; rump also yellow. *Wings black with a conspicuous yellow band on coverts and another on flight feathers*; black tail has yellow at base. ♀ much duller, lacking black on head; *wings and tail as in* ♂. Grayish olive above with some dusky back streaking and yellowish rump. Olive yellowish below, yellower on belly. Siskins most often occur in small groups, and forage from high in trees down to the ground; they are often quite tame. Distinctive flight strongly undulating. Pretty song a long-continued and irregular twittering, sometimes given by several ♂♂ at once. Call a pretty "tseeu," often given in flight. Classified in the family Fringillidae.

SAFFRON FINCH
CANÁRIO-DA-TERRA

ORANGE-FRONTED YELLOWFINCH
CANÁRIO-DO-AMAZONAS

GRASSLAND YELLOWFINCH
CANÁRIO-TIPIU

STRIPE-TAILED YELLOWFINCH
CANÁRIO-RASTEIRO

HOODED SISKIN
PINTASSILGO

ARREMON sparrows are *boldly patterned, mainly ground-dwelling* emberizid finches that are *inconspicuous inhabitants of woodland and forest.*

PECTORAL SPARROW Arremon taciturnus   15 cm | 6"

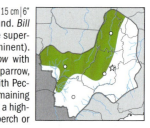

Uncommon and rather local in undergrowth of forest and woodland. *Bill black.* Head black with a gray median crown streak and long white superciliary; otherwise *olive above,* shoulders yellow (often quite prominent). White below with a *black pectoral band.* ♀ duller, *buffier below* with pectoral band restricted to sides. Both sexes of Saffron-billed Sparrow, regardless of race, always have a mostly yellow bill; it overlaps with Pectoral only marginally. Hops on or near the ground in pairs, usually remaining in dense cover and hard to see; generally not with mixed flocks. Song a high-pitched, almost hissing "zitip, tzeee-tzeee-tzeee," given from a low perch or sometimes the ground.

SAFFRON-BILLED SPARROW Arremon flavirostris   15.5-16 cm | 6-6.25"

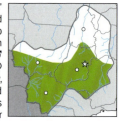

Uncommon to locally fairly common in undergrowth of deciduous woodland and forest. Bill mostly orange-yellow. ♂ in Mato Grosso and Mato Grosso do Sul (**A**) has a *mostly gray mantle* with olive on wing-coverts; head black with a white postocular stripe. Mostly white below with a *black pectoral band.* ♂ in Goiás and Minas Gerais (**B**) is *olive-backed* with only upper back gray. ♀ like respective ♂ but duller and washed with buff below. Pectoral Sparrow, which can overlap with Saffron-billed locally, differs in its all-black bill and longer white superciliary (starting at bill, not from eye). Saffron-billed's unobtrusive behavior is much as in Pectoral Sparrow. Song a similar spitting "tsit, tsee-tsi-tsi, tseép-seép-tseép."

🇧🇷 The recently described **São Francisco Sparrow** (*A. franciscanus*) occurs in undergrowth of deciduous woodland in São Francisco valley of N Minas Gerais; it is not known to occur with either Pectoral or Saffron-billed Sparrow. Bill mostly yellow (only culmen is black) and black pectoral band interrupted (only on sides) in both sexes.

RUFOUS-COLLARED SPARROW Zonotrichia capensis   14-15 cm | 5.5-6"

Uncommon and local in open and semiopen shrubby habitats and often around houses in our area. Though a familiar and common bird in many parts of South America, not especially numerous here and scarce or absent in Pantanal. Slightly bushy-crested. *Head gray with black striping and with contrasting and conspicuous rufous hindneck collar.* Above rufescent brown with black striping on back; wings with two whitish bars. Below whitish with *black patch on sides of breast.* Juvenile duller and streakier, especially below, but showing at least a trace of the rufous collar. Often tame, hopping and feeding on the ground; usually in pairs or small family groups. Song typically consists of 1-2 pretty slurred whistles followed by a trill: "tee-teeoo, treeeee." Call a "chip" or "chink." Very popular as a cagebird in much of Brazil.

HOUSE SPARROW Passer domesticus   15 cm | 6"

Introduced and locally common in and around towns and cities, and in agricultural terrain. Though it remains absent away from human habitation, the familiar House Sparrow continues to spread and increase in our area. ♂ has *gray crown, chestnut nape, white cheeks and sides of neck,* and *black bib;* back brown streaked black, rump plain gray. Below whitish. The pattern is soiled in many birds. More nondescript ♀ has *yellow on bill.* Grayish brown above with blackish streaking and a *pale superciliary;* below dingy whitish. A strict commensal of man, House Sparrows are noisy, aggressive, and social. In many areas they nest under the shade awnings of gas stations. Native to Eurasia, and classified in the family Passeridae.

COMMON WAXBILL Estrilda astrild   11.5 cm | 4.5"

Introduced and now locally not uncommon in and around various towns and cities in SE of our area; apparently still spreading. An unmistakable small finch with a *long tail* and *vivid red bill.* Brown above, vermiculated darker, with red *streak through eye.* Below pale buff vermiculated dusky on sides and breast; red stain on midbelly. Waxbills feed on the ground in small groups, usually monospecific but sometimes together with seedeaters; they seem to favor lawns. Native to tropical Africa, and classified in the family Estrildidae.

PECTORAL SPARROW
TICO-TICO-DE-BICO-PRETO

SAFFRON-BILLED SPARROW
TICO-TICO-DE-BICO-AMARELO

RUFOUS-COLLARED SPARROW
TICO-TICO

HOUSE SPARROW
PARDAL

COMMON WAXBILL
BICO-DE-LACRE

AMERICAN BLACKBIRDS (Icteridae) are a diverse group of passerine birds found only in the New World. Their coloration varies, though many are mainly or entirely black; all have pointed bills. Many are gregarious, notably certain caciques and oropendolas.

BAYWING *Agelaioides badius* 18.5 cm | 7.25"

Locally common in semiopen and agricultural areas and light woodland in SW (including Pantanal) and E of our area. In SW (illustrated) *ashy brownish gray with ocular area blackish, rufous wings*. In E, *notably paler and buffier*. Dull-plumaged, but no similar bird shows the *contrasting rufous wings*; cf. juvenile Screaming Cowbird (a brood parasite on Baywing), which briefly resembles it. Almost always in small flocks, sometimes feeding with Shiny Cowbirds, mainly on ground. Unlike true cowbirds, Baywings are *not* brood parasites; pairs construct their own nests or sometimes appropriate nests of other species. Helpers are often present, helping to defend against the Screaming Cowbirds that parasitize them. Song, often given in concert by flock members, a series of whistled notes on varying pitches, usually sounding off key. Formerly called the Bay-winged Cowbird.

SCREAMING COWBIRD *Molothrus rufoaxillaris* 19 cm | 7.5"

Uncommon in open areas and light woodland; *spreading northward. Lustrous black with a slight greenish blue sheen*. Juvenile *very different*, resembling the adult Baywing, but quickly molting into adult plumage. Resembles ♂ Shiny Cowbird, but *bill shorter and stouter* (imparting a snub-nosed look) and lacking Shiny's violet gloss. Behavior similar to Shiny Cowbird, and sometimes with it. Rarely in large flocks and, though mainly feeding on the ground, only infrequently assembles around cattle. A brood parasite on the Baywing in most of its range, here Screamings have begun to parasitize the Chopi Blackbird. ♂'s distinctive song a brief, explosive "zhhleeee," often given as it fluffs neck feathers.

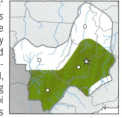

SHINY COWBIRD *Molothrus bonariensis* 19-21.5 cm | 7.5-8.5"

Widespread and often common in semiopen and agricultural terrain. ♂ *glossy purplish black*, gloss strongest on head and forecrown. ♀ *dark chocolate brown*, very slightly paler below. Juvenile resembles ♀ but paler below with blurry dark streaking. Various blackbirds are larger and less glossy than the ♂ Shiny; they have longer, more pointed bills. Cf. Screaming Cowbird. Usually in flocks, when the sexual dimorphism is a helpful identification point (other similar blackbirds show *no* dimorphism). Forages mainly on ground, often holding its tail cocked; frequently attends cattle. The species is a brood parasite on various songbirds. Song a musical twittering, often given by several ♂♂ together, their neck feathers ruffled.

GIANT COWBIRD *Molothrus oryzivorus* ♂ 37 cm | 14.5" ♀ 32 cm | 12.5"

Uncommon and local in forest borders and adjacent open areas. Iris usually orange. ♂ *glossy purplish black. Conspicuous ruff on neck* imparts a small-headed effect. ♀ sootier; ruff less evident. *Even the smaller ♀ is much larger than other cowbirds*. Flying birds can be recognized by their *distinctively undulating flight* (several quick flaps alternating with a dip when wings are closed). Otherwise the species is mainly seen on the ground, sometimes in small groups, occasionally visiting feeders at ecolodges or accompanying cattle. ♀♀ linger around nesting colonies of oropendolas and caciques, on which the species is a brood parasite. Notably quiet.

CHOPI BLACKBIRD *Gnorimopsar chopi* 23-24 cm | 9-9.5"

Familiar and usually common in semiopen and agricultural terrain throughout much of our area. Fairly short bill has an *inconspicuous but diagnostic diagonal groove on lower mandible*. Glossy black; *feathers of crown and nape narrow and pointed*, imparting a hackled effect. Cf. ♂ Shiny Cowbird (which has a much more glossy sheen) and Screaming Cowbird (with shorter bill). Chopis are often tame and conspicuous: they parade about on the ground and on lawns, perch in shade trees around buildings, and often roost in palms. They do not associate with cattle. A variety of loud and musical calls is given, sometimes as a medley by a flock. They call even during the heat of day, and also in the predawn. Vocalizations include a repeated single "peer," "pur-peer" (or "chopi"), or "chup," these sometimes preceding a rich gurgled song.

VU Forbes's Blackbird (*Curaeus forbesi*) is a very rare and local resident around marshes in N Minas Gerais. Acutely endangered, this blackbird resembles the Chopi but has an even longer and more pointed bill, and is a sootier black. Its mouth lining is *red*, readily seen in singing birds.

BAYWING
ASA-DE-TELHA

SCREAMING COWBIRD
CHUPIM-PICUMÃ

SHINY COWBIRD
CHUPIM

GIANT COWBIRD
IRAÚNA-GRANDE

CHOPI BLACKBIRD
PÁSSARO-PRETO, GRAÚNA

YELLOW-RUMPED MARSHBIRD *Pseudoleistes guirahuro* 23.5 cm | 9.25"
Uncommon to locally fairly common in grasslands and fields near water and marshes in S of our area. Fairly long pointed bill. *Brown above* with *yellow rump* and yellow on lesser wing-coverts (both visible especially in flight). Throat and chest dark brown, *lower underparts bright yellow*. A conspicuous bird that forages principally on the ground adjacent to the marshes where it nests and roosts. Usually occurs in small flocks, sometimes associating with other blackbirds such as Chopis. Often noisy, with a loud and rich gurgling song and other calls that will often draw attention to a flock flying past, or to birds that are hidden in tall vegetation.

BOBOLINK *Dolichonyx oryzivorus* 18 cm | 7"
Uncommon boreal migrant (occurring mostly as a transient, Sep-Apr) in marshes, ricefields, and damp pastures with tall grass; recorded mainly in Pantanal. *Pointed tail feathers*. ♀ and nonbreeding ♂ streaked buffy brown and blackish above with *bold buff and dusky head striping*. Below *yellowish buff* with blackish streaking on flanks. Breeding plumage ♂ (rarely seen here, only just prior to its departure for the N; attained as buff feather tips wear off) mostly *black* with a *golden buff nuchal collar, broad white scapulars*, and *large white rump patch*. ♀ White-browed Blackbird lacks the "spiky" tail feathers and has a longer bill; it generally shows some pink tinge below. Bobolinks are gregarious birds that can occur in large flocks on their wintering grounds, which lie mainly further south. Feeding birds often associate with *Sporophila* seedeaters or other seed-eating birds. In some ares they consume much rice. Call a distinctive metallic "pink," often given in flight; ♂'s bubbly song is not heard here. Breeds in North America.

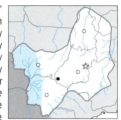

WHITE-BROWED BLACKBIRD *Sturnella superciliaris* 18 cm | 7"
Locally common in grasslands and lusher pastures. ♂ black above with a *prominent white postocular stripe*; bend of wing red (obvious in flight). *Throat and breast bright red*, belly black. Fresh-plumaged birds have their feathers edged brownish, obscuring the black and red though *the white postocular is always obvious*. ♀ streaked blackish and buff above with a long buff superciliary and crown stripe. Buff below, streaked dusky across chest and on belly, *variably suffused pink on breast*. Gregarious; even when nesting often found in loose colonies, though each ♂ does defend a small territory. Regularly forages around cattle. Frequent in agricultural areas, seeming to have adapted to fertilized pastures better than most other birds. ♂'s song is given in flight, a buzz followed by a series of "chuk" notes, the buzz as it ascends, the terminal notes as it drops back into the grass.

AGELASTICUS and ***CHRYSOMUS*** blackbirds are rather small, *mainly or entirely black* icterids with *sharply pointed bills* that are found *in or near marshes*. The two species, now in separate genera, were formerly placed in the genus *Agelaius*.

UNICOLORED BLACKBIRD *Agelasticus cyanopus* 19 cm | 7.5"
Locally common in marshes and margins of shallow ponds and lakes; most numerous in Pantanal. ♂ *uniform glossy black*. ♀ streaked blackish and brown above, with a yellowish superciliary, blackish lores, and cheeks; *wings blackish edged rufous-chestnut*. Below oily yellow streaked dusky. Cf. Chopi Blackbird (not associated with water). Usually in pairs, at most small groups. Strictly tied to the vicinity of water, where it walks on floating vegetation and clambers about in nearby reeds, grass, and low shrubbery. Song a loud and ringing "tchew-tchew-tchew-tchew."

CHESTNUT-CAPPED BLACKBIRD *Chrysomus ruficapillus* 18.5 cm | 7.25"
Rather uncommon and local in and around marshes and other wet situations, foraging in adjacent open and agricultural areas. ♂ glossy black with *crown and throat chestnut*. Nondescript ♀ dark olivaceous brown above with obscure blackish streaking; *foreneck buffy fawn*, below dull olivaceous. ♂ chestnut is so dark that except in ideal light it can be hard to discern; at such times the duller ♀ can actually be easier to identify. A gregarious bird (more so than the Unicolored) that occurs in flocks year-round and nests colonially. It rarely forages very far from water. Song a drawn-out, descending "chree-chree-chrrrr." Also gives sweet, canary-like whistles and trills.

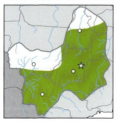

YELLOW-RUMPED MARSHBIRD
CHUPIM-DO-BREJO

BOBOLINK
TRISTE-PIA

WHITE-BROWED BLACKBIRD
POLÍCIA-INGLESA-DO-SUL

UNICOLORED BLACKBIRD
CARRETÃO-DO-BREJO

CHESTNUT-CAPPED BLACKBIRD
GARIBALDI

SCARLET-HEADED BLACKBIRD *Amblyramphus holosericeus* 24 cm | 9.5"
Uncommon and local in extensive marshes in Pantanal. Stunning and unmistakable. Bill long, slender, and sharply pointed (almost appearing upswept). Black with *brilliant scarlet head, neck, and breast*; thighs also scarlet. Juvenile all black, but it soon acquires red on throat and breast, this gradually spreading onto the head. Strictly confined to reedbeds, even there seeming thinly spread, with pairs holding large territories; gathers in small groups when not breeding, then sometimes associating with other icterids. Breeding ♂♂ are often conspicuous, perching atop reeds or on adjacent shrubs. Loud, ringing song a clear and melodic "cleer-cleer-clur, clululu."

TROUPIALS are beautiful *mainly bright orange* icterids that range in woodland and scrub. The three allopatric species (one in Venezuela) were formerly often considered conspecific.

ORANGE-BACKED TROUPIAL *Icterus croconotus* 23-23.5 cm | 9-9.25"
Uncommon to fairly common in gallery woodland and borders; most numerous in the Pantanal. Bright orange with a *black face and bib* and *small bare blue ocular area*; wings (aside from an orange epaulet and white patch on secondaries) and tail also black. Nearly unmistakable; cf. only the Campo Troupial (occurring to E). *In its range Orange-backed Troupial is the only bright orange bird of its size.* Found singly or in pairs, usually moving independently of mixed flocks; generally conspicuous. Often pirates the nests of Yellow-rumped Cacique and other species such as thornbirds and Great Kiskadees. Song a series of loud, rich musical phrases, each most often two-parted, e.g., "tree-tur" or "cheer-to," often with a softer introductory note; this is repeated slowly, sometimes with long intervening pauses.

CAMPO TROUPIAL *Icterus jamacaii* 23-23.5 cm | 9-9.25"
Uncommon to fairly common in scrub, deciduous woodland, and borders in NE of our area. Resembles Orange-backed Troupial but has an *entirely black head*, no bare blue skin behind eye, black bib with a jagged lower edge, and more black on back. *In its range, Campo Troupial is the only bright orange bird of its size.* Behavior much like Orange-backed Troupial, as is its simple whistled song, which is often long-continued. Often tagged as a cagebird, and this has resulted in a reduction in its numbers in many areas.

EPAULET ORIOLE *Icterus cayanensis* 20.5-21 cm | 8-8.25"
Widespread and locally fairly common to common in woodland, forest borders, and clearings with scattered trees. In Pantanal (**A**) black with a *chestnut epaulet*. In most of our area (**B**) similar but *epaulet more tawny-orange* and thighs yellow. There is nothing really very similar in our area, though note that the chestnut epaulet is so dark that it can be hard to discern in poor light. Usually found singly or in pairs, though nonbreeders often range in groups, occasionally of up to 20 or more birds. Regularly visits flowering trees. Agile, often twitching its rather long tail. Song a leisurely series of whistled notes and phrases with intervening pauses, "sweet... peeyr... pyur... peeyr... pyur... kreer... wrrrt..." or "purit... skeeyr... tyi-pur-twee-twee...." Sometimes one will include mimicry of other birds.

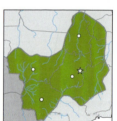

GOLDEN-WINGED CACIQUE *Cacicus chrysopterus* ♂ 20.5 cm | 8"; ♀ 18 cm | 7"
A *small* cacique found locally in deciduous and gallery woodland in *SW of our area*. Bill bluish horn; iris can be either pale blue or yellow. *Black with yellow wing-coverts and rump*. Can look *shaggy-crested*. Yellow-rumped Cacique is larger with yellow extending to basal tail and on its crissum. Sympatric Epaulet Orioles have chestnut on their wing-coverts. Ranges in pairs and small groups; a much quieter and less conspicuous bird than the Yellow-rumped Cacique, and a solitary nester. Song a ringing "gloo-gloo-gleéyu-gleéyu," sometimes given as a duet by members of a pair. Call a more nasal "wreyur."

SCARLET-HEADED BLACKBIRD
CARDEAL-DO-BANHADO

ORANGE-BACKED TROUPIAL
JOÃO-PINTO

CAMPO TROUPIAL
CORRUPIÃO

EPAULET ORIOLE
ENCONTRO

GOLDEN-WINGED CACIQUE
SOLDADO

CACIQUES are fairly large, mainly black icterids, most species *with some red or yellow on wings, rump, or tail*. They are noisy birds that attract attention. *Nesting is colonial*, ♀♀ constructing purse-shaped nests and suspending them from the outer limbs of trees out in the open. ♂♂ are notably larger than ♀♀. The Golden-winged Cacique is depicted on the previous plate.

SOLITARY CACIQUE *Cacicus solitarius*   ♂ 27 cm | 10.5" ♀ 23 cm | 9"
Uncommon to fairly common in dense undergrowth in clearings and riparian woodland; most numerous in the Pantanal. Bill greenish ivory; *iris dark. All black*. Red-rumped Cacique's red rump can be hidden on perched birds (but even then, note its blue eye). Cf. ♂ Unicolored Blackbird (also all black, but smaller and with a dark bill). Skulks in tangled undergrowth, often in pairs, probing into dead leaves and other suspended debris, sometimes hanging upside-down; at times will ascend to feed, more in the open, at flowering trees. Not colonial. Has a varied repertoire of loud and strident calls, including a penetrating "keeyoh keeyoh keeyoh" or "kyoong kyoong kyoong," a fast "chochochochochocho," and a repeated "kway." Also gives a nasal "wheeeah-ah" (like a donacobius), less often a loud "popp!"

RED-RUMPED CACIQUE *Cacicus haemorrhous*   ♂ 28 cm | 11" ♀ 23 cm | 9"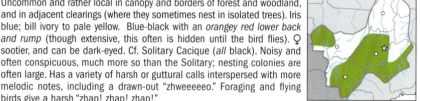
Uncommon and rather local in canopy and borders of forest and woodland, and in adjacent clearings (where they sometimes nest in isolated trees). Iris blue; bill ivory to pale yellow. Blue-black with an *orangey red lower back and rump* (though extensive, this often is hidden until the bird flies). ♀ sootier, and can be dark-eyed. Cf. Solitary Cacique (*all* black). Noisy and often conspicuous, much more so than the Solitary; nesting colonies are often large. Has a variety of harsh or guttural calls interspersed with more melodic notes, including a drawn-out "zhweeeeeo." Foraging and flying birds give a harsh "zhap! zhap! zhap!"

✓ **YELLOW-RUMPED CACIQUE** *Cacicus cela*   ♂ 28 cm | 11" ♀ 24.5 cm | 9.5"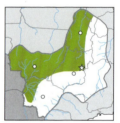
Fairly common to common in canopy and borders of forest and woodland, and in adjacent clearings; most numerous northward. Iris blue; bill ivory yellow to pale grayish. Glossy black with a *large yellow patch on wing-coverts; rump, crissum and base of tail also yellow*. ♀ sootier, and can be dark-eyed. The smaller Golden-winged Cacique lacks yellow on its tail and crissum; it occurs only in the far south. Crested Oropendola has yellow only on lateral tail feathers. A flamboyant bird that attracts attention, Yellow-rumped Cacique nests in small colonies, usually in the semiopen, sometimes with oropendolas. It feeds mainly on fruit, also on insects and at flowering trees. Has a wide repertoire of loud calls, some harsh and slashing, others more melodic; imitations of other species are sometimes incorporated. ♂♂ sing mainly around nesting colonies, leaning forward and ruffling their rump feathers.

✓ **CRESTED OROPENDOLA** *Psarocolius decumanus* ♂ 46.5 cm | 18.25" ♀ 37 cm | 14.5"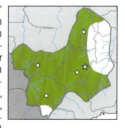
Widespread and often common in borders of forest and woodland and in clearings with scattered tall trees. Iris pale blue; bill ivory to pale greenish yellow. Glossy black with dark chestnut rump and crissum; *tail bright yellow with blackish central feathers*. Some birds have a scatter of yellow or white feathers in the black. The only oropendola in our area, and as such not likely to be confused; compare to the caciques, which are shorter-tailed (this especially evident in flight). Conspicuous and gregarious birds, Crested Oropendolas often range with other icterids but usually nest apart, in an isolated tree. In flight looks stretched-out, with steady rowing wingbeats. The displaying ♂ leans forward with fluttering wings and raised tail (sometimes appearing almost to fall off its perch!), uttering a loud gurgling, slashing song that accelerates. Both sexes often give a loud "chak" call.

Olive Oropendola (*P. yuracares*) occurs in small numbers in forest and edge in the far north of our area (e.g., Ilha do Bananal). Olive is a very large oropendola with an orange-tipped black bill and conspicuous bare pink cheeks; its foreparts are dusky-olive, rearparts mostly bright chestnut, and tail mainly yellow.

References

Antas, Paulo de Tarso Zuquim and Haroldo Palo Jr.
Pantanal – Guia de Aves,
2004. Rio de Janeiro: SESC.
http//www.avespantanal.com.br
[website in Portuguese].

Ayres, José Márcio et al.
Os Corredores Ecológicos das Florestas Tropicais do Brasil,
2005. Belém: Sociedade Civil Mamirauá.

Buzzetti, Dante and Silvestre Silva.
Nurturing New Life – Nests of Brazilian Birds,
2005. São Paulo: Editora Terceiro Nome.

Byers, Clive.
A Photographic Guide to Birds of Southern Brazil.
2008. London: New Holland Publishers.

CBRO (Brazilian Ornithological Records Committee).
Checklist of the Birds of Brazil.
www.cbro.org.br
[bilingual website].

Coelho, Henrique R. P. et al.
Guia de Campo Bonito Field Guide,
2005. Bonito: Guias de Campo.

Colombini, Fabio and Martha Argel.
Maravilhas do Brasil: Aves [Wonders of Brazil: Birds, bilingual edition], 2005.
São Paulo, SP: Escrituras.

Endrigo, Edson.
Aves – Chapada dos Veadeiros,
2008. São Paulo:
Aves e Fotos Editora.

—. *Aves do Cerrado*,
2008. São Paulo:
Aves e Fotos Editora.

Erize, Francisco, Jorge R. Rodriguez Mata, and Maurice Rumboll.
Birds of South America, Non-Passerines,
2006. Princeton: Princeton University Press.

Forum on Brazilian Birds
[OrnitoBr, Lista Brasileira de Ornitologia].
http://br.groups.yahoo.com/group/ornitobr
[website in Portuguese].

Frisch, Johan Dalgas and Christian Dalgas Frisch.
Aves Brasileiras e Plantas que as Atraem,
2005. São Paulo:
Dalgas Ecoltec.

Hughes, B. et al.
Action Plan for the Conservation of the Brazilian Merganser (Mergus octosetaceus), 2006.
Brasília: Ibama
[Threatened Species Series, 3].

Palo, Haroldo Jr.
Pantanal [bilingual edition],
1992. Rio de Janeiro:
Cor/Ação Editora.

Philips Guides: Brazil – Pantanal & Bonito,
2000. São Paulo:
Editora Horizonte.

Philips Guides: Brazil – National Parks,
2002. São Paulo:
Editora Horizonte.

Projeto Arara Azul.
[Hyacinth Macaw Project].
http://www.projetoararaazul.org.br/Arara
[website in Portuguese].

Ridgely, R.S. and G. Tudor.
The Birds of South America,
vol. 1, 1989. Austin:
University of Texas Press.

—. *The Birds of South America*,
vol. 2, 1994.
Austin: University of Texas Press.

—. *The Songbirds of South America*, 2009.
Austin: University of Texas Press.

Sick, Helmut.
Birds in Brazil, 1993.
Princeton: Princeton University Press.

—. *Ornitologia Brasileira*, 1997.
Rio de Janeiro: Nova Fronteira.

Sigrist, Tomas.
Aves do Brasil: Uma Visão Artística,
2006. São Paulo:
Avis Brasilis Editora.

—. *Guia de Campo Avifauna Brasileira* [Field Guide to the Birds of Brazil, bilingual edition], 2009. Vinhedo:
Avis Brasilis Editora.

Sousa, Deodato.
Todas as Aves do Brasil,
1998. Feira de Santana, Bahia: Editora Dall.

Syposz, Wieslaw Jan.
Aves do Pantanal e do Cerrado de Mato Grosso [Birds of Pantanal and Cerrado of Mato Grosso, bilingual edition], 2005.
Cuiabá: Carlini & Caniato Editorial.

—. *Chapada dos Guimarães: Jardim do cerrado* [The Cerrado Garden, bilingual edition],
2007. Cuiabá:
Carlini & Caniato Editorial.

WikiAves.
The Encyclopedia of Brazilian Birds.
www.wikiaves.com.br
[bilingual website].

Wheatley, Nigel.
Where to Watch Birds in South America, 1995. Princeton:
Princeton University Press.

Index of English Names

Illustrated species appear in **boldface** type. Species accounts are opposite illustrations.

Amazon
  Mealy, 122
  Orange-winged, **123**
  Turquoise-fronted, **123**
Anhinga, **39**
Ani
  Greater, **127**
  Smooth-billed, **127**
Antbird
  Bananal, **205**
  Band-tailed, **201**
  Black-faced, 202
  Black-throated, **205**
  Mato Grosso, **205**
  Southern Chestnut-tailed, 204
  Stripe-backed, **203**
  White-browed, **203**
Antpipit, Southern, **223**
Antpitta, White-browed, 206
Antshrike
  Amazonian, **199**
  Barred, **197**
  Bolivian Slaty, 198
  Glossy, **197**
  Great, **197**
  Planalto Slaty, **199**
  Rufous-winged, **197**
  Silvery-cheeked, 196
  Variable, **199**
Antthrush
  Black-faced, **207**
  Rufous-capped, **207**
Antvireo, Plain, **199**
Antwren
  Amazonian Streaked, **201**
  Black-bellied, **203**
  Black-capped, **201**
  Large-billed, **201**
  Rufous-winged, 200
  Rusty-backed, **203**
  Southern White-fringed, **203**
  White-flanked, **201**
Araçari
  Chestnut-eared, **167**
  Lettered, **167**
  Red-necked, **167**
Attila
  Rufous-tailed, **235**
  White-eyed, **235**

Bananaquit, **275**
Baywing, **309**
Becard
  Black-capped, 242
  Chestnut-crowned, **243**
  Crested, **245**
  Green-backed, **243**
  Pink-throated, **245**
  White-winged, **243**
Bellbird, Bare-throated, **251**
Bittern
  Cory's Least, 48
  Least, **49**
  Pinnated, **51**
Blackbird
  Chestnut-capped, **311**
  Chopi, **309**
  Forbes's, 308
  Scarlet-headed, **313**
  Unicolored, **311**
  White-browed, **311**

Bobolink, **311**
Buzzard-Eagle, Black-chested, **79**

Cacholote
  Caatinga, 184
  Gray-crested, **185**
Cacique
  Golden-winged, **313**
  Red-rumped, **315**
  Solitary, **315**
  Yellow-rumped, **315**
Canastero, Cipó, **191**
Caracara
  Black, **85**
  Chimango, 84
  Red-throated, **85**
  Southern Crested, **85**
  Yellow-headed, **85**
Cardinal
  Crimson-fronted, **289**
  Red-crested, **289**
  Yellow-billed, **289**
Casiornis
  Ash-throated, **235**
  Rufous, **235**
Chachalaca, Chaco, **37**
Chat, Rose-breasted, **273**
Conebill, Chestnut-vented, **275**
Coquette
  Dot-eared, **149**
  Frilled, **149**
Cormorant, Neotropic, **39**
Cotinga
  Pompadour, 250
  Swallow-tailed, **251**
Cowbird
  Giant, **309**
  Screaming, **309**
  Shiny, **309**
Crake
  Ash-throated, **91**
  Gray-breasted, **91**
  Ocellated, **91**
  Rufous-faced, **91**
  Rufous-sided, **91**
  Russet-crowned, **91**
  Uniform, **93**
  Crescentchest, Collared, **207**
Cuckoo
  Ash-colored, **125**
  Dark-billed, **125**
  Guira, **129**
  Little, **127**
  Pavonine, **129**
  Pearly-breasted, **125**
  Pheasant, **129**
  Squirrel, **127**
  Striped, **129**
  Yellow-billed, **125**
Curassow, Bare-faced, **37**

Dacnis, Blue, **275**
Donacobius, Black-capped, **261**
Doradito, Crested, **217**
Dove
  Blue Ground, **113**
  Blue-eyed Ground, **111**
  Eared, **113**

  Gray-fronted, **113**
  Long-tailed Ground, **111**
  Picui Ground, **111**
  Plain-breasted Ground, **111**
  Ruddy Ground, **111**
  Scaled, **111**
  White-tipped, **113**
Duck
  Black-bellied Whistling, **43**
  Comb, **41**
  Fulvous Whistling, **43**
  Masked, **47**
  Muscovy, **41**
  White-faced Whistling, **43**
Eagle
  Crested, **83**
  Crowned Solitary, **79**
  Harpy, **83**
Egret
  Cattle, **53**
  Great, **53**
  Snowy, **53**
Elaenia
  Forest, **213**
  Gray, **213**
  Greenish, **213**
  Highland, **209**
  Large, **209**
  Lesser, **211**
  Olivaceous, 210
  Plain-crested, **211**
  Small-billed, **211**
  White-crested, **211**
  Yellow-bellied, **209**
Emerald
  Glittering-bellied, **151**
  Glittering-throated, **151**
  Sapphire-spangled, 150
  Versicolored, **151**
Euphonia
  Chestnut-bellied, **279**
  Golden-rumped, **279**
  Purple-throated, **279**
  Rufous-bellied, 278
  Thick-billed, **279**
  Violaceous, **279**
Falcon
  Aplomado, **89**
  Barred Forest, **87**
  Bat, **89**
  Collared Forest, **87**
  Laughing, **87**
  Orange-breasted, **89**
  Peregrine, **89**
Finch
  Black-capped Warbling, **301**
  Black-masked, **301**
  Blue, **295**
  Cinereous Warbling, **301**
  Coal-crested, **301**
  Gray Pileated, **301**
  Great Pampa, **303**
  Pale-throated Serra, **303**
  Red Pileated, **301**
  Saffron, **305**
Fire-Eye
  White-backed, **205**
  White-shouldered, **205**
Firewood-Gatherer, **191**

Flatbill
  Ochre-lored, **223**
  Yellow-olive, **223**
  Zimmer's, 222
Flicker, Campo, **173**
Flycatcher
  Alder, 224
  Boat-billed, **237**
  Bran-colored, **225**
  Brown-crested, **233**
  Cliff, **235**
  Crowned Slaty, **239**
  Dusky-capped, **233**
  Euler's, **225**
  Fork-tailed, **241**
  Fuscous, **225**
  Gray-hooded, **221**
  Ochre-bellied, 220
  Piratic, **239**
  Rusty-margined, **237**
  Sepia-capped, **221**
  Short-crested, **233**
  Social, **237**
  Southern Scrub, **211**
  Streaked, **239**
  Sulphury, **241**
  Swainson's, **233**
  Variegated, **239**
  Vermilion, **225**
Foliage-Gleaner
  Buff-fronted, **193**
  Henna-capped, **195**
  Ochre-breasted, **193**
  Planalto, **193**
  White-eyed, **195**
Fruitcrow
  Bare-necked, **253**
  Purple-throated, **253**
  Red-ruffed, **253**
Gallinule
  Azure, **95**
  Purple, **95**
  Spot-flanked, **95**
Gnatcatcher, Masked, **265**
Gnateater, Rufous, **207**
Gnatwren, Long-billed, **265**
Godwit, Hudsonian, **101**
Goldenthroat, White-tailed, **153**
Goose, Orinoco, **43**
Grassfinch
  Lesser, **303**
  Wedge-tailed, **303**
Grassquit
  Blue-black, **295**
  Dull-colored, **295**
  Sooty, **295**
Grebe
  Least, **47**
  Pied-billed, **47**
Greenlet
  Ashy-headed, **269**
  Buff-cheeked, **269**
  Gray-eyed, **269**
Grosbeak
  Black-backed, **293**
  Glaucous-blue, **293**
  Ultramarine, **293**
Guan
  Chestnut-bellied, **37**
  Common Piping, **37**

317

Rusty-margined, **37**
Spix's, 36
White-browed, 36
White-crested, 36
Harrier, Long-winged, **69**
Hawk
 Bicolored, **67**
 Black-collared, **77**
 Crane, **71**
 Gray, **73**
 Great Black, **79**
 Harris's, **77**
 Roadside, **73**
 Rufous-thighed, **67**
 Savanna, **77**
 Short-tailed, **73**
 Swainson's, **75**
 Tiny, **67**
 White, **71**
 White-tailed, **75**
 Zone-tailed, **75**
Hawk-Eagle
 Black, **81**
 Black-and-white, **81**
 Ornate, **81**
Hermit
 Buff-bellied, **145**
 Cinnamon-throated, **145**
 Planalto, **145**
 Reddish, **145**
 Rufous-breasted, **145**
Heron
 Agami, **51**
 Black-crowned Night, **49**
 Boat-billed, **49**
 Capped, **55**
 Cocoi, **55**
 Fasciated Tiger, **51**
 Little Blue, **53**
 Rufescent Tiger, **51**
 Striated, **49**
 Whistling, **55**
 Zigzag, **49**
Hoatzin, **39**
Honeycreeper
 Purple, **275**
 Red-legged, **275**
Hornero
 Band-tailed, **185**
 Pale-legged, **185**
 Rufous, **185**
Hummingbird
 Sombre, **147**
 Swallow-tailed, **147**
Ibis
 Bare-faced, **57**
 Buff-necked, **55**
 Green, **57**
 Plumbeous, **57**
 White-faced, **57**
Inezia
 Amazonian, 214
 Plain, **215**
Jabiru, **59**
Jacamar
 Brown, **161**
 Paradise, **161**
 Rufous-tailed, **161**
Jacana, Wattled, **97**
Jay
 Curl-crested, **255**
 Plush-crested, **255**
 Purplish, **255**
 White-naped, **255**

318

Kestrel, American, **89**
Kingbird
 Eastern, **241**
 Tropical, **241**
 White-throated, **241**
Kingfisher
 Amazon, **159**
 American Pygmy, **159**
 Green, **159**
 Green-and-rufous, **159**
 Ringed, **159**
Kiskadee
 Great, **237**
 Lesser, **237**
Kite
 Double-toothed, **65**
 Gray-headed, **71**
 Hook-billed, **69**
 Mississippi, **65**
 Pearl, **63**
 Plumbeous, **65**
 Rufous-thighed, **65**
 Snail, **69**
 Swallow-tailed, **63**
 White-tailed, **63**
Lapwing, Southern, **99**
Leaftosser, Rufous-breasted, **195**
Limpkin, **57**
Macaw
 Blue-and-yellow, **115**
 Blue-winged, **117**
 Golden-collared, **117**
 Hyacinth, **115**
 Red-and-green, **115**
 Red-bellied, **115**
 Red-shouldered, **117**
Manakin
 Band-tailed, **249**
 Blue, **247**
 Blue-backed, **247**
 Fiery-capped, **249**
 Flame-crested, **247**
 Helmeted, **247**
 Pin-tailed, **249**
 Red-headed, **249**
 White-bearded, **249**
Mango, Black-throated, **147**
Marshbird, Yellow-rumped, **311**
Martin
 Brown-chested, **257**
 Gray-breasted, **257**
 Purple, **257**
 Sand, **257**
Merganser, Brazilian, **47**
Miner, Campo, **185**
Mockingbird
 Chalk-browed, **261**
 White-banded, **261**
Monjita
 Gray, **227**
 White, **227**
 White-rumped, **227**
Moorhen, Common, **95**
Motmot
 Blue-crowned, **161**
 Rufous-capped, **161**
Mourner, Grayish, **235**
Myiobius
 Black-tailed, **223**
 Yellow-rumped, **223**
Nighthawk
 Band-tailed, **137**

Common, **137**
Least, **137**
Lesser, **137**
Nacunda, **139**
Plain-tailed, 136
Sand-colored, 136
Short-tailed, **137**
Nightjar
 Band-winged, 140
 Ladder-tailed, **139**
 Little, **141**
 Rufous, **141**
 Scissor-tailed, **139**
 Sickle-winged, **141**
 Spot-tailed, **141**
 White-winged, **141**
Nothura
 Lesser, **35**
 Spotted, **35**
 White-bellied, **35**
Nunbird
 Black-fronted, **165**
 White-fronted, **165**
Nunlet
 Rufous-capped, 162
 Rusty-breasted, **163**
Oriole, Epaulet, **313**
Oropendola
 Crested, **315**
 Olive, 314
Osprey, **63**
Owl
 Amazonian Pygmy, 132
 Barn, **133**
 Black-banded, **135**
 Buff-fronted, **133**
 Burrowing, **133**
 Crested, **135**
 Ferruginous Pygmy, **133**
 Great Horned, **131**
 Mottled, **135**
 Short-eared, **131**
 Spectacled, **135**
 Striped, **131**
 Stygian, **131**
 Tawny-bellied Screech, 132
 Tropical Screech, **133**
Palmcreeper, Point-tailed, **191**
Parakeet
 Black-hooded, **110**
 Blaze-winged, **121**
 Blue-crowned, **117**
 Caatinga, 118
 Dusky-headed, **116**
 Golden-capped, **119**
 Green-cheeked, **121**
 Jandaya, **119**
 Monk, **119**
 Peach-fronted, **119**
 Pfrimer's, **121**
 White-eyed, **117**
 Yellow-chevroned, **121**
Parrot
 Blue-headed, **123**
 Scaly-headed, **123**
 Yellow-faced, **123**
Parrotlet, Blue-winged, **121**
Parula, Tropical, **269**
Pauraque, **139**
Peppershrike, Rufous-browed, **269**
Pewee, Tropical, **225**
Phalarope, Wilson's, **105**
Piculet
 Bar-breasted, **171**

Rusty-necked, 170
Spotted, 170
White-barred, 170
White-wedged, **171**
Pigeon
 Pale-vented, **109**
 Picazuro, **109**
 Plumbeous, **109**
 Scaled, **109**
Piha, Screaming, **251**
Pintail, White-cheeked, **45**
Pipit
 Ochre-breasted, **265**
 Yellowish, **265**
Piprites, Wing-barred, 244
Plover
 American Golden, **101**
 Collared, **99**
 Pied, **99**
Pochard
 Rosy-billed, 44
 Southern, **45**
Potoo
 Common, **135**
 Great, **135**
Puffbird
 Caatinga, **163**
 Chaco, **163**
 Pied, 164
 Spotted, **165**
 Striolated, **163**
 Swallow-winged, **165**
 White-eared, **163**
 White-necked, **165**
Pygmy-Tyrant
 Rufous-sided, **217**
 Tawny-crowned, **217**
Quail-Dove, Ruddy, **113**
Rail
 Blackish, **93**
 Giant Wood, **93**
 Gray-necked Wood, **93**
 Spotted, **93**
Reedfinch, Long-tailed, **303**
Rhea, Greater, **31**
Sabrewing, Gray-breasted, **147**
Saltator
 Black-throated, **291**
 Buff-throated, **291**
 Golden-billed, **291**
 Grayish, **291**
 Green-winged, **291**
Sandpiper
 Baird's, 104
 Buff-breasted, **101**
 Least, **105**
 Pectoral, **105**
 Solitary, **103**
 Spotted, **105**
 Stilt, **105**
 Upland, **101**
 White-rumped, **105**
Sapphire
 Gilded, **153**
 Rufous-throated, 152
 White-chinned, **151**
Schiffornis
 Greenish, **247**
 Thrush-like, 246
Screamer
 Horned, **41**
 Southern, **41**
Scythebill, Red-billed, **183**

Seedeater
  Black-and-tawny, 298
  Black-bellied, **299**
  Capped, **297**
  Chestnut, **299**
  Dark-throated, **299**
  Double-collared, **297**
  Entre Ríos, 298
  Hooded, **295**
  Lined, **295**
  Marsh, **299**
  Plumbeous, **297**
  Rufous-rumped, **299**
  Rusty-collared, **297**
  Tawny-bellied, **299**
  White-bellied, **297**
  Yellow-bellied, **295**
Seedfinch
  Great-billed, **293**
  Lesser, **293**
Seriema, Red-legged, **31**
Sharpbill, **251**
Shrike-Tanager, White-winged, 284
Sirystes, Eastern, **233**
Siskin, Hooded, **305**
Skimmer, Black, **107**
Snipe
  Giant, **103**
  South American, **103**
Spadebill, White-throated, **223**
Sparrow
  Grassland, **303**
  House, **307**
  Pectoral, **307**
  Rufous-collared, **307**
  Saffron-billed, **307**
  São Francisco, 306
  Yellow-browed, **303**
Spinetail
  Araguaia, 186
  Chotoy, **189**
  Cinereous-breasted, **187**
  Gray-headed, **189**
  Ochre-cheeked, **189**
  Pale-breasted, **187**
  Pallid, **189**
  Red-shouldered, 188
  Rufous-capped, **187**
  Rusty-backed, **189**
  Sooty-fronted, **187**
  Spix's, **187**
  White-lored, **187**
  Yellow-chinned, **189**
Spoonbill, Roseate, **59**
Starthroat
  Blue-tufted, **153**
  Long-billed, **153**
  Stripe-breasted, **153**
Stilt
  Black-necked, **99**
  White-backed, **99**
Stork
  Maguari, **59**
  Wood, **59**
Streamcreeper, Sharp-tailed, **195**
Suiriri
  Campo, **209**
  Chapada, 208
Sunbittern, **97**
Sungem, Horned, **149**
Sungrebe, **97**

Swallow
  American Cliff, **257**
  Bank, see Sand Martin
  Barn, **257**
  Blue-and-white, **259**
  Southern Rough-winged, **259**
  Tawny-headed, **259**
  White-banded, 258
  White-rumped, **259**
  White-winged, **259**
Swan, Coscoroba, 40
Swift
  Ashy-tailed, **143**
  Biscutate, **143**
  Gray-rumped, 142
  Great Dusky, **143**
  Lesser Swallow-tailed, 142
  Neotropical Palm, **143**
  Short-tailed, **143**
  White-collared, **143**
Tachuri
  Bearded, **217**
  Gray-backed, **217**
Tanager
  Bay-headed, **277**
  Black-faced, **287**
  Black-goggled, **283**
  Blue-necked, **277**
  Burnished-buff, **277**
  Cinnamon, **287**
  Cone-billed, **287**
  Fawn-breasted, **281**
  Flame-crested, **285**
  Gilt-edged, 276
  Gray-headed, **283**
  Guira, **273**
  Hooded, **273**
  Lowland Hepatic, **283**
  Magpie, **289**
  Masked, **277**
  Orange-headed, **273**
  Palm, **281**
  Paradise, 276
  Ruby-crowned, **285**
  Rufous-headed, **273**
  Sayaca, **281**
  Scarlet-throated, **289**
  Shrike-like, **287**
  Silver-beaked, **283**
  Swallow, **281**
  Turquoise, **277**
  White-lined, **285**
  White-rumped, **287**
  White-shouldered, **285**
Tapaculo, Brasilia, **207**
Teal
  Brazilian, **45**
  Ringed, **45**
Tern
  Common, **107**
  Large-billed, **107**
  Yellow-billed, **107**
Thornbird
  Greater, **191**
  Rufous-fronted, **191**
Thrush
  Cocoa, **267**
  Creamy-bellied, **267**
  Eastern Slaty, **267**
  Pale-breasted, **267**
  Rufous-bellied, **267**
  White-necked, 266
Tinamou
  Brown, 32
  Dwarf, **35**

Gray, **33**
  Little, **33**
  Red-winged, **35**
  Small-billed, **33**
  Tataupa, **33**
  Undulated, **33**
Tityra
  Black-crowned, **245**
  Black-tailed, **245**
  Masked, **245**
Tody-Flycatcher
  Common, **219**
  Rusty-fronted, **219**
  Spotted, **219**
  Yellow-lored, **219**
Tody-Tyrant
  Pearly-vented, **219**
  Stripe-necked, **219**
Topaz, Ruby, **149**
Toucan
  Channel-billed, **169**
  Green-billed, **169**
  Toco, **169**
  White-throated, 168
Toucanet, Gould's, **167**
Trogon
  Amazonian White-tailed, **155**
  Black-tailed, **155**
  Black-throated, **157**
  Blue-crowned, **157**
  Collared, **157**
  Surucua, **155**
  Violaceous, **157**
Troupial
  Campo, **313**
  Orange-backed, **313**
Tyrannulet
  Minas Gerais, 220
  Mottle-cheeked, **221**
  Mouse-colored, **215**
  Planalto, **213**
  Reiser's, **213**
  River, 214
  Sooty, **215**
  Southern Beardless, **215**
  White-bellied, **215**
  White-crested, **215**
  White-lored, 212
  Yellow, **221**
Tyrant
  Amazonian Black, **231**
  Black-backed Water, **229**
  Blue-billed Black, **231**
  Cattle, **227**
  Cinereous, **231**
  Cock-tailed, **229**
  Crested Black, **231**
  Hudson's Black, 230
  Long-tailed, **233**
  Masked Water, **229**
  Riverside, **231**
  São Francisco Black, **231**
  Sharp-tailed Grass, **217**
  Southern Bristle, **221**
  Spectacled, **229**
  Streamer-tailed, **227**
  Velvety Black, 230
  White-headed Marsh, **229**
  White-winged Black, 230
  Yellow-browed, **239**
Tyrant-Manakin
  Dwarf, 242
  Pale-bellied, **243**
Veery, 266
Violetear, White-vented, **147**

Vireo, Red-eyed, **269**
Vulture
  Black, **61**
  Greater Yellow-headed, 60
  King, **61**
  Lesser Yellow-headed, **61**
  Turkey, **61**
Warbler
  Flavescent, **271**
  Golden-crowned, **271**
  White-bellied, **271**
  White-rimmed, **271**
  White-striped, **271**
Waxbill, Common, **307**
Woodcreeper
  Black-banded, **181**
  Buff-throated, **181**
  Great Rufous, **183**
  Lesser, **179**
  Long-billed, **183**
  Moustached, **181**
  Narrow-billed, **179**
  Olivaceous, **179**
  Plain-brown, **179**
  Planalto, **181**
  Scaled, 178
  Straight-billed, **179**
  Striped, 178
  White-throated, **181**
Woodnymph
  Fork-tailed, **151**
  Violet-capped, **151**
Woodpecker
  Blond-crested, **175**
  Checkered, **173**
  Chestnut, 174
  Cream-backed, 176
  Cream-colored, **175**
  Crimson-crested, **177**
  Golden-green, **171**
  Green-barred, **173**
  Kaempfer's, **175**
  Lineated, **177**
  Little, **171**
  Pale-crested, **175**
  Red-necked, **177**
  Red-stained, **171**
  Ringed, 174
  White, **175**
  White-fronted, **173**
  White-spotted, 170
  Yellow-fronted, 172
  Yellow-throated, 170
  Yellow-tufted, **173**
Woodstar, Amethyst, **149**
Wren
  Buff-breasted, **263**
  Fawn-breasted, **263**
  Grass, **265**
  Moustached, **263**
  Southern House, **263**
  Thrush-like, **263**
Xenops
  Great, **195**
  Plain, **193**
  Streaked, **193**
Xenopsaris, White-naped, **243**
Yellowfinch
  Grassland, **305**
  Orange-fronted, **305**
  Stripe-tailed, **305**
Yellowlegs
  Greater, **103**
  Lesser, **103**
Yellowthroat, Southern, **271**

Index of Scientific Names

Accipiter
  bicolor, 66
  erythronemius, 66
  superciliosus, 66
Accipitridae, 62
Actitis macularius, 104
Aegolius harrisii, 132
Agamia agami, 50
Agelaioides badius, 308
Agelasticus cyanopus, 310
Ajaia ajaja, 58
Alcedinidae, 158
Alectrurus tricolor, 228
Alipiopsitta xanthops, 122
Alopochelidon fucata, 258
Amaurolimnas concolor, 92
Amazilia
  fimbriata, 150
  lactea, 150
  versicolor, 150
Amazona
  aestiva, 122
  amazonica, 122
  farinosa, 122
Amazonetta brasiliensis, 44
Amblyramphus holosericeus, 312
Ammodramus
  aurifrons, 302
  humeralis, 302
Anas bahamensis, 44
Anatidae, 40
Anhima cornuta, 40
Anhimidae, 40
Anhinga anhinga, 38
Anhingidae, 38
Anodorhynchus hyacinthinus, 114
Anthracothorax nigricollis, 146
Anthus
  lutescens, 264
  nattereri, 264
Antilophia galeata, 246
Anumbius annumbi, 190
Aphantochroa cirrhochloris, 146
Apodidae, 142
Ara
  ararauna, 114
  chloropterus, 114
Aramides
  cajanea, 92
  ypecaha, 92
Aramus guarauna, 56
Aratinga
  acuticaudata, 116
  aurea, 118
  auricapillus, 118
  cactorum, 118
  jandaya, 118
  leucophthalma, 116
  weddellii, 116
Ardea
  alba, 52
  cocoi, 54
Ardeidae, 48
Arremon
  flavirostris, 306
  franciscanus, 306
  taciturnus, 306
Arundinicola leucocephala, 228
Asio
  clamator, 130
  flammeus, 130
  stygius, 130
Asthenes luizae, 190
Athene cunicularia, 132
Atticora fasciata, 258
Attila
  bolivianus, 234
  phoenicurus, 234

Automolus leucophthalmus, 194

Bartramia longicauda, 100
Baryphthengus ruficapillus, 160
Basileuterus
  culicivorus, 270
  flaveolus, 270
  hypoleucus, 270
  leucoblepharus, 270
  leucophrys, 270
Berlepschia rikeri, 190
Botaurus pinnatus, 50
Brachygalba lugubris, 160
Brotogeris chiriri, 120
Bubo virginianus, 130
Bubulcus ibis, 52
Bucconidae, 162
Bucco tamatia, 164
Busarellus nigricollis, 76
Buteo
  albicaudatus, 74
  albonotatus, 74
  brachyurus, 72
  magnirostris, 72
  nitidus, 72
  swainsoni, 74
Buteogallus
  meridionalis, 76
  urubitinga, 78
Butorides striatus, 48

Cacicus
  cela, 314
  chrysopterus, 312
  haemorrhous, 314
  solitarius, 314
Cairina moschata, 40
Calidris
  bairdii, 104
  fuscicollis, 104
  himantopus, 104
  melanotos, 104
  minutilla, 104
Calliphlox amethystina, 148
Callonetta leucophrys, 44
Campephilus
  leucopogon, 176
  melanoleucus, 176
  rubricollis, 176
Camptostoma obsoletum, 214
Campylopterus largipennis, 146
Campylorhamphus
  trochilirostris, 182
Campylorhynchus turdinus, 262
Caprimulgidae, 136
Caprimulgus
  longirostris, 140
  maculicaudus, 140
  parvulus, 140
  rufus, 140
Capsiempis flaveola, 220
Caracara plancus, 84
Cardinalidae, 288
Carduelis magellanica, 304
Cariama cristata, 30
Cariamidae, 30
Casiornis
  fuscus, 234
  rufus, 234
Cathartes
  aura, 60
  burrovianus, 60
  melambrotus, 60
Cathartidae, 60
Catharus fuscescens, 266
Celeus
  elegans, 174
  flavescens, 174
  flavus, 174

lugubris, 174
  obrieni, 174
  torquatus, 174
Cercomacra
  ferdinandi, 204
  melanaria, 204
Certhiaxis cinnamomeus, 188
Chaetura
  brachyura, 142
  cinereiventris, 142
  meridionalis, 142
Charadriidae, 98
Charadrius collaris, 98
Charitospiza eucosma, 300
Chauna torquata, 40
Chelidoptera tenebrosa, 164
Chiroxiphia
  caudata, 246
  pareola, 246
Chloroceryle
  aenea, 158
  amazona, 158
  americana, 158
  inda, 158
Chlorostilbon aureoventris, 150
Chondrohierax uncinatus, 68
Chordeiles
  acutipennis, 136
  minor, 136
  pusillus, 136
  rupestris, 136
Chrysolampis mosquitus, 148
Chrysomus ruficapillus, 310
Chrysoptilus melanochloros, 172
Ciconia maguari, 58
Ciconiidae, 58
Circus buffoni, 68
Cissopis leverianus, 288
Cistothorus platensis, 264
Claravis pretiosa, 112
Cnemotriccus fuscatus, 224
Coccycua minuta, 126
Coccyzus
  americanus, 124
  euleri, 124
  melacoryphus, 124
Cochlearius cochlearius, 48
Coereba flaveola, 274
Colaptes campestris, 172
Colibri serrirostris, 146
Colonia colonus, 232
Columbidae, 108
Columbina
  cyanopis, 110
  minuta, 110
  picui, 110
  talpacoti, 110
Compsothraupis loricata, 288
Conirostrum speciosum, 274
Conopophaga lineata, 206
Conopophagidae, 206
Contopus cinereus, 224
Coragyps atratus, 60
Corvidae, 254
Coryphaspiza melanotis, 300
Coryphospingus
  cucullatus, 300
  pileatus, 300
Corythopis delalandi, 222
Coscoroba coscoroba, 40
Cotingidae, 250
Cracidae, 34
Cranioleuca
  pallida, 188
  semicinerea, 188
  vulpina, 188
Crax fasciolata, 36
Crotophaga
  ani, 126
  major, 126

Crypturellus
  obsoletus, 32
  parvirostris, 32
  soui, 32
  tataupa, 32
  undulatus, 32
Cuculidae, 124
Culicivora caudacuta, 216
Curaeus forbesi, 308
Cyanerpes
  caeruleus, 274
  cyaneus, 274
Cyanocompsa brissonii, 292
Cyanocorax
  chrysops, 254
  cristatellus, 254
  cyanomelas, 254
  cyanopogon, 254
Cyanoloxia glaucocaerulea, 292
Cyclarhis gujanensis, 268
Cypseloides senex, 142
Cypsnagra hirundinacea, 286

Dacnis cayana, 274
Daptrius ater, 84
Dendrocincla fuliginosa, 178
Dendrocolaptes
  picumnus, 180
  platyrostris, 180
Dendrocolaptidae, 178
Dendrocygna
  autumnalis, 42
  bicolor, 42
  viduata, 42
Diopsittaca nobilis, 116
Dolichonyx oryzivorus, 310
Donacobius atricapilla, 260
Donacospiza albifrons, 302
Dromococcyx
  pavoninus, 128
  phasianellus, 128
Dryocopus lineatus, 176
Dysithamnus mentalis, 198

Egretta
  caerulea, 52
  thula, 52
Elaenia
  albiceps, 210
  chiriquensis, 210
  cristata, 210
  flavogaster, 208
  mesoleuca, 210
  obscura, 208
  parvirostris, 210
  spectabilis, 208
Elanoides forficatus, 62
Elanus leucurus, 62
Eleothreptus
  anomalus, 140
  candicans, 140
Emberizidae, 292
Emberizoides
  herbicola, 302
  ypiranganus, 302
Embernagra
  longicauda, 302
  platensis, 302
Empidonax alnorum, 224
Empidonomus varius, 238
Estrilda astrild, 304
Eucometis penicillata, 282
Eupetomena macroura, 146
Euphonia
  chlorotica, 278
  cyanocephala, 278
  laniirostris, 278
  pectoralis, 278
  rufiventris, 278
  violacea, 278

Eurypyga helias, 96
Euscarthmus
 meloryphus, 216
 rufomarginatus, 216

Falco
 deiroleucus, 88
 femoralis, 88
 peregrinus, 88
 rufigularis, 88
 sparverius, 88
Falconidae, 84
Fluvicola
 albiventer, 228
 nengeta, 228
Formicariidae, 206
Formicarius
 analis, 206
 colma, 206
Formicivora
 grisea, 202
 melanogaster, 202
 rufa, 202
Forpus xanthopterygius, 120
Furnariidae, 184
Furnarius
 figulus, 184
 leucopus, 184
 rufus, 184

Galbula
 dea, 160
 ruficauda, 160
Galbulidae, 160
Gallinago
 paraguaiae, 102
 undulata, 102
Gallinula chloropus, 94
Gampsonyx swainsonii, 62
Geositta poeciloptera, 184
Geothlypis velata, 270
Geotrygon montana, 112
Geranoaetus melanoleucus, 78
Geranospiza caerulescens, 70
Glaucidium
 brasilianum, 132
 hardyii, 132
Glaucis hirsuta, 144
Gnorimopsar chopi, 308
Granatellus pelzelni, 272
Griseotyrannus
 aurantioatrocristatus, 238
Gubernetes yetapa, 226
Guira guira, 128
Gyalophylax hellmayri, 188
Gymnoderus foetidus, 252

Harpagus
 bidentatus, 64
 diodon, 64
Harpia harpyja, 82
Harpyhaliaetus coronatus, 78
Heliactin bilopha, 148
Heliomaster
 furcifer, 152
 longirostris, 152
 squamosus, 152
Heliornis fulica, 96
Hemithraupis
 guira, 272
 ruficapilla, 272
Hemitriccus
 margaritaceiventer, 218
 striaticollis, 218
Herpetotheres cachinnans, 86
Herpsilochmus
 atricapillus, 200
 longirostris, 200
 rufimarginatus, 200
Heterocercus linteatus, 246
Himantopus
 melanurus, 98
 mexicanus, 98
Hirundinea ferruginea, 234

Hirundinidae, 256
Hirundo rustica, 256
Hoploxypterus cayanus, 98
Hydropsalis
 climacocerca, 138
 torquata, 138
Hylocharis
 chrysura, 152
 cyanus, 150
 sapphirina, 150
Hylocryptus rectirostris, 194
Hylopezus ochroleucus, 206
Hylophilus
 amaurocephalus, 268
 muscicapinus, 268
 pectoralis, 268
Hymenops perspicillatus, 228
Hypocnemoides maculicauda, 200

Ibycter americanus, 84
Icteridae, 308
Icterus
 cayanensis, 312
 croconotus, 312
 jamacaii, 312
Ictinia
 mississippiensis, 64
 plumbea, 64
Ilicura militaris, 248
Inezia
 inornata, 214
 subflava, 214
Ixobrychus exilis, 48

Jabiru mycteria, 58
Jacana jacana, 96

Knipolegus
 aterrimus, 230
 cyanirostris, 230
 franciscanus, 230
 hudsoni, 230
 lophotes, 230
 nigerrimus, 230
 orenocensis, 230
 poecilocercus, 230
 striaticeps, 230

Lanio versicolor, 284
Laridae, 106
Laterallus exilis, 90
 melanophaius, 90
 viridis, 90
 xenopterus, 90
Lathrotriccus euleri, 224
Legatus leucophaius, 238
Lepidocolaptes
 angustirostris, 178
 squamatus, 178
Leptodon cayanensis, 70
Leptopogon amaurocephalus, 220
Leptotila
 rufaxilla, 112
 verreauxi, 112
Leuconerpes candidus, 174
Leucopternis albicollis, 70
Limosa haemastica, 100
Lipaugus vociferans, 250
Lochmias nematura, 194
Lophornis
 gouldii, 148
 magnificus, 148
Lophostrix cristata, 134
Lurocalis semitorquatus, 136

Machaeropterus pyrocephalus, 248
Machetornis rixosa, 226
Manacus manacus, 248
Megaceryle torquata, 158
Megarynchus pitangua, 236
Megascops
 choliba, 132
 watsonii, 132

Megaxenops parnaguae, 194
Melanerpes
 cactorum, 172
 cruentatus, 172
 flavifrons, 172
Melanopareia torquata, 206
Melanopareiidae, 206
Mergus octosetaceus, 46
Mesembrinibis cayennensis, 56
Micrastur
 ruficollis, 86
 semitorquatus, 86
Micrococcyx cinereus, 124
Micropygia schomburgkii, 90
Milvago
 chimachima, 84
 chimango, 84
Mimidae, 260
Mimus
 saturninus, 260
 triurus, 260
Mionectes
 oleagineus, 220
 rufiventris, 220
Molothrus
 bonariensis, 308
 oryzivorus, 308
 rufoaxillaris, 308
Momotidae, 160
Momotus momota, 160
Monasa
 morphoeus, 164
 nigrifrons, 164
Morphnus guianensis, 82
Motacillidae, 264
Mycteria americana, 58
Myiarchus
 ferox, 232
 swainsoni, 232
 tuberculifer, 232
 tyrannulus, 232
Myiobius
 atricaudus, 222
 mastacalis, 222
Myiodynastes maculatus, 238
Myiopagis
 caniceps, 212
 gaimardii, 212
 viridicata, 212
Myiophobus fasciatus, 224
Myiopsitta monachus, 118
Myiozetetes
 cayanensis, 236
 similis, 236
Myrmeciza
 atrothorax, 204
 hemimelaena, 204
Myrmoborus
 leucophrys, 202
 myotherinus, 202
Myrmorchilus strigilatus, 202
Myrmotherula
 axillaris, 200
 multostriata, 200

Nandayus nenday, 118
Nasica longirostris, 182
Nemosia pileata, 272
Neochen jubata, 42
Neopelma pallescens, 242
Neothraupis fasciata, 286
Netta
 erythrophthalma, 44
 peposaca, 44
Nomonyx dominicus, 46
Nonnula
 rubecula, 162
 ruficapilla, 162
Notharchus
 hyperrhynchus, 164
 tectus, 164
Nothura
 boraquira, 34
 maculosa, 34
 minor, 34

Notiochelidon cyanoleuca, 258
Nyctibiidae, 134
Nyctibius
 grandis, 134
 griseus, 134
Nycticorax nycticorax, 48
Nyctidromus albicollis, 138
Nyctiprogne
 leucopyga, 136
 vielliardi, 136
Nystalus
 chacuru, 162
 maculatus, 162
 striatipectus, 162
 striolatus, 162

Opisthocomus hoazin, 38
Ornithion inerme, 212
Ortalis canicollis, 36
Orthopsittaca manilata, 114
Oryzoborus
 angolensis, 292
 maximiliani, 292
Oxyruncus cristatus, 250

Pachyramphus
 castaneus, 242
 marginatus, 242
 minor, 244
 polychopterus, 242
 validus, 244
 viridis, 242
Pandion haliaetus, 62
Panyptila cayennensis, 142
Parabuteo unicinctus, 76
Pardirallus
 maculatus, 92
 nigricans, 92
Paroaria
 baeri, 288
 capitata, 288
 coronata, 288
Parula pitiayumi, 268
Parulidae, 270
Passer domesticus, 306
Patagioenas
 cayennensis, 108
 picazuro, 108
 plumbea, 108
 speciosa, 108
Penelope
 jacquacu, 36
 jacucaca, 36
 ochrogaster, 36
 pileata, 36
 superciliaris, 36
Petrochelidon pyrrhonota, 256
Phacellodomus
 ruber, 190
 rufifrons, 190
Phaeomyias murina, 214
Phaethornis
 nattereri, 144
 pretrei, 144
 ruber, 144
 subochraceus, 144
Phaetusa simplex, 106
Phalocrocoracidae, 38
Phalacrocorax brasilianus, 38
Phalaropus tricolor, 104
Pheucticus aureoventris, 292
Phibalura flavirostris, 250
Philohydor lictor, 236
Philydor
 lichtensteini, 192
 rufum, 192
Phimosus infuscatus, 56
Phyllomyias
 fasciatus, 212
 reiseri, 212
Phylloscartes
 roquettei, 220
 ventralis, 220

321

Piaya cayana, 126
Picidae, 170
Piculus
  chrysochloros, 170
  flavigula, 170
Picumnus
  albosquamatus, 170
  aurifrons, 170
  cirratus, 170
  fuscus, 170
  pygmaeus, 170
Pilherodius pileatus, 54
Pionus
  maximiliani, 122
  menstruus, 122
Pipile pipile, 36
Pipra
  fasciicauda, 248
  rubrocapilla, 248
Pipraeidea melanonota, 280
Pipridae, 246
Piprites chloris, 244
Piranga flava, 282
Pitangus sulphuratus, 236
Platyrinchus mystaceus, 222
Plegadis chihi, 56
Pluvialis dominica, 100
Podager nacunda, 138
Podicipedidae, 46
Podilymbus podiceps, 46
Poecilotriccus latirostris, 218
Pogonotriccus eximius, 220
Polioptila dumicola, 264
Polioptilidae, 264
Polystictus
  pectoralis, 216
  superciliaris, 216
Polytmus guainumbi, 152
Poospiza
  cinerea, 300
  melanoleuca, 300
Porphyriops melanops, 94
Porphyrospiza caerulescens, 294
Porphyrula
  flavirostris, 94
  martinica, 94
Porzana albicollis, 90
Primolius
  auricollis, 116
  maracana, 116
Procnias nudicollis, 250
Progne
  chalybea, 256
  subis, 256
  tapera, 256
Psarocolius
  decumanus, 314
  yuracares, 314
Pseudocolopteryx sclateri, 216
Pseudoleistes guirahuro, 310
Pseudoseisura
  cristata, 184
  unirufa, 184
Psittacidae, 114
Pteroglossus
  bitorquatus, 166
  castanotis, 166
  inscriptus, 166
Pulsatrix perspicillata, 134
Pyriglena
  leuconota, 204
  leucoptera, 204
Pyrocephalus rubinus, 224
Pyroderus scutatus, 252
Pyrrhura
  devillei, 120
  molinae, 120
  pfrimeri, 120

Querula purpurea, 252

Rallidae, 90

Ramphastidae, 166
Ramphastos
  dicolorus, 168
  toco, 168
  tucanus, 168
  vitellinus, 168
Ramphocaenus melanurus, 264
Ramphocelus carbo, 282
Rhea americana, 30
Rheidae, 30
Rhynchothraupis mesoleuca, 286
Rhynchotus rufescens, 34
Rhytipterna simplex, 234
Riparia riparia, 256
Rostrhamus sociabilis, 68
Rynchops niger, 106

Sakesphorus
  cristatus, 196
  luctuosus, 196
Saltator
  atricollis, 290
  aurantiirostris, 290
  coerulescens, 290
  maximus, 290
  similis, 290
Sarcoramphus papa, 60
Sarkidiornis melanotos, 40
Satrapa icterophrys, 238
Scardafella squammata, 110
Schiffornis
  turdinus, 246
  virescens, 246
Schistochlamys
  melanopis, 286
  ruficapillus, 286
Schoeniophylax phryganophilus, 188
Scleruridae, 184
Sclerurus scansor, 194
Scolopacidae, 100
Scytalopus novacapitalis, 206
Selenidera gouldii, 166
Serpophaga
  hypoleuca, 214
  munda, 214
  nigricans, 214
  subcristata, 214
Sicalis
  citrina, 304
  columbiana, 304
  flaveola, 304
  luteola, 304
Sirystes sibilator, 232
Sittasomus griseicapillus, 178
Spizaetus
  melanoleucus, 80
  ornatus, 80
  tyrannus, 80
Sporophila
  bouvreuil, 296
  caerulescens, 296
  cinnamomea, 298
  collaris, 296
  hypochroma, 298
  hypoxantha, 298
  leucoptera, 296
  lineola, 294
  melanogaster, 298
  melanops, 294
  nigricollis, 294
  nigrorufa, 298
  palustris, 298
  plumbea, 296
  ruficollis, 298
  zelichi, 298
Stelgidopteryx ruficollis, 258
Sterna hirundo, 106
Sternula superciliaris, 106
Streptoprocne
  biscutata, 142
  zonaris, 142

Strigidae, 130
Strix
  huhula, 134
  virgata, 134
Sturnella superciliaris, 310
Sublegatus modestus, 210
Suiriri
  affinis, 208
  islerorum, 208
Synallaxis
  albescens, 186
  albilora, 186
  frontalis, 186
  hypospodia, 186
  ruficapilla, 186
  scutata, 188
  simoni, 186
  spixi, 186
Syndactyla dimidiata, 192
Syrigma sibilatrix, 54

Tachornis squamata, 142
Tachybaptus dominicus, 46
Tachycineta
  albiventer, 258
  leucorrhoa, 258
Tachyphonus
  coronatus, 284
  cristatus, 284
  luctuosus, 284
  rufus, 284
Tangara
  cayana, 276
  chilensis, 276
  cyanicollis, 276
  cyanoventris, 276
  gyrola, 276
  mexicana, 276
  nigrocincta, 276
Taoniscus nanus, 34
Tapera naevia, 128
Taraba major, 196
Tersina viridis, 280
Thalurania
  furcata, 150
  glaucopis, 150
Thamnophilidae, 196
Thamnophilus
  amazonicus, 198
  caerulescens, 198
  doliatus, 196
  pelzelni, 198
  sticturus, 198
  torquatus, 196
Theristicus
  caerulescens, 56
  caudatus, 54
Thlypopsis sordida, 272
Thraupidae, 272
Thraupis
  palmarum, 280
  sayaca, 280
Threskiornithidae, 54
Thryothorus
  genibarbis, 262
  guarayanus, 262
  leucotis, 262
Tiaris
  fuliginosus, 294
  obscurus, 294
Tigrisoma
  fasciatum, 50
  lineatum, 50
Tinamidae, 32
Tinamus tao, 32
Tityra
  cayana, 244
  inquisitor, 244
  semifasciata, 244
Tityridae, 244
Todirostrum
  cinereum, 218
  maculatum, 218
  poliocephalum, 218

Tolmomyias
  assimilis, 222
  flaviventris, 222
  sulphurescens, 222
Trichothraupis melanops, 282
Tringa
  flavipes, 102
  melanoleuca, 102
  solitaria, 102
Trochilidae, 144
Troglodytes musculus, 262
Troglodytidae, 262
Trogon
  collaris, 156
  curucui, 156
  melanurus, 154
  rufus, 156
  surrucura, 154
  violaceus, 156
  viridis, 154
Trogonidae, 154
Tryngites subruficollis, 100
Turdidae, 266
Turdus
  albicollis, 266
  amaurochalinus, 266
  fumigatus, 266
  leucomelas, 266
  rufiventris, 266
  subalaris, 266
Tyranneutes stolzmanni, 242
Tyrannidae, 208
Tyrannopsis sulphurea, 240
Tyrannus
  albogularis, 240
  melancholicus, 240
  savanna, 240
  tyrannus, 240
Tyto alba, 132

Uropelia campestris, 110

Vanellus chilensis, 98
Veniliornis
  affinis, 170
  mixtus, 172
  passerinus, 170
  spilogaster, 170
Vireonidae, 268
Vireo olivaceus, 268
Volatinia jacarina, 294

Xenops
  minutus, 192
  rutilans, 192
Xenopsaris albinucha, 242
Xiphocolaptes
  albicollis, 180
  falcirostris, 180
  major, 182
Xipholena
  punicea, 250
Xiphorhynchus
  fuscus, 178
  guttatus, 180
  obsoletus, 178
  picus, 178
Xolmis
  cinereus, 226
  irupero, 226
  velatus, 226

Zebrilus undulatus, 48
Zenaida auriculata, 112
Zonotrichia capensis, 306